Fundamentals and Advances in Metal Matrix Composites

The scope of this book covers the fundamental background of metal matrix composites (MMCs), their processing and fabrication, testing and characterization, exploration of materials for MMCs and green MMCs, and advancements in all aspects of fabrication, testing, and applications. Development or fabrication of MMCs with evaluation of mechanical and tribological properties as well as machinability evaluation, optimization of fabrication process, and machining operations are covered.

Features:

- Covers advanced processing strategies and machining studies for composite materials
- Discusses representative volume element-based FEM modelling approaches and sustainability
- Sheds light on advancements in MMC application, fabrication, and testing
- Reviews green MMCs and sustainability in MMCs development
- Includes case studies and intelligent modelling methodologies

This book is aimed at graduate students, researchers, and professionals in micro/nanoscience and technology, mechanical engineering, industrial engineering, metallurgy, and composites.

Advanced Materials Processing and Manufacturing

Series Editor: Kapil Gupta

The CRC Press Series in Advanced Materials Processing and Manufacturing covers the complete spectrum of materials and manufacturing technology, including fundamental principles, theoretical background, and advancements. Considering the accelerated importance of advances for producing quality products for a wide range of applications, the titles in this series reflect the state-of-the-art in understanding and engineering the materials processing and manufacturing operations. Technological advancements for enhancement of product quality, process productivity, and sustainability, are on special focus including processing for all materials and novel processes. This series aims to foster knowledge enrichment on conventional and modern machining processes. Micro-manufacturing technologies such as micro-machining, micro-forming, and micro-joining, and Hybrid manufacturing, additive manufacturing, near net shape manufacturing, and ultra-precision finishing techniques are also covered.

Advanced Materials Characterization
Basic Principles, Novel Applications, and Future Directions
Ch Sateesh Kumar, M. Muralidhar Singh and Ram Krishna

Thin-Films for Machining Difficult-to-Cut Materials
Challenges, Applications, and Future Prospects
Ch Sateesh Kumar and Filipe Daniel Fernandes

Advanced Materials Processing and Manufacturing
Research, Technology, and Applications
Amogelang Sylvester Bolokang and Maria Ntsoaki Mathabathe

Advanced Joining Technologies
Edited by Manjaiah M, Shivraman Thapliyal and Adepu Kumar

Nanofinishing of Materials for Advanced Industrial Applications
Edited by Faiz Iqbal, Dilshad Ahmad Khan and Zafar Alam

Environmentally Benign Machining
Edited by Şenol Bayraktar and Sunil Pathak

Post-Processing of Parts and Components Fabricated by Fused Deposition Modeling
Techniques and Advancements
Edited by Vinayak R. Malik, Vivek Tiwary and Arunkumar P.

Forming and Machining of Polymers, Ceramics, and Composites
Edited by Matruprasad Rout and Kishore Debnath

Fundamentals and Advances in Metal Matrix Composites
Edited by Tharmaraj Ramakrishnan and PM Gopal

For more information about this series, please visit: www.routledge.com/Advanced-Materials-Processing-and-Manufacturing/book-series/CRCAMPM

Fundamentals and Advances in Metal Matrix Composites

Edited by
Tharmaraj Ramakrishnan and PM Gopal

CRC Press
Taylor & Francis Group
Boca Raton London

CRC Press is an imprint of the
Taylor & Francis Group, an **informa** business

Designed cover image: TK

First edition published 2025
by CRC Press
2385 NW Executive Center Drive, Suite 320, Boca Raton FL 33431

and by CRC Press
4 Park Square, Milton Park, Abingdon, Oxon, OX14 4RN

CRC Press is an imprint of Taylor & Francis Group, LLC

ISBN: 978-1-032-80452-1 (hbk)
ISBN: 978-1-032-81407-0 (pbk)
ISBN: 978-1-003-49971-8 (ebk)

DOI: 10.1201/9781003499718

Typeset in Times
by Apex CoVantage, LLC

Contents

Preface.....ix
About the Editors.....xi
Contributors.....xiii

Chapter 1 Processing, Properties and Applications of Metal Matrix Composites.....1

Jarubula Ranga Raya Chowdary, Abdul Sattar Kajavali, and Veeravalli Ramakoteswara Rao

Chapter 2 Development of Metal Matrix Composites Through Liquid State Technique.....15

Senthil Murugan Subramanian and Subhaschandra Kattimani

Chapter 3 Exploring Liquid Metallurgy Processes for Advanced Metal Matrix Composite.....35

Abdur Rahman Mohammed Habeebullah, Sirajudeen Nizeeruddin, and Sathickbasha Katharbasha

Chapter 4 Development of Metal Matrix Composites Through Solid-State Techniques.....51

Saravanan Rathinasamy, Omkumar Meenakshisundaram, Sathish Thanikodi, and Jothi Arunachalam Solairaju

Chapter 5 Additive Manufacturing of Metal Matrix Composites.....66

Ved Prakash Dubey, Shivam Sharma, Mateusz Kopec, and Zbigniew Kowalewski

Chapter 6 A Review on Selective Laser Melting of Aluminium Metal Matrix Composites.....79

Kumar Ramar, Gedion Habtay Gebremicheal, Ghebremlak Andemariam, Lewhat Michael Weldegebriel, Hurui Tesfalem Woldu, Vijay Ananth Suyamburajan, and Ajith Arul Daniel

Chapter 7 Investigation on Properties of Metal Matrix Composites Fabricated Through Additive Manufacturing.....90

Aravindh Raj Babu Rudrakotti and Velmurugan Chandrasekaran

Chapter 8 Advancement in Fabrication of Metal Matrix Composites.....108

Sudhagar Sukkasamy, Vijayakumar Palanivel, Suresh Subramanian, Yessian Robinson, and Duraisamy Velmurugan

Chapter 9 Fabrication of Intelligent Metal Matrix Composites and Their Applications.....121

Saravanan Rathinasamy, Omkumar Meenakshisundaram, Sathish Thanikodi, and Jothi Arunachalam Solairaju

Chapter 10 Investigation of the Effects of Process Parameters on Metal Matrix Composites of A356 with SiC on Mechanical and Tribological Characteristics 138

Padmanabhan Sambandam, Giridharan Krishnan, Deepak James Raj, Vinod Kumar Thangaraj, and Chakravarthi Gurijala

Chapter 11 Mechanical Characteristics of Hybrid Metal Matrix Composites Fabricated through Liquid State Technique 152

Alphonse Bovas Herbert Bejaxhin, Selvaraj Hari Krishna Raj, Pathinettampadian Gurusamy, and Nandagopan Ramanan

Chapter 12 Investigation on Properties of Friction Stir Welded Metal Matrix Composites 169

Ganapathy Srinivasan Ramachandran, Rajaravi Chinnathambi, Selvam Murugan, and Palani Subbiah Kone

Chapter 13 Industry 4.0 Based Approaches for Fabrication, Processing and Testing of Metal Matrix Composites 181

Sneha Haresh Dhoria, Yellinedi Madhavi, and Kota Sobha

Chapter 14 Modern Machining of Metal Matrix Composites 196

Thavasilingam Kannan, Thiagarajan Srinivasan, Naga Chandrika Kondiboyana Kumari, Madhan Mohankumar, and Santhosh Sivaraj

Chapter 15 Conventional Machining of Metal Matrix Composites 208

Meby Selvaraj and Navasingh Rajesh Jesudoss Hynes

Chapter 16 Investigation of Electrical Discharge Machining on Powder Metallurgy Aluminium Metal Matrix Composites 222

Alphonse Bovas Herbert Bejaxhin, Selvaraj Hari Krishna Raj, Pathinettampadian Gurusamy, and Nandagopan Ramanan

Chapter 17 Joining of Metal Matrix Composites 240

Shenbaga Velu Pitchumani, Venkatachalam Gopalan, Vignesh Nagarajan Jawahar, Navasingh Rajesh Jesudoss Hynes, and Raja Thiyagarajan

Chapter 18 Advanced Forming Methods and Processing Condition of Metal Matrix Composites 255

Sharath Peramenahalli Chikkegowda, Shrishail Basaprabhu Sollapur, and Bharathkumar Bidare Rangaswamy

Chapter 19 Sustainable Approaches for Metal Matrix Composites Fabrication 267

Arunachalam Saravanakumar, Dodda Ravikanth, Tamil Selvan Manickam, and Manickam Ramesh

Chapter 20 New Insights into Metal Matrix Composite Recycling and Reclamation................280

Raja Thiyagarajan, Shenbaga Velu Pitchumani, Senthil Kumar Marikannan, Venkatachalam Gopalan, and Sakthi Selvarasu

Index..291

Preface

Metal matrix composites (MMCs) represent a class of materials that combine the desirable properties of metals with the distinct advantages of reinforcements. Compared to their monolithic counterparts, these composites offer enhanced mechanical properties, superior thermal stability, and improved wear resistance. MMCs are increasingly finding applications across a wide range of industries due to their ability to meet the demanding performance criteria required in these fields. The edited book *Fundamentals and Advances in Metal Matrix Composites* aims to comprehensively understand this dynamic field, addressing both foundational concepts and the latest advancements.

The book begins with an overview of the fundamental principles, processing methods, properties, and applications of MMCs. This sets the stage for a detailed exploration of specific techniques for developing MMCs, starting with liquid state techniques and delving into the intricacies of liquid metallurgy processes. The focus then shifts to the solid-state techniques, offering insights into alternative methods for fabricating MMCs. The growing importance of additive manufacturing is thoroughly explored, covering the basics and the investigation of properties fabricated through these innovative methods.

Advancements in fabrication techniques and the development of intelligent MMCs are also discussed, emphasizing their modern applications. A detailed investigation of process parameters and their effects on the mechanical and tribological characteristics of specific MMCs is presented. The mechanical properties of hybrid MMCs, fabricated through liquid state techniques, are scrutinized, followed by an in-depth analysis of the properties of friction stir welded MMCs.

The transformative impact of Industry 4.0 on MMC fabrication, processing, and testing is explored, highlighting the integration of digital technologies. Various machining techniques are addressed in detail, from modern and conventional approaches to the MMCs. Challenges and innovations in joining and forming methods of MMCs are thoroughly examined. Sustainable fabrication methods and the recycling of MMCs are also discussed, reflecting the growing emphasis on sustainability in materials science.

Each chapter is contributed by experts in the field, ensuring that the content is both authoritative and current. This book is intended for researchers, engineers, and students seeking an in-depth understanding of MMCs. It aims to serve as a valuable resource, bridging the gap between fundamental concepts and advanced research in this exciting field. We hope that this book will inspire further innovation and exploration, contributing to the advancement of MMCs.

Editors
Tharmaraj Ramakrishnan
PM Gopal

About the Editors

Tharmaraj Ramakrishnan is Assistant Professor in the Department of Mechanical Engineering at SRM Institute of Science and Technology, Ramapuram Campus, Chennai, India. Tharmaraj completed his PhD in the Department of Mechanical Engineering at the National Institute of Technology Warangal, India. In the past, Tharmaraj worked as Post Doctoral Research Fellow at the Institute of Fundamental Technological Research, Polish Academy of Sciences, Warsaw, Poland. His area of research interests are metal matrix composites, friction welding, and powder metal forming.

PM Gopal is Assistant Professor in the Centre for Material Science, Department of Mechanical Engineering, Karpagam Academy of Higher Education, Coimbatore. He completed his master's in Production Engineering and PhD in Mechanical Engineering (in the area of MMC) at Anna University Regional Campus Coimbatore under Anna University Chennai. His areas of interest include MMC, additive manufacturing, and waste management. He has published more than 40 articles and ten book chapters in the area of composite fabrication, property analysis, and machining.

Contributors

Ghebremlak Andemariam
Mai-Nefhi College of Engineering and Technology
Asmara, Eritrea

Saravanakumar Arunachalam
Sophisticated Testing and Instrumentation Centre (AU-STIC), Centre of Excellence Advanced Materials Synthesis (CoE- AMS)
Department of Mechanical Engineering
Alliance School of Applied Engineering
Alliance University, Bengaluru, India

Alphonse Bovas Herbert Bejaxhin
Saveetha School of Engineering, Saveetha Institute of Medical and Technical Sciences
Saveetha University
Chennai, Tamil Nadu, India

Velmurugan Chandrasekaran
Indian Institute of Information Technology Tiruchirappalli
Tamil Nadu, India

Sharath Peramenahalli Chikkegowda
Jain (Deemed to be University)
Bengaluru, Karnataka, India

Rajaravi Chinnathambi
Hindhusthan College of Engineering and Technology
Coimbatore, Tamil Nadu, India

Jarubula Ranga Raya Chowdary
Annamalai University Faculty of Engineering and Technology
Chidambaram, Tamil Nadu, India

Ajith Arul Daniel
VELS Institute of Science, Technology and Advanced Studies
Chennai, Tamil Nadu, India

Ved Prakash Dubey
Institute of Fundamental Technological Research
Polish Academy of Sciences
Warsaw, Poland

Gedion Habtay Gebremicheal
Mai-Nefhi College of Engineering and Technology
Asmara, Eritrea

Venkatachalam Gopalan
Vellore Institute of Technology
Chennai, Tamil Nadu, India

Chakravarthi Gurijala
Easwari Engineering College
Chennai, Tamil Nadu, India

Abdur Rahman Mohammed Habeebullah
B.S. Abdur Rahman Crescent Institute of Science and Technology
Chennai, Tamil Nadu, India

Sneha Haresh Dhoria
RVR & JC College of Engineering
Guntur, Andhra Pradesh, India

Selvaraj Hari Krishna Raj
Vel Tech Rangarajan Dr Sagunthala R&D Institute of Science and Technology
Chennai, Tamil Nadu, India

Deepak James Raj
Sathyabama Institute of Science and Technology
Chennai, Tamil Nadu, India

Abdul Sattar Kajavali
Annamalai University Faculty of Engineering and Technology
Chidambaram, India

Thavasilingam Kannan
Easwari Engineering College
Chennai, Tamil Nadu, India

Sathickbasha Katharbasha
B.S. Abdur Rahman Crescent Institute of Science and Technology
Chennai, Tamil Nadu, India

Subhaschandra Kattimani
National Institute of Technology
Surathkal, Karnataka, India

Naga Chandrika Kondiboyana Kumari
Easwari Engineering College
Chennai, Tamil Nadu, India

Mateusz Kopec
Institute of Fundamental Technological Research
Polish Academy of Sciences
Warsaw, Poland

Zbigniew Kowalewski
Institute of Fundamental Technological Research
Polish Academy of Sciences
Warsaw, Poland

Giridharan Krishnan
Easwari Engineering College
Chennai, Tamil Nadu, India

Yellinedi Madhavi
RVR & JC College of Engineering
Guntur, Andhra Pradesh, India

Tamil Selvan Manickam
Dhanalakshmi Srinivasan College of Engineering
Coimbatore, Tamil Nadu, India

Senthil Kumar Marikannan
Vellore Institute of Technology
Chennai, Tamil Nadu, India

Omkumar Meenakshisundaram
College of Engineering Guindy
Anna University
Chennai, Tamil Nadu, India

Madhan Mohankumar
Easwari Engineering College
Chennai, Tamil Nadu, India

Selvam Murugan
Vel Tech Multi Tech Dr Rangarajan Dr Sakunthala Engineering College
Chennai, Tamil Nadu, India

Vignesh Nagarajan Jawahar
Mepco Schlenk Engineering College
Sivakasi, Tamil Nadu, India

Sirajudeen Nizeeruddin
B.S. Abdur Rahman Crescent Institute of Science and Technology
Chennai, Tamil Nadu, India

Gurusamy Pathinettampadian
Chennai Institute of Technology
Chennai, Tamil Nadu, India

Vijayakumar Palanivel
Nehru Institute of Technology (Autonomous)
Coimbatore, Tamil Nadu, India

Shenbaga Velu Pitchumani
Vellore Institute of Technology
Chennai, Tamil Nadu, India

Thiyagarajan Raja
Kongunadu College of Engineering and Technology
Trichy, Tamil Nadu, India

Nandagopan Ramanan
Sri Jayaram Institute of Engineering and Technology
Chennai, Tamil Nadu, India

Kumar Ramar
Mai-Nefhi College of Engineering and Technology
Asmara, Eritrea

Manickam Ramesh
KIT-Kalaignar Karunanidhi Institute of Technology
Coimbatore, Tamil Nadu, India

Bharathkumar Bidare Rangaswamy
Jain College of Engineering and Technology
Hubballi, Karnataka, India

Navasingh Rajesh Jesudoss Hynes
Mepco Schlenk Engineering College
Sivakasi, Tamil Nadu, India
and
Opole University of Technology
Opole, Poland

Veeravalli Ramakoteswara Rao
RVR & JC College of Engineering
Guntur, Andhra Pradesh, India

Saravanan Rathinasamy
Savetha School of Engineering
Saveetha Institute of Medical and Technical Sciences
Saveetha University
Chennai, Tamil Nadu, India

Dodda Ravikanth
KSRM College of Engineering
Kadapa, Andhra Pradesh, India

Yessian Robinson
Erode Sengunthar Engineering College
Tamil Nadu, India

Aravindh Raj Babu Rudrakotti
Indian Institute of Information Technology
Tiruchirappalli, Tamil Nadu, India

Padmanabhan Sambandam
Vel Tech Rangarajan Dr Sagunthala R&D Institute of Science and Technology
Chennai, Tamil Nadu, India

Meby Selvaraj
AAA College of Engineering and Technology
Sivakasi, Tamil Nadu, India

Sakthi Selvarasu
Arifa Institute of Technology
Esanoor, Tamil Nadu, India

Shivam Sharma
CMEMS, School of Engineering
University of Minho, Campus de Azurém
Guimarães, Portugal

Santhosh Sivaraj
Easwari Engineering College
Chennai, Tamil Nadu, India

Kota Sobha
RVR & JC College of Engineering
Guntur, Andhra Pradesh, India

Jothi Arunachalam Solairaju
Savetha School of Engineering
Saveetha Institute of Medical and Technical Sciences
Saveetha University
Chennai, Tamil Nadu India

Shrishail Basaprabhu Sollapur
Jain (Deemed to be University)
Bengaluru, Karnataka, India

Thiagarajan Srinivasan
Easwari Engineering College
Chennai, Tamil Nadu, India

Ganapathy Srinivasan Ramachandran
Vel Tech Multi Tech Dr Rangarajan Dr Sakunthala Engineering College
Chennai, Tamil Nadu, India

Palani Subbiah Kone
Sri Venkateswara College of Engineering & Technology
Thirupachur, Chennai, Tamil Nadu, India

Senthil Murugan Subramanian
Rajalakshmi Engineering College
Chennai, Tamil Nadu, India

Suresh Subramanian
Erode Sengunthar Engineering College
Tamil Nadu, India

Sudhagar Sukkasamy
KIT-Kalaignar Karunanidhi Institute of Technology
Coimbatore, Tamil Nadu, India

Vijay Ananth Suyamburajan
VELS Institute of Science Technology and Advanced Studies
Chennai, Tamil Nadu, India

Vinod Kumar Thangaraj
Vels Institute of Science, Technology & Advanced Studies (VISTAS)
Chennai, Tamil Nadu, India

Sathish Thanikodi
Savetha School of Engineering
Saveetha Institute of Medical and Technical Sciences
Saveetha University
Chennai, Tamil Nadu, India

Duraisamy Velmurugan
Muthayammal Engineering College
Rasipuram, Tamil Nadu, India

Lewhat Michael Weldegebriel
Mai-Nefhi College of Engineering and Technology
Asmara, Eritrea

Hurui Tesfalem Woldu
Mai-Nefhi College of Engineering and Technology
Asmara, Eritrea

1 Processing, Properties and Applications of Metal Matrix Composites

Jarubula Ranga Raya Chowdary, Abdul Sattar Kajavali, and Veeravalli Ramakoteswara Rao

1.1 INTRODUCTION OF METAL MATRIX COMPOSITES

Composites are made by combining different materials which offer enhanced strength, corrosion resistance, flexibility, durability, and improved strength-to-weight ratios. These versatile materials have been used in sectors like medicine, construction, automotive, aerospace, and more. By adding stronger or stiffer phases like fibres or particles, their properties can be further improved [1]. Aluminium matrix composites, used in aircraft and automobiles, are valued for their fuel efficiency and noise reduction. Although composites have existed since ancient times, they became commercially available after World War II, offering unique properties by combining existing materials.

The demand for lightweight, strong, and stiff materials has led to increased research on innovative MMCs, such as aluminium metal matrix composites (AlMMCs) reinforced with materials like SiC or Al_2O_3, Graphite, Graphene, Flyash, etc., which have been successfully used in automotive applications for improved wear resistance. MMCs were a groundbreaking technological advancement in military systems, initially driven by performance enhancement [2]. Despite initial challenges, they gained momentum in the late 1970s due to increased emphasis on cost-effectiveness and risk awareness.

Traditional materials have been employed in industries; however, they are becoming less and less useful because of their limitations in strength, mechanical qualities, and thermal expansion [3–5]. Because of this, there is an increasing need for novel materials to satisfy the changing requirements of cutting-edge technological applications. By fusing the great elasticity and toughness of ceramics with the strength and ductility of metals [6]. MMCs provide a good substitute when compared to the matrix material; the properties of these composites are drastically changed by the addition of additional phases [7].

MMCs have garnered significant research interest due to their unique combination of matrix metal and reinforcing materials, offering material scientists flexibility in design. These versatile materials can be tailored for various applications. Key design factors for MMCs are shown in Figure 1.1.

1.2 KEY FEATURES OF MMC

Improved mechanical properties: MMCs offer enhanced stiffness, strength, hardness, wear resistance, and stability, thanks to the interaction between the matrix and reinforcement.

Lightweight potential: MMCs can be lighter than unreinforced metals while maintaining or exceeding mechanical performance, making them ideal for applications where weight reduction is essential, such as aerospace and transportation.

Tailorable characteristics: MMCs can be customized by adjusting reinforcement type, weight, and allocation to achieve specific properties for different applications.

DOI: 10.1201/9781003499718-1

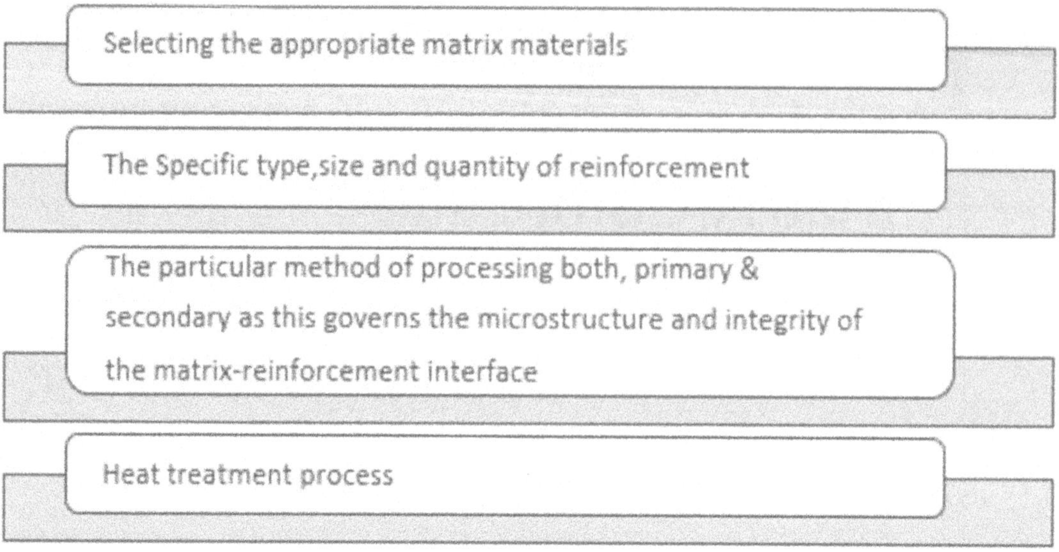

FIGURE 1.1 Factors while designing metal matrix composites.

1.3 EVOLUTION OF METAL MATRIX COMPOSITES

MMCs have seen significant advancements over the years, driven by the requirement for materials with enhanced mechanical, thermal, and other specialized properties. Here is a brief overview of the key stages in the evolution of MMC:

- **Early Developments (1950s–1970s):** In the 1950s and 1960s, investigators reinforced aluminium matrices with silicon carbide and boron, laying the foundation for MMCs. By improving metal's characteristics and functionality, these initiatives paved the way for developments in material engineering and production.
- **Pioneering Research (1970s–1980s):** The 1970s and 1980s saw significant advancements in MMCs, with researchers focusing on optimizing properties and processing techniques. Aluminium became a preferred matrix material due to its lightweight and ease of processing. This period established key principles of MMC design, shaping their development and industrial applications.
- **Advancements in Processing Techniques (1990s):** The 1990s saw major advancements in MMC processing, with methods like powder metallurgy, stir-casting, and infiltration enhancing reinforcement distribution and bonding. Powder metallurgy became a key technique, enabling superior mechanical properties and tailored characteristics, expanding MMC applications across industries, and driving further innovation in material science.
- **Diversification of Matrix and Reinforcement Materials (2000s–2010s):** In the 2000s and 2010s, MMC's research expanded to include matrix materials like magnesium and titanium, aiming to tailor properties such as corrosion resistance, strength, and weight. Advances in reinforcements like ceramic fibres and nanoparticles further enhanced mechanical and thermal performance, meeting diverse industrial demands.
- **Application-Specific Development (2010s–Present):** Recent trends emphasize customizing MMCs for industries like electronics, automotive, and aerospace. Research has optimized their structure and composition, improving performance and reliability to address sector-specific needs and expand applications.

- **Advanced Manufacturing Technologies (Present–Future):** Technologies like additive manufacturing (3D printing) offer precision and flexibility for intricate MMC structures, enhancing performance in applications such as aerospace. These advancements promise expanded use and innovation in MMC design and production.
- **Environmental and Sustainable Considerations:** Rising environmental awareness and demand for sustainable materials are reshaping MMC development. Efforts focus on creating more effective, recyclable MMCs with reduced environmental impact using innovative raw materials and manufacturing techniques.
- Material science's emphasis on sustainability draws attention to MMCs as a way to address environmental issues. MMCs promote a sustainable future through the use of recycled materials, renewable resources, and effective manufacturing techniques. In order to satisfy business demands while balancing performance and environmental responsibility, researchers are working to create MMCs with specific features.

1.4 IMPORTANCE OF METAL MATRIX COMPOSITES

Table 1.1 shows the key properties of commonly used matrix metals. MMCs with copper, steel, and nickel alloys are ideal for tooling, industrial, and electronic packaging applications. In contrast, MMCs with titanium, magnesium, and aluminium alloys are well suited for lightweight structural uses in automotive, aerospace, and defence [7–9]. Various reinforcements used in composite preparation are shown in Figure 1.2a and 1.2b.

TABLE 1.1
Properties of Common Base Materials

Base alloy	Density (g/cc)	Modulus (Gpa)	Yield strength (MPa)
Magnesium	1.738	42–47	70–400
Aluminium	2.7	68–82	30–500
Titanium	4.506	90–120	240–900
Nickel	8.9	150–245	300–1900
Copper	8.96	112–148	300–500

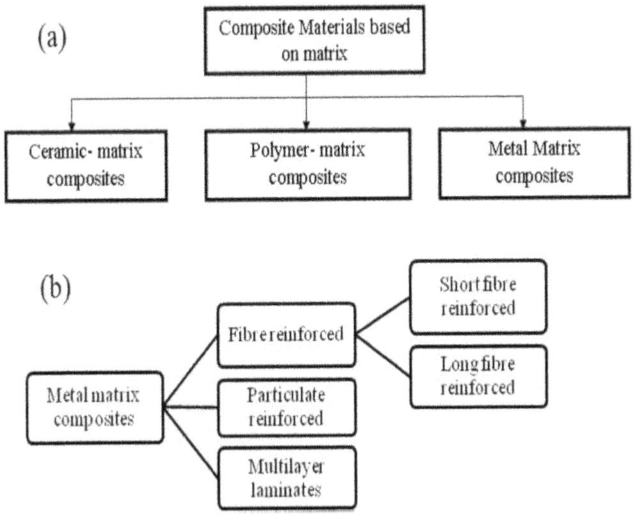

FIGURE 1.2 **a)** Composites based on matrix **b)** Metal matrix composites types.

1.5 CLASSIFICATION OF MMC's

MMC classifications depend on the reinforcement type, which includes materials like silicon carbide (SiC), alumina (Al_2O_3), boron carbide (B_4C), and carbon fibres. Common base metals include cobalt (Co), aluminium (Al), nickel (Ni), titanium (Ti), and copper (Cu). These metals are chosen for their properties and compatibility with reinforcements to create enhanced composites for various applications.

MMCs are classified into two types based on reinforcement:

Continuous Reinforcement MMCs: Fibers are aligned in one direction, improving mechanical properties along the fibre.

Discontinuous Reinforcement MMCs: Reinforcements like fibres, flakes, or particles are scattered, providing isotropic properties with consistent mechanical characteristics in all directions.

1.6 FABRICATION PROCESS

Several fabrication techniques for metal matrix composites (MMCs) offer distinct advantages, as shown in Figure 1.3.

1.6.1 Liquid Phase Processing

In this method, reinforcement is added while the base metal is molten, making it ideal for producing large quantities, especially for low-melting metals like Al and Mg. It is divided into four main categories, with spray processes shown in Figure 1.4. Stir-casting is a popular method for MMC production, where the metal is cast in a semi-solid state. Continuous agitation improves material properties by preventing viscosity rise and promoting spherical particle formation [10–12].

Agitation ensures even distribution of reinforcement throughout the matrix, preventing agglomeration or floating. This technique helps disperse ceramic particles, even though they are not easily

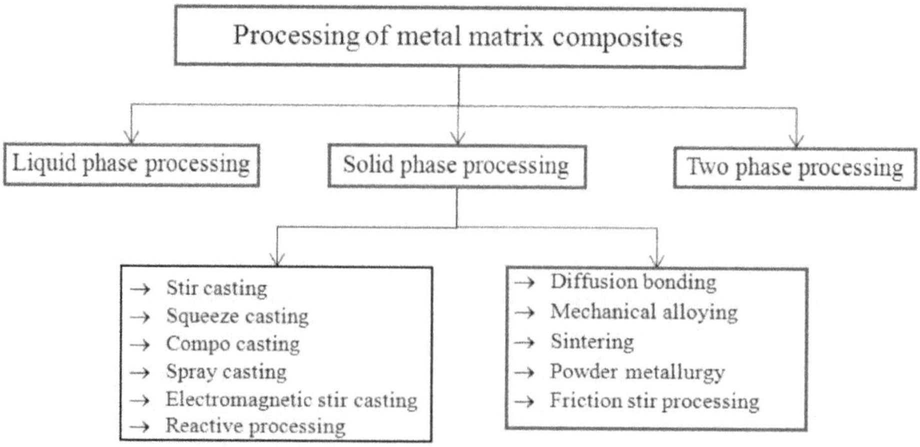

FIGURE 1.3 Various fabrication techniques.

wetted by the base metal. Continuous mixing enhances bonding and wetting—especially with longer agitation—and reduces viscosity, allowing for easier casting. It also minimizes shrinkage cavities by processing the slurry at semi-solid conditions, close to a solid state [13].

1.6.2 Solid State Processing

Solid-state processing process is key for adding high amounts of reinforcement to the base metal, with powder metallurgy being the primary technique, as shown in Figure 1.5.

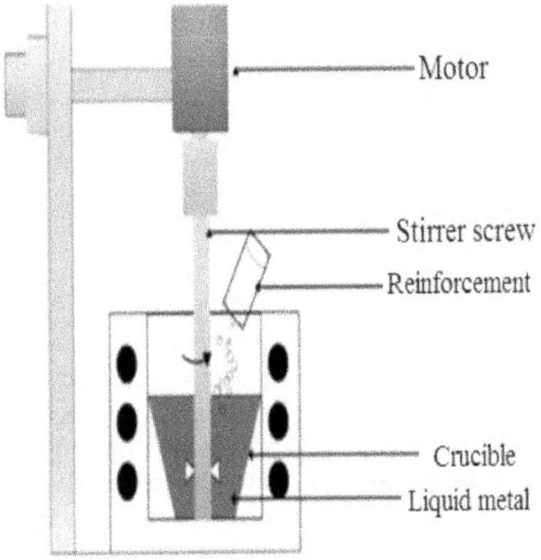

FIGURE 1.4 Liquid phase processing. Example-stir casting.

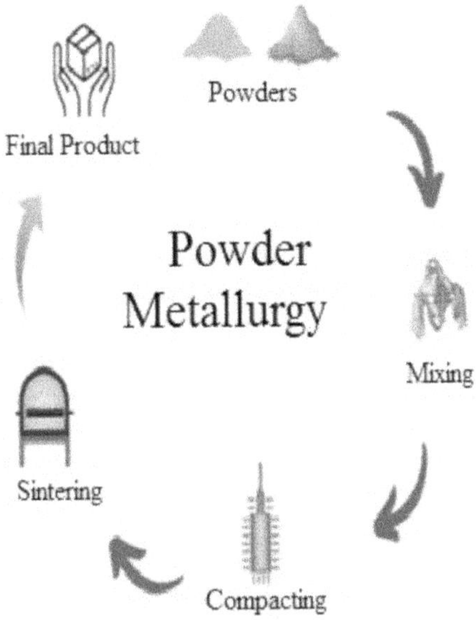

FIGURE 1.5 Example of solid state processing (powder metallurgy process).

Reinforcements are mixed with powdered metals and solidified at room or elevated temperatures, followed by sintering. Secondary processes like rolling, forging, and extrusion can also be used, though these methods can be expensive, limiting their use to specific applications and preventing bulk production. Additionally, high tooling costs further restrict the use of powder metallurgy.

Gaseous-state processing, like chemical vapor deposition and physical vapor deposition, controls composite composition by depositing vapour onto substrates. Liquid-state fabrication is preferred due to lower cost, flexibility, and wider shapes, but issues like poor parameter control persist [14, 15].

Metal matrix composites (MMCs) exhibit distinctive mechanical and deformation behaviour due to their combination of metal matrices and reinforcements, enhancing strength and stiffness and causing elastic deformation under stress. Reinforcements, like fibres or particles, enhance plastic deformation resistance in metal materials. The distribution, size, and weight of reinforcements significantly influence deformation behaviour. Reinforcements affect crack initiation and propagation, enhancing fracture toughness. Thermal expansion mismatch can induce residual stresses, impacting composite deformation and mechanical performance. Understanding these interactions is crucial for high-performance applications.

1.7 INTRODUCTION OF AL6082 MMC

Aluminium 6082 is a reliable alloy used in aerospace, military, and marine industries due to its tensile strength and corrosion resistance [16]. However, its effectiveness in naval conditions is limited due to reduced hardness and wear resistance. Reinforcing agents like $ZrSiO_4$ and TiC can improve their performance, opening up opportunities for improved applications in maritime environments. Despite its drawbacks, aluminium 6082 is essential for high-stress applications like cranes, bridges, and transportation parts due to its superior wear and decay conflict and uniform grain structure [16].

Aluminium 6082 is a crucial alloy for high-stress applications like cranes, bridges, and trusses due to its toughness and durability against corrosion. Its superior wear and decay resistance and uniform grain structure are due to its high silicon content and low copper content. Al6082's composition includes various reinforcements like silicon carbide, boron carbide, graphite, fly ash, and ceramic particulates [17]. The composition of Al6082 is given in Table 1.2

The different advanced methodologies in the fabrication of Al6082 are selective laser melting (SLM), friction stir welding (FSW), cryogenic machining, high-pressure torsion, equal channel angular pressing (ECAP), and Laser Beam Welding (LBW).

1.7.1 OBJECTIVES

The major objective of this chapter is to provide a detailed and comprehensive analysis of metal matrix composites (MMCs) made of Al6082, covering their applications, mechanical properties, and production techniques.

TABLE 1.2
Composition of Al6082

Alloying elements	Mg	Si	Mn	Fe	Cr	Cu	Zn	Ti	Al
Content (wt%)	0.6–1.2	0.7–1.3	0.4–1.0	0–0.5	0–0.25	0–0.1	0–0.2	0–0.1	Remaining

To Examine Fabrication Procedures: It explores the benefits, drawbacks, and effects of fabrication methods like casting, powder metallurgy, and solid-state processing.

To examine Mechanical Properties: The review examines the mechanical properties of Al6082 MMCs, including tensile strength, hardness, toughness, and fatigue behaviour, considering factors such as volume fraction, processing conditions, and reinforcement type.

Investigate Microstructural Features: It investigates the microstructure of Al6082 MMCs, including reinforcement distribution, grain structure, and interfacial interactions between the reinforcement and matrix.

1.7.2 Comparison of Aluminium MMCs with Other MMCs

Aluminium metal matrix composites (MMCs) are lighter, corrosion-resistant, and have a better strength-to-weight ratio compared to other materials like magnesium and copper. They are easier to process, cost-effective, and suitable for various industries. Al MMCs also offer superior fatigue resistance, wear resistance, and lower environmental impact compared to other materials. They can be engineered for improved stability and performance over time, making them a preferred choice in various applications. Al MMCs are widely used in automotive, aerospace, and electronics, with established manufacturing processes and a mature market.

1.8 LITERATURE REVIEW ON DIFFERENT PROPERTIES OF MMC

Thimmarayan et al. studied the impact of particle volume on SiC composites, finding improved tensile, yield strength, and elasticity but reduced ductility [18]. Sabbagh et al. study the hot rolling process of aluminium metal matrix composites, focusing on stir-cast composites with AA6061 and AA 6082 alloys and fine SiCp particulates [19]. They use optimal casting techniques and material pre-treatments to ensure successful particle insertion and distribution. The research shows that improved casting quality correlates with improved rolled product quality, with subsequent hot rolling enhancing mechanical properties.

Sharma et al. conducted a study on Al6082 and graphite powder-reinforced ceramics and found that the addition of particulate-reinforced ceramic reduced wear rate, influenced by factors like load, sliding velocity, and sliding distance, using stir casting techniques [20]. Sharma et al. study explores the impact of graphite particles on Al6082 MMC composites' dry sliding wear characteristics [21]. They found that adding graphite reduced micro-hardness and macro-hardness by 11.11% and 10.44%, respectively. The study highlights the importance of understanding graphite's influence on wear behaviour for practical applications. Kaushik et al. conducted a study on the wear behaviour of stir cast Al 6082–SiC–Gr MMC, capered with Al 6082–SiC composites and base alloy [22]. They found that AlMMCs outperformed alloy and Al–SiC composites, with improvements of 16.4% and 27% in cast and T6 heat-treated grit sizes.

Sharma et al. developed Hybrid AlMMCs by reinforcing AA6082 alloy with varying weight percentages of ball-milled ceramic particulates like Si3N4 and graphite [23]. The composites were examined for microstructures and mechanical properties, showing enhanced hardness and ultimate tensile strength but a decrease in % elongation with higher weight percentages in the aluminium matrix. Puneeth et al. evaluate the fracture hardiness of Al6082-based Al_2O_3/Al_2SiO_5 composites using the stir casting technique [24]. They analyze Al_2O_3 wt%, stirring speed, and stir time's impact on mechanical properties, finding Al_2O_3 wt% as the most influential factor. Kumar et al. modified Al 6082 composites with coconut shell ash (CSA) and zirconium oxide (ZrO_2) particles [25]. The stir-casting process resulted in a lubricating effect, decreasing hardness with 10% CSA. However, the combination of CSA and ZrO_2 reduced impact strength. Sekar et al.'s study on A6082 MMC prepared through a hybrid stir and squeeze casting process found a significant enhancement in Charpy impact strength, ultimate tensile strength, and hardness [26]. Kumar et al. conducted a study on the combination of Al 6082 with 5% B_4C, using fly ash as reinforcement [27]. They fabricated specimens through stir casting and conducted wear tests under various conditions. The Al/5%B_4C sample showed the lowest wear value. Jayachandra et al. studied the fabrication of A6082 hybrid composites using stir and squeeze casting, adjusting ZrO_2 content while maintaining SiC [28]. They found that incorporating 1 wt% ZrO_2 increased Charpy Impact strength by 60%, and ultimate tensile strength by 33% with 0.5 wt% ZrO_2. However, composites with 1 and 1.5 wt% ZrO_2 showed reduced UTS and brittle breakdown. Welded composites with 1 wt% SiC and 1 wt% ZrO_2 showed a 17% hardness increase.

Samal et al. fabricated composites using AA 6082 alloy and red mud particles as reinforcement, resulting in reduced wear and increased sliding distance in their stir casting tests [29]. Santecchia et al. studied AA 6082(T6) coated with durable anodized alumina using stir casting [30]. The wear test showed a maximum friction coefficient of 0.93 for the GHA/Si3N4 tribosystem. The GHA50-based Si3N4 tribo-couple maintained the highest value of about 9 mm^3/Nm compared to the bearing steel ball, indicating its superior wear performance. Shalini et al. conducted a stir-casting study using Al 6082 as the base alloy and Y_2O_3 + Gr with different reinforcements [31]. The tests were conducted at a sliding speed of 1 m/s and various loads ranging from 20 to 60 N. The samples with 4% graphite and 6% Y_2O_3 had the least wear, measuring 0.002 mm^3/m at 60 N 50% less than the AA 6082's 0.004 mm^3/m at 60 N loads. Rathi et al. investigated AA 6082 and MoS_2 composites and showed that increasing the weight fraction of MoS_2 improved the wear property, with composites fabricated via stir-casting [32]. Singh et al. conducted wear tests on al 6082 and heat-treated 6082 (T6) under various conditions [33]. They optimized these variables using the Taguchi process. Results showed an increase in wear performance up to 800 m sliding distance, with load being the primary factor influencing wear loss in both composites. Zhu et al. conducted a study on SiCnp/Al6082 composites using squeeze and stir-casting methods [34]. The results showed a uniform distribution of SiCnp in Al6082, leading to a 30.55% grain refinement compared to the alloy without SiCnp. The composite's density was also enhanced post-SC process.

Tan et al. explore the development of hybrid composite brake discs, containing an A357/SiC aluminium metal matrix composite top layer and an Al6082 Al alloy bottom base [35]. The discs show remarkable wear and friction performance during full-scale dynamometer braking experiments in city rail vehicle operations. Key factors include even distribution of SiC particles and secondary particles, minimization of crack initiation, and improved mechanical properties. The lightweight discs are high technology readiness, promising alternatives to heavy iron and steel brake discs.

Manu Samet et al. studied the wear properties of an AA6082 composite reinforced with hybrid ceramics, using varying corn cob ash weight percentages [36]. They found that sliding velocity, sliding distance, and applied load significantly affect the wear rate. Kumar et al. successfully crafted Al6082 bimodal hybrid composites using a modified stir-squeeze cast method [37]. Mechanical alloying ensured uniform Al particle coating, enhancing reinforcement wettability. Morphological evaluation revealed homogeneous SiC particle dispersion, improving compressive strength and ultimate strength. However, impact strength decreased due to reduced ductility with reinforcing phases. Maximum mechanical property improvement was observed with 3 wt% SiC and 0.5 wt% TiO_2. Baghel et al. used a stir-casting technique to create advanced Al6082 composites with varying proportions of multi-walled carbon nanotubes (MWCNTs) [38]. The composites underwent a heat treatment cycle to enhance mechanical strength. The 0.9 wt% MWCNTs/Al6082 nanocomposites showed significant improvements in ultimate tensile strength, yield strength, and hardness.

Baghel et al. optimized the stir-casting process for Al6082 composites reinforced with multi-walled carbon nanotubes using a multi-objective optimization framework [39]. The study examined output parameters like UTS, hardness, compressive strength, and MWCNT distribution uniformity, using neutrosophic sets and TOPSIS methods. The optimal combination was identified using neutrosophic grey relational analysis. Sivalingam et al. studied the machinability and energy consumption of Al6082-based hybrid composites, fabricated using stir casting and incorporating boron carbide and molybdenum reinforcements [40].

Chakravarthy et al.'s conducted study on AA6082 strengthening using RHA and B_4C reinforcements revealed that Al-100%, RHA-5%, and B_4C-5% compositions yielded superior mechanical properties [41]. Mishra et al. improved aluminium 6082 yield, tensile strength, and hardness using 8%, 9%, and 10% SiC reinforcement, with maximum tensile strength at 9% SiC and decreased hardness with increased reinforcement [42]. Somayaji et al. studied the impact of adding graphite particles to Al6082, evaluating mechanical and tribological qualities through wear, tensile, hardness, and microstructural tests [43]. Results showed that graphite significantly increases tensile strength and wear resistance, while decreasing hardness.

1.8.1 Effect of Mechanical Properties

The graph in Figure 1.6 shows the hardness and tensile strength of Al6082 with varying weight percentages of different reinforcements, leading to the following conclusions.

The data shows that different reinforcement types affect the hardness of composite materials, with each type having varying baseline hardness values. The weight percentage of reinforcement generally enhances the composite material's hardness. Engineers can compare the performance of different reinforcement types at the same weight percentage to determine the optimal combination for specific applications. The data can also be used to optimize the composition of MMCs for hardness-critical applications, such as high wear resistance or improved surface hardness.

1.8.2 Applications of AL6082

Aluminium 6082 is a corrosion-resistant alloy used in aerospace, marine, automobile, construction, electronics, and general engineering. Its lightweight nature makes it ideal for structural components, while its low hardness makes it suitable for harsh marine conditions. However, it is limited by brittleness and corrosion issues, while copper MMCs are used for high-conductivity applications but face limitations in weight and brittleness. The comparison is given in Table 1.3

The choice of MMC material depends on the application: Aluminium MMCs offer balanced performance, Magnesium MMCs excel in weight reduction, and Copper MMCs are ideal for high electrical conductivity or wear resistance.

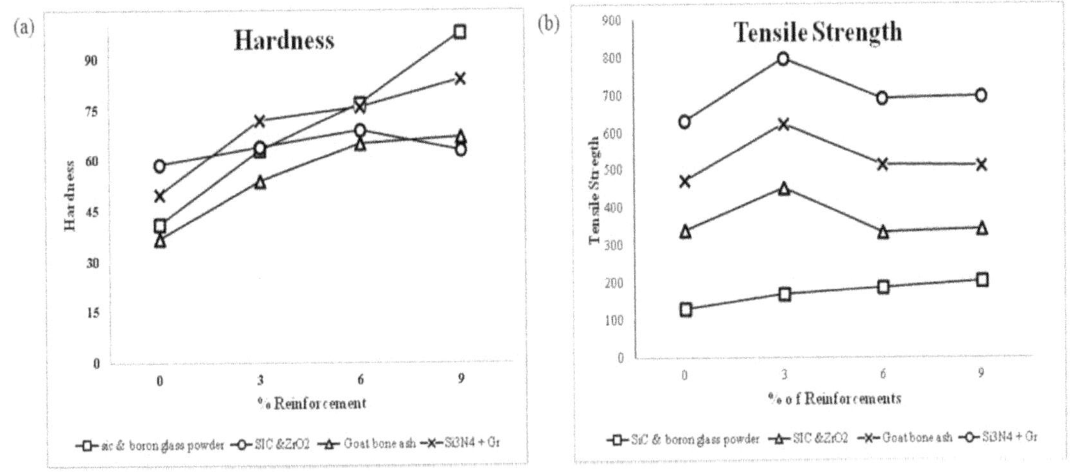

FIGURE 1.6 a) Hardness and b) Tensile strength of various reinforcements with Al 6082 as base metal.

TABLE 1.3
Comparison of Al MMC with Magnesium and Copper MMC

Property	Aluminium MMCs	Magnesium MMCs	Copper MMCs
Density	Low	Very Low	High
Strength	High	High (Strength-to-weight ratio highest)	Very High
Ductility	Moderate	Low	Low
Corrosion Resistance	Good (alloy dependent)	Poor	Moderate
Fabrication	Relatively easy	More challenging	Similar to Aluminium MMCs
Applications	Pistons, Engines, Aerospace	Niche aerospace, transportation	Electrical contacts, heat sinks

1.8.3 MMCs under Different Environments

Metal matrix composites (MMCs) exhibit varying behaviours under various environmental conditions, with tribological performance influenced by reinforcement material, operating conditions, and environmental factors. Erosive wear behaviour and wear resistance are affected by a slurry environment, with weight loss increasing in corrosive environments. Understanding MMCs' behaviour is crucial for optimal performance in practical applications.

1.9 CHALLENGES AND FUTURE DIRECTIONS

Aluminium 6082 Metal Matrix Composites (MMCs) encounter several challenges that can hinder their widespread use and performance in industrial applications. One major issue is the high manufacturing costs associated with the production of MMCs. These composites often require specialized equipment and processes, such as stir casting or powder metallurgy, which can be expensive. Additionally, the use of advanced reinforcement materials like carbon nanotubes, graphene, or high-performance ceramics can further increase production costs due to their limited availability and high prices.

Another significant challenge is achieving homogeneous reinforcement dispersion throughout the matrix. For MMCs to perform optimally, the reinforcing phases (such as ceramic particles, fibres, or whiskers) must be evenly distributed within the metal matrix. Inconsistent distribution can lead to localized areas of weakness, reducing the overall effectiveness of the composite. This is particularly critical in load-bearing applications where the performance of the material depends on how well the reinforcement can carry the applied stress.

Interfacial bonding between the matrix and the reinforcement is also a crucial factor in enhancing the mechanical properties of MMCs. If the bonding is weak or imperfect, the load transfer between the matrix and reinforcement will be inefficient, resulting in reduced strength and stiffness. Strong interfacial bonding is necessary to ensure that the reinforcement effectively contributes to the composite's overall mechanical performance. Poor bonding can also lead to issues such as delamination or interfacial failure under stress, which can significantly compromise the material's integrity.

Furthermore, processing-induced flaws, such as porosity or voids, can develop during the fabrication of MMCs. These defects can occur due to air entrapment, incomplete consolidation of the materials, or improper cooling. Porosity and voids reduce the material's density and strength, affecting both its mechanical and thermal properties. Addressing these flaws is essential for ensuring the reliability and consistency of the material, especially in applications where structural integrity is critical.

To improve the performance and broaden the application of Al6082 MMCs, ongoing research and development are focused on optimizing manufacturing processes, improving material quality, and developing cost-effective solutions for reinforcement dispersion and bonding.

1.9.1 Future Directions

Advanced Manufacturing Techniques: Advanced manufacturing techniques are becoming more and more necessary to produce Al6082 MMCs at a reasonable price. While investigating novel approaches like automated systems and hybrid processes, researchers concentrate on refining established techniques like liquid metal infiltration, casting, and powder metallurgy. These developments increase commercial viability for a greater variety of applications, decrease waste, and boost production efficiency.

Exploration of Novel Reinforcement Materials: To improve Al6082 MMCs, researchers are looking for additional reinforcement materials other than silicon carbide (SiC) and alumina. Investigating nanomaterials that can enhance mechanical characteristics, thermal stability, and electrical conductivity—such as graphene, carbon nanotubes (CNTs), and nanoceramics—will open up new avenues for high-performance applications.

The Novel applications of Al6082 metal matrix composites (MMCs) include enhancing hardness, wear resistance, and thermal properties for advanced engineering applications. These composites, reinforced with materials like waste glass powder, show improved mechanical, physical, and thermal properties, making them suitable for structural applications [44]. Al6082 alloy, when reinforced with hard ceramic particles like Si_3N_4, exhibits increased strength and stiffness, making it ideal for strong structural applications [45]. Additionally, hybrid Al MMCs, incorporating Zirconium di Oxide (ZrO_2) and eggshell particles as reinforcements, demonstrate enhanced tensile and compressive strength, crucial for automotive applications. The versatility and potential of Al6082 MMCs in meeting stringent specifications for various industries highlight their promising future in diverse industrial and commercial applications.

Understanding Microstructural Evolution: Research on microstructural evolution in Al6082 MMCs is crucial for optimizing properties and performance. Advanced characterization techniques like electron microscopy, XRD, and spectroscopy are used to study this evolution at various stages.

Integration of Nanotechnology: By improving Al6082 MMCs' strength, hardness, and thermal conductivity, nanotechnology can increase their performance and range of applications in electronics, automotive, and aerospace.

Additive Manufacturing Techniques: Additive manufacturing techniques like 3D printing and laser melting are being used to fabricate complex Al6082 MMC components, enabling rapid prototyping, on-demand production, and customization for specific applications.

1.10 SUMMARY

In summary, MMCs based on aluminium 6082 offer better mechanical strength and tribological performance than traditional alloys, marking a significant breakthrough in materials science. They are perfect for sectors including engineering, automotive, and aerospace because of their adaptability. By demonstrating their improved qualities through reinforcing phases, this chapter emphasizes their significance in engineering applications. High performance is provided by these composites, guaranteeing dependability and longevity in vital industries. Furthermore, they provide sustainable and energy-efficient answers to contemporary problems, with continuous research aimed at refining microstructure and processing methods to promote novel innovations.

REFERENCES

[1] Miracle, D.B. and Donaldson, S.L., 2001. *Composites, ASM Handbook.* ASM International, vol. 21, p. 1201.

[2] Trzepieciński, T., Najm, S.M., Sbayti, M., Belhadjsalah, H., Szpunar, M. and Lemu, H.G., 2021. New advances and future possibilities in forming technology of hybrid metal–polymer composites used in aerospace applications. *Journal of Composites Science*, *5*(8), p. 217.

[3] Soboyejo, W.O. and Srivatsan, T.S., 2006. *Advanced Structural Materials: Properties, Design Optimization, and Applications.* CRC Press, p. 528.

[4] Armbrister, C.E., Okoli, O.I. and Shanbhag, S., 2015. Micromechanics predictions for two-phased nanocomposites and three-phased multiscale composites: A review. *Journal of Reinforced Plastics and Composites*, *34*(8), pp. 605–623.

[5] Hooker, J.A. and Doorbar, P.J., 2000. Metal matrix composites for aeroengines. *Materials Science and Technology*, *16*(7–8), pp. 725–731.

[6] Trojanová, Z., Drozd, Z., Kúdela, S., Száraz, Z. and Lukáč, P., 2007. Strengthening in Mg–Li matrix composites. *Composites Science and Technology*, *67*(9), pp. 1965–1973.

[7] Adebisi, A.A., Maleque, M.A. and Rahman, M.M., 2011. Metal matrix composite brake rotor: Historical development and product life cycle analysis. *International Journal of Automotive and Mechanical Engineering*, *4*, pp. 471–480.

[8] Jayalakshmi, S. and Gupta, M., 2015. *Metallic Amorphous Alloy Reinforcements in Light Metal Matrices.* Cham: Springer International Publishing.

[9] Jayalakshmi, S., Singh, R.A. and Gupta, M., 2018. Metallic glasses as potential reinforcements in Al and Mg matrices: A review. *Technologies*, *6*(2), p. 40.

[10] Kaczmar, J.W., Pietrzak, K. and Włosiński, W., 2000. The production and application of metal matrix composite materials. *Journal of Materials Processing Technology*, *106*(1–3), pp. 58–67.

[11] Miracle, D.B., 2005. Metal matrix composites—from science to technological significance. *Composites Science and Technology*, *65*(15–16), pp. 2526–2540.

[12] Yadav, P., Ranjan, A., Kumar, H., Mishra, A. and Yoon, J., 2021. A contemporary review of aluminium MMC developed through stir-casting route. *Materials*, *14*(21), p. 6386.

[13] Flemings, M.C., 1981. Continuous production of strip by rheocasting. *Final Report,* August 16, 1978–September 30, 1981 (No. DOE/CS/40310-T1). Cambridge: Massachusetts Institute of Technology, Materials Processing Center.

[14] Miracle, D.B., 2005. Metal matrix composites—from science to technological significance. *Composites Science and Technology*, *65*(15–16), pp. 2526–2540.

[15] Harrigan Jr, W.C., 1998. Commercial processing of metal matrix composites. *Materials Science and Engineering: A*, *244*(1), pp. 75–79.

[16] Pasupulla, A.P., Amornphimoltham, P., Halabo, T.T., Abebe, H. and Ramakrishna, M.M., 2022, July. Microstructure and characterization of Al 6082 alloy with addition of $ZrSiO_4$ and TiC reinforcements. *AIP Conference Proceedings*, *2473*, p. 020002.

[17] Sharma, P., Khanduja, D. and Sharma, S., 2016. Dry sliding wear investigation of Al6082/Gr metal matrix composites by response surface methodology. *Journal of Materials Research and Technology*, *5*(1), pp. 29–36.

[18] Thimmarayan, R. and Thanigaiyarasu, G., 2010. Effect of particle size, forging and ageing on the mechanical fatigue characteristics of Al6082/SiCp metal matrix composites. *The International Journal of Advanced Manufacturing Technology*, *48*, pp. 625–632.

[19] El-Sabbagh, A., Soliman, M., Taha, M. and Palkowski, H., 2012. Hot rolling behaviour of stir-cast Al 6061 and Al 6082 alloys—SiC fine particulates reinforced composites. *Journal of Materials Processing Technology*, *212*(2), pp. 497–508.

[20] Sharma, P., Khanduja, D. and Sharma, S., 2015. Production of hybrid composite by a novel process and its physical comparison with single reinforced composites. *Materials Today: Proceedings*, *2*(4–5), pp. 2698–2707.

[21] Sharma, P., Khanduja, D. and Sharma, S., 2016. Dry sliding wear investigation of Al6082/Gr metal matrix composites by response surface methodology. *Journal of Materials Research and Technology*, *5*(1), pp. 29–36.

[22] Kaushik, N.C. and Rao, R.N., 2016. Effect of grit size on two body abrasive wear of Al 6082 hybrid composites produced by stir casting method. *Tribology International*, *102*, pp. 52–60.

[23] Sharma, P., Sharma, S. and Khanduja, D., 2017. Production and characterization of AA6082-(Si_3N_4+Gr) stir cast hybrid composites. *Particulate Science and Technology*, *35*(2), pp. 158–165.

[24] Puneeth, N., Satheesh, J. and Naveen, G.J., 2019. Processing and evaluation of Al6082 based hybrid matrix composites for fracture toughness using taguchi method. *Procedia Structural Integrity*, *14*, pp. 53–59.

[25] Kumar, K.R., Pridhar, T. and Balaji, V.S., 2018. Mechanical properties and characterization of zirconium oxide (ZrO_2) and coconut shell ash (CSA) reinforced aluminium (Al 6082) matrix hybrid composite. *Journal of Alloys and Compounds*, *765*, pp. 171–179.

[26] Sekar, K., Jayachandra, G. and Aravindan, S., 2018. Mechanical and welding properties of A6082-SiC-ZrO_2 hybrid composite fabricated by stir and squeeze casting. *Materials Today: Proceedings*, *5*(9), pp. 20268–20277.

[27] Kumar, R.A., Sait, A.N. and Subramanian, K., 2018. Evaluation of wear behaviour for Al/B_4C/fly ash composites by stir casting process. *Journal of Scientific & Industrial Research*, *77*, pp. 282–287.

[28] Jayachandra, G., Venkata Sandeep, J. and Sekar, K., 2019. Mechanical and welding properties of A6082–SiC–ZrO_2 hybrid composite fabricated by stir and squeeze casting. In *Recent Advances in Material Sciences. Lecture Notes on Multidisciplinary Industrial Engineering*, Singapore: Springer, pp. 77–92.

[29] Samal, P., Mandava, R.K. and Vundavilli, P.R., 2020. Dry sliding wear behavior of Al 6082 metal matrix composites reinforced with red mud particles. *SN Applied Sciences*, *2*(2), p. 313.

[30] Santecchia, E., Cabibbo, M., Hamouda, A.M.S., Musharavati, F., Popelka, A. and Spigarelli, S., 2020. Dry sliding tribological properties of a hard anodized AA6082 aluminum alloy. *Metals*, *10*(2), p. 207.

[31] Kumar, T.S., Shalini, S. and Kumar, K.K., 2020. Effect of friction stir processing and hybrid reinforcement on wear behaviour of AA6082 alloy composite. *Materials Research Express*, *7*(2), p. 026507.

[32] Singh, H., Singh, K., Vardhan, S. and Mohan, S., 2021. Study on the wear performance of AA 6061 and AA 6082 based metal matrix composites. *Materials Today: Proceedings*, *43*, pp. 660–664.

[33] Zhu, J., Jiang, W., Li, G., Guan, F., Yu, Y. and Fan, Z., 2020. Microstructure and mechanical properties of SiCnp/Al6082 aluminum matrix composites prepared by squeeze casting combined with stir casting. *Journal of Materials Processing Technology*, *283*, p. 116699.

[34] Tan, D., Xia, S., Yob, A., Yang, K., Yan, S., Givord, M. and Liang, D., 2022. Evaluation of the wear resistance of aluminium-based hybrid composite brake discs under relevant city rail environments. *Materials & Design*, *215*, p. 110504.

[35] Manu, S., Radhika, N., Sidvilash, V. and Mohanraj, T., 2022. Investigation on the mechanical and wear behaviour of Al-6082-BN-B_4C-corn cob ash hybrid composites. *Tribology in Industry*, *44*(2), p. 294.

[36] Kumar, S., Dixit, G. and Dixit, S., 2023. Investigating microstructural and mechanical behavior of stir-squeeze cast TiO_2–SiC/Al6082 bimodal hybrid composites. *Journal of Materials Research and Technology*, *26*, pp. 150–163.

[37] Baghel, M. and Krishna, C.M., 2023. Microstructural evolution and mechanical properties of multiwalled carbon nanotubes/Al6082 nanocomposites through optimized T6 heat treatment process. *Proceedings of the Institution of Mechanical Engineers, Part E: Journal of Process Mechanical Engineering*, *237*(4), pp. 1145–1157.

[38] Baghel, M. and Krishna, C.M., 2023. Multi-objective optimization of mechanical and microstructural characteristics in a stir casting process of MWCNTs/Al6082 composites using neutrosophic TOPSIS and GRA. *Journal of the Chinese Institute of Engineers*, *46*(4), pp. 345–354.

[39] Sivalingam, V., Zhou, Q., Ross, N.S., Tiwari, S., Selvam, B., Sun, J. and Gupta, M.K., 2023. Identification of tool wear and surface morphology measurements in sustainable milling of Al 6082 hybrid metal matrix composite. *Journal of Materials Research and Technology*, *27*, pp. 7570–7581.

[40] Chakravarthy, M.P. and Rao, D.S., 2022. Evaluation of mechanical properties of aluminium alloy (AA 6082) reinforced with rice husk ash (RHA) and boron carbide (B_4C) hybrid metal matrix composites using stir casting method. *Materials Today: Proceedings*, *66*, pp. 580–586.

[41] Mishra, D. and Nanda, A.K., 2023. Experimental investigation on mechanical properties of stir casted aluminum SiC metal matrix composites. *Materials Today: Proceedings*, *74*, pp. 1023–1027.

[42] Puneeth, N., Satheesh, J., Naveen, G.J., Mohan, N., Bhandari, L. and Kiran, G., 2020. Development and investigation on elastic plastic fracture analysis of Al6082 reinforced alumina/aluminium silicate MMCs using taguchi approach. *Materials Today: Proceedings*, *27*, pp. 2243–2248.

[43] Somayaji, A., Nagaral, M., Anjinappa, C., Alkahtani, M.Q., Billady, R.K., Kumar, N., Auradi, V., Islam, S., Chowdary, J.R.R., Razak, A. and Khan, M.A., 2023. Influence of graphite particles on the mechanical and wear characterization of Al6082 alloy composites. *ACS Omega*, *8*(30), pp. 26828–26836.

[44] Wubieneh, T.A. and Tegegne, S.T., 2022. Fabrication and characterization of aluminum (Al-6061) matrix composite reinforced with waste glass for engineering applications. *Journal of Nanomaterials, 2022*, pp. 1–8, 409750.
[45] Nturanabo, F., Masu, L. and Kirabira, J.B., 2019. Novel applications of aluminium metal matrix composites. In *Aluminium Alloys and Composites*, IntechOpen, pp. 1–24.

2 Development of Metal Matrix Composites Through Liquid State Technique

Senthil Murugan Subramanian and Subhaschandra Kattimani

2.1 INTRODUCTION

The SCP is a prominent liquid-state technique vital for fabricating a wide range of MMCs tailored to diverse industrial applications [1]. As a liquid metallurgy or mechanical alloying process, SCP involves the mechanical stirring of reinforcing particulates into a molten metal matrix [2]. This method excels in achieving optimal bonding and uniform mixing between the matrix and ceramic reinforcements while promoting essential chemical interactions between them. In industrial settings, SCP is frequently scaled up to accommodate large production volumes. Automation and advanced equipment are often incorporated to enhance efficiency and ensure consistency in the manufacturing process. MMCs produced via SCP are widely utilized in sectors like automotive and aerospace, owing to their superior mechanical and thermal properties. Notably, hybrid matrix composites—those with multiple reinforcements—often demonstrate improved characteristics compared to composites with single reinforcements [3]. Non-ferrous alloys are commonly used in SCP due to their compatibility with reinforcements, meeting various industrial demands. Composites, being multi-phase materials, can be customized by carefully controlling process parameters and managing the addition of reinforcements. However, despite its advantages, SCP has certain limitations. Challenges such as achieving a homogeneous mixture and improving wettability during MMC processing remain significant hurdles [4]. Figure 2.1 illustrates the typical procedure for processing castings, including metals, alloys, and composites.

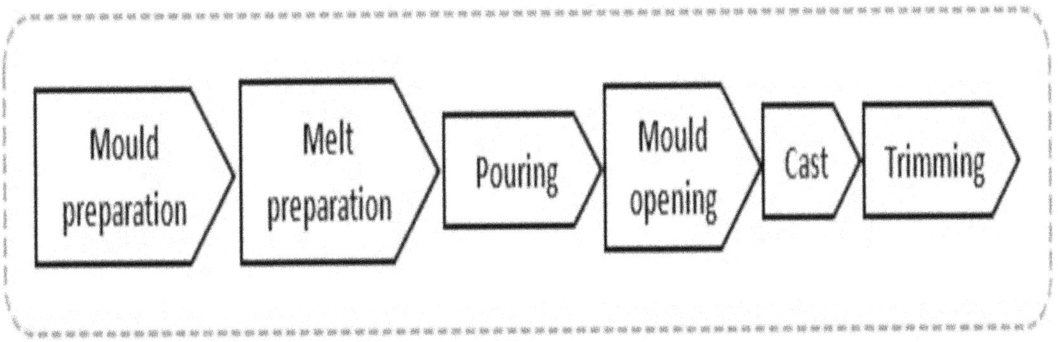

FIGURE 2.1 Typical casting preparation procedure.

DOI: 10.1201/9781003499718-2

2.2 SC ARRANGEMENT AND PROCEDURE

The SCP is a versatile and sustainable manufacturing method (Figure 2.2a) for producing MMCs with tailored properties to meet diverse industrial demands. Figure 2.2b illustrates the production steps of MMCs via SCP. The selection of materials, process parameters, and post-processing techniques is guided by the specific requirements of the final product. SCP typically utilizes either an induction furnace or a crucible furnace to melt the base metal. The choice of furnace depends on the type and quantity of the metal being processed. The crucible, which holds the molten metal, must be made from materials capable of withstanding high temperatures and chemical reactions, such as graphite or ceramic. A key element of SCP is the stirring mechanism, which ensures uniform dispersion of reinforcing particles within the molten metal. This is typically achieved using a mechanical stirrer or impeller with different designs. Critical parameters such as stirring speed and duration must be carefully controlled to achieve a homogenous mixture and optimize the composite's properties.

Reinforcing particles, such as ceramic or metallic particles of the desired size and shape, are added to the molten metal during the stirring process. The process is conducted in an inert gas atmosphere to prevent oxidation, especially with reactive metals, typically using argon or nitrogen. A temperature control system regulates the molten metal temperature, which is managed through the furnace control system. Finally, a mould or die shapes and solidifies the molten composite into the desired form. Adherence to safety protocols and appropriate personal perspective equipment (PPE) are essential when working with molten metals and high-temperature equipment. The process begins with weighing and preparing the base metal and reinforcing particles according to the required composition. The base metal is melted in a crucible or induction furnace until it reaches a molten state, enabling effective mixing. Pre-prepared reinforcing particles are then introduced into the molten metal, followed by simultaneous stirring at a controlled speed and duration to ensure uniform distribution. The molten composite is then poured into the prepared mould or die, allowed to cool, and solidified into the final shape. Controlled cooling ensures the desired microstructure, and additional post-processing, such as heat treatment, machining, or surface finishing, may be performed to achieve specific properties and dimensions.

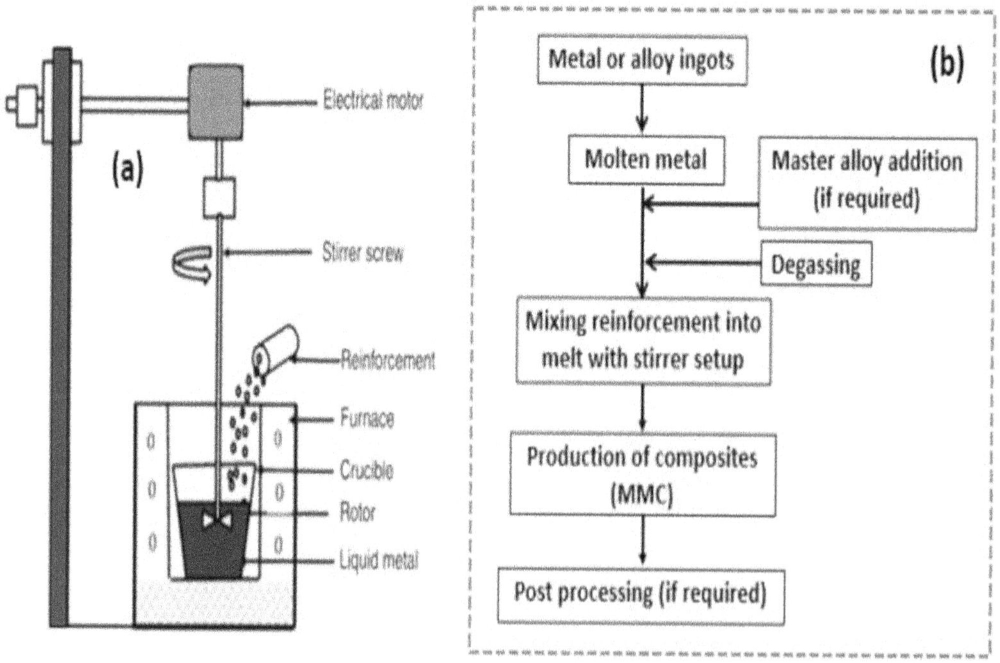

FIGURE 2.2 a) Stir casting arrangement, b) MMC processing steps.

2.3 MMCs

The MMCs are materials where a lightweight metal matrix, such as aluminium, titanium, or magnesium, is reinforced with particles, fibres, or whiskers to enhance properties (Figure 2.3). During the SCP, a mechanical stirrer or impeller agitates the molten metal to achieve a uniform distribution of reinforcing particles and prevent settling [5]. Stirring is crucial for producing a homogeneous mixture. The choice of metal alloy as the matrix material depends on the desired properties of the final composite, while reinforcements typically include ceramics, carbides, borides, or other high-strength materials [6]. Figure 2.4 illustrates various processing routes for MMCs, while Figure 2.5

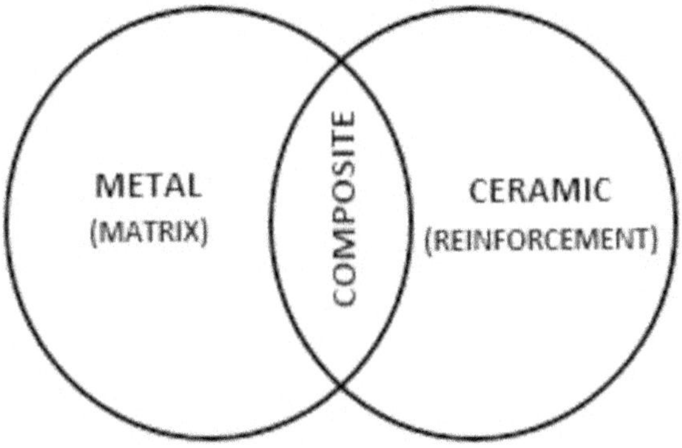

FIGURE 2.3 MMCs concept.

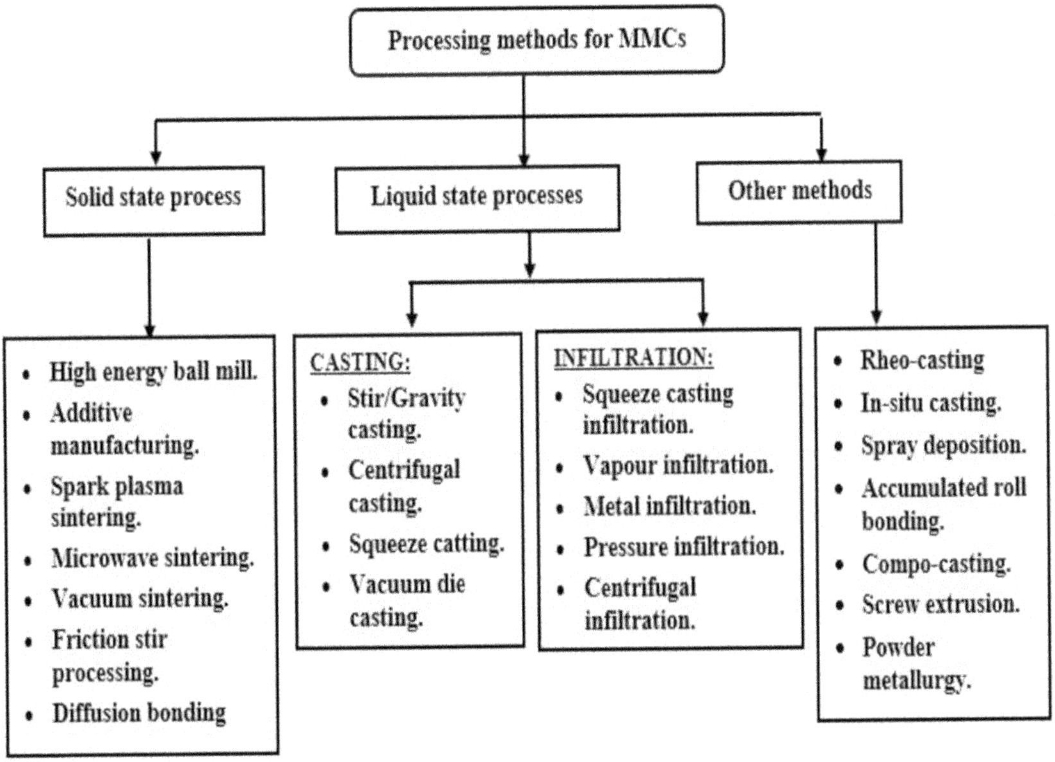

FIGURE 2.4 MMCs processing methods.

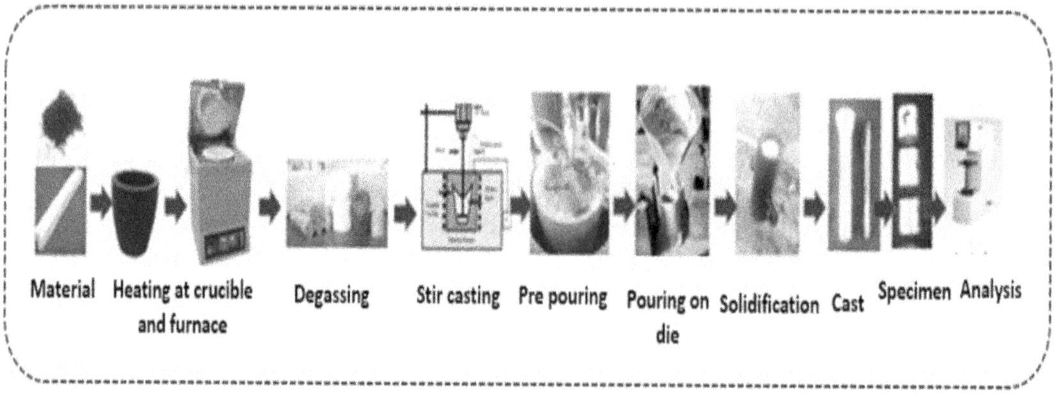

FIGURE 2.5 Stir casting arrangement for MMC production and analysis.

shows the SCP arrangement for MMC development. SCP offers a versatile route for tailoring composite properties by optimizing materials and process parameters.

Common reinforcing materials for MMCs include boron carbide (B_4C), silicon carbide (SiC), alumina (Al_2O_3), fly ash, graphite, jute ash, pumice, and carbonized coal particles [7–10]. The combination of a metal matrix and reinforcement enhances physical, thermal, and mechanical properties (such as strength, stiffness, and hardness) and wear resistance [11]. Lightweight metals like aluminium and magnesium are commonly used as matrices, and the addition of lightweight reinforcements further improves the strength-to-weight ratio, making MMCs ideal for weight-sensitive applications. Improved thermal conductivity makes certain MMCs suitable for heat dissipation, while their superior wear resistance supports high-wear applications. Tailoring the type, size, and volume fraction of reinforcements allows customization for specific needs, including enhanced corrosion resistance. MMCs are widely used in aerospace, automotive, electronics, and sports industries, with applications ranging from structural components and brake rotors to heat sinks. Manufacturing methods include stir casting, powder metallurgy, and infiltration, with SCP being versatile and economical. Key challenges include achieving uniform reinforcement distribution, ensuring strong matrix-reinforcement bonding, and managing thermal expansion mismatches. Quality control, such as non-destructive testing and adherence to standards, is critical for ensuring MMC integrity. The SCP setup and processing parameters—matrix material, reinforcement type, stirring conditions, and cooling rates—significantly influence the microstructure and performance of the resulting MMC.

2.3.1 Development of Aluminium Matrix Composites (AMCs)

The development of AMCs using the SCP requires careful attention to material selection, process parameters, and post-casting treatments to achieve the desired properties. The process must be tailored to the specific application and performance requirements of the composite material. AMCs can be classified into two types: in-situ and ex-situ composites, with the steps for in-situ composites illustrated in Figure 2.6. SCP is a versatile and economical technique widely used for AMCs due to the favourable properties of aluminium alloys. The outline of the general development process, which allows for the production of AMCs with enhanced mechanical, thermal, and physical properties suited for various industrial applications, are *1) Matrix Material Selection:* Choose a suitable aluminium alloy based on requirements such as strength, corrosion resistance, and thermal conductivity. *2) Reinforcement Selection:* Select appropriate reinforcing materials, commonly ceramics like SiC or Al_2O_3, based on the desired properties. Prepare the reinforcement in the required form (particles or fibres) and size, and pre-treat if necessary to improve compatibility with the aluminium matrix. *3) Melting the Matrix:* Preheat the aluminium alloy in an induction or crucible furnace to

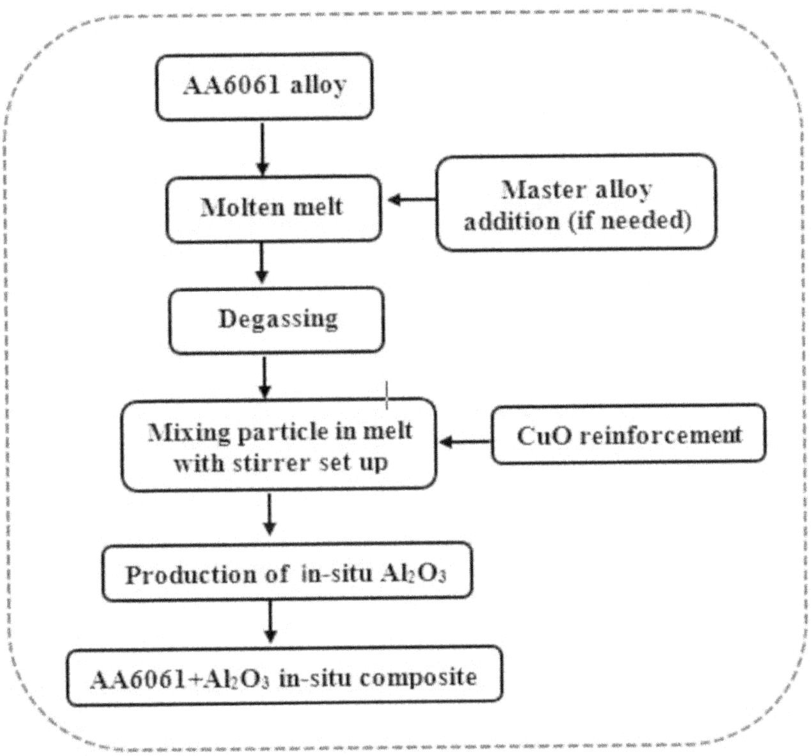

FIGURE 2.6 Production of aluminium matrix in-situ composite.

its melting temperature, ensuring it is fully molten. *4) Reinforcement Addition:* Gradually introduce the pre-treated reinforcing materials into the molten aluminium while stirring. The stirring ensures uniform distribution of the reinforcements within the matrix.

The stirring mechanism in SCP ensures a uniform distribution of reinforcing particles within the molten aluminium alloy. Activating the stirrer creates a homogeneous mixture, preventing particle settling and improving the mechanical properties of the composite. The stirred molten metal is then poured into a prepared mould or die, using methods such as gravity or pressure casting, and cooled in a controlled manner to achieve the desired microstructure. Post-casting treatments, such as extrusion, hot isostatic pressing (HIP), or heat treatment, may be applied to enhance the composite's microstructure and mechanical properties. Microscopic analyses using optical or scanning electron microscopy (SEM) provide insights into particle distribution and composite quality. Mechanical testing, including tensile, hardness, and impact tests, evaluates the material's performance for specific applications. Quality control measures, including defect inspection and adherence to standards, ensure consistency and reliability. Raj Tilank et al. developed an aluminium matrix composite using a computerized SCP setup, incorporating car engine scraps and petrochemical alumina catalysts as reinforcements [12]. Figures 2.7a and 2.7b show the microstructure of AMC reinforced with graphite via stir-gravity casting [13]. In the microstructure, the phase with a white region signifies the primary aluminium, the dark blocky phases represent the primary silicon, and the extreme black phase represents graphite. The particle distribution of graphite in gravity casting exhibited a high degree of uniformity. The majority of graphite particles are adhered to primary silicon particles. Graphite enhances silicon fineness and alpha aluminium, improving uniformity and eliminating the dendritic shape observed in LM30 alloys. The composite achieved a Brinell hardness of 125.3 BHN. Figures 2.7c and 2.7d display the die and the prepared MMC, respectively.

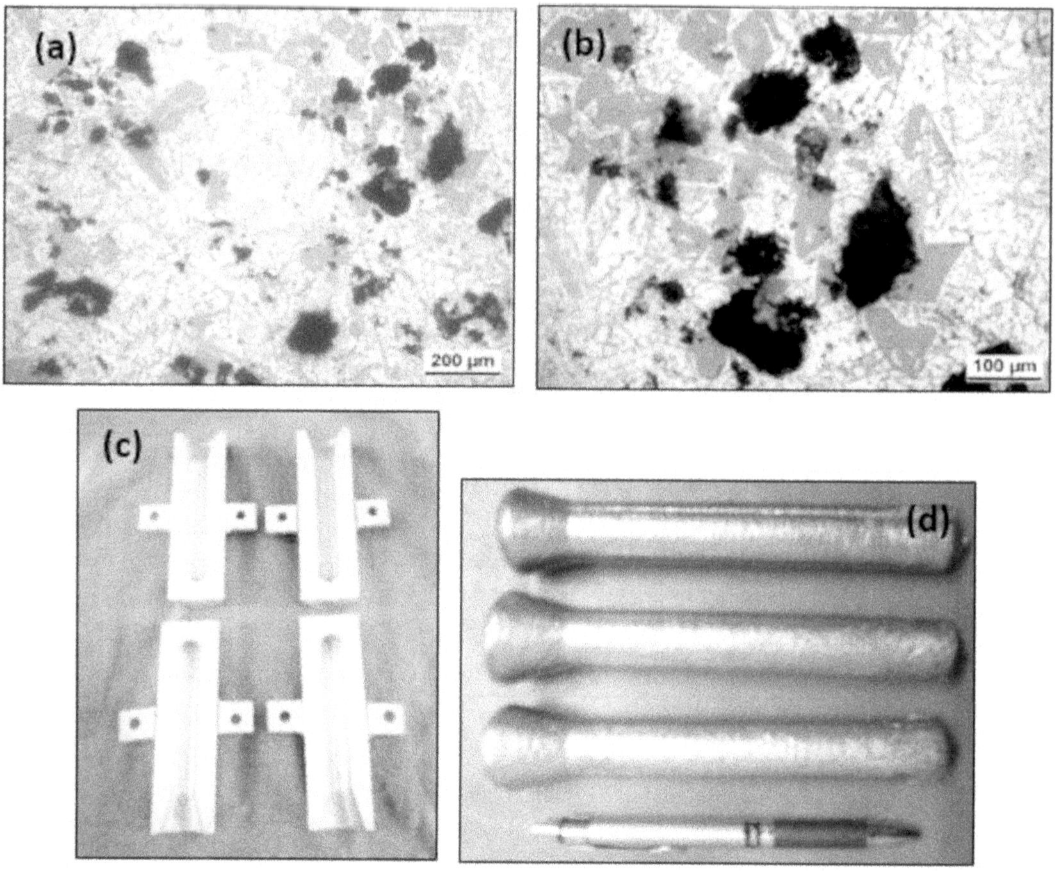

FIGURE 2.7 **a)** Microstructure of Al+Graphite MMC, **b)** accumulation of graphite with silicon, **c)** die used for composite casting, **d)** prepared Al+Graphite MMC through gravity casting.

2.3.2 Development of Magnesium Matrix Composites

The development of magnesium matrix composites (MgMCs) through the stir-casting process involves several essential steps. Magnesium alloys, known for their lightweight properties, are further enhanced by incorporating reinforcing materials, improving strength, wear resistance, and overall performance. The process outline is almost similar to the AMCs's outline, i) selection of a suitable magnesium alloy based on application-specific requirements, such as strength, corrosion resistance, and weight; ii) usage of reinforcements like CNTs, SiC, B_4C, graphene, GNPs, Al_2O_3, or metallic glass to improve mechanical, chemical, and wear properties. Pretreat reinforcing particles to enhance compatibility with the magnesium matrix, iii) preheating the magnesium alloy in an induction or crucible furnace to its melting temperature; iv) gradually introduce the pre-prepared reinforcement particles into the molten magnesium under controlled conditions while stirring to ensure uniform distribution and prevent settling; v) pour the stirred molten composite into a mould using gravity or pressure casting and cool it in a controlled manner to achieve the desired microstructure; vi) apply extrusion or heat treatment to further enhance (post-casting) mechanical properties and refine the microstructure. Microscopy inspection methods can help examine the particle distribution and composite quality. Similarly, performing mechanical tests (e.g., tensile, hardness, impact) is required to assess application suitability. Implement quality control measures throughout the process to ensure consistency

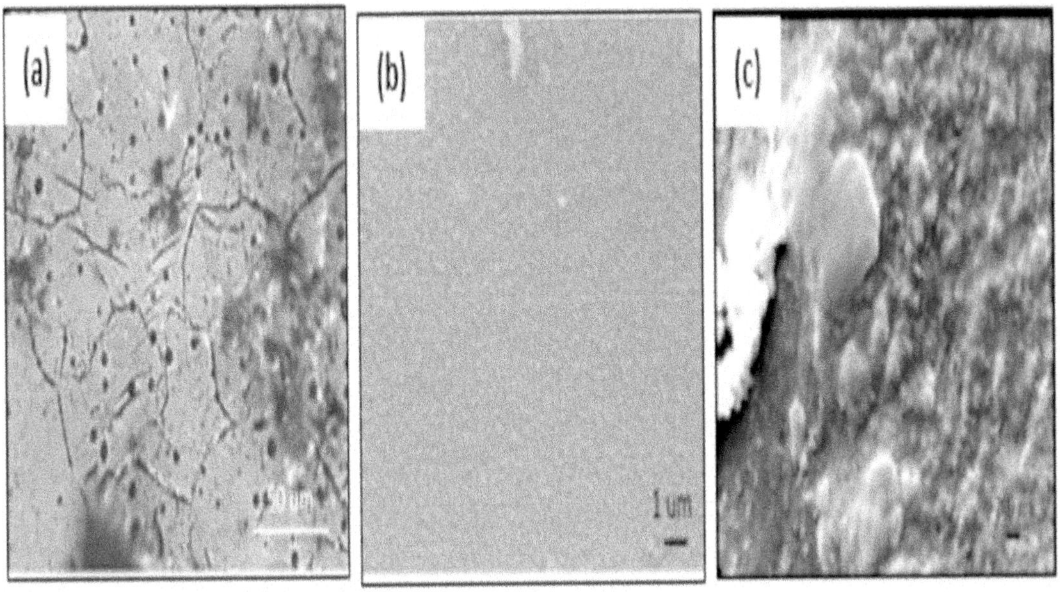

FIGURE 2.8 SEM images of Mg-CAN composite developed through SCP [16]: **a)** in situ composite production of CeO_2, **b)** nanoparticle production in Mg-CAN composite, **c)** micro size particle Mg-CAN composite.

and reliability in the production of MgMCs. This includes checking for defects, monitoring process parameters, and adhering to specified standards. Magnesium's high reactivity necessitates the use of inert gases like argon or fluxing agents to minimize oxidation [14]. Reinforcement compatibility must be carefully considered for optimal performance. Studies, such as those by Bhaskar Chandra Kandpal et al. and Singh H et al. highlight MgMCs for biomedical applications [15, 16]. For example, CeO_2 reinforcement was synthesized in situ using ceric ammonium nitrate and SCP [17]. Figure 2.8 illustrates SEM images of Mg-CAN composites developed through SCP.

2.3.3 Development of FGMs

The FGMs are an innovative class of materials in which properties vary continuously without any discontinuity across the material. The SCP process plays a pivotal role in the development of metal composite FGMs. The FGM concept was introduced in Japan in 1984 during the spaceplane project, particularly for use as a thermal barrier material. Designed to endure extreme conditions, FGMs can withstand surface temperatures of up to 2000 K and a temperature gradient of 1000 K across a 10 mm section. While conventional materials fail under such extreme environments, FGMs offer the capability to meet these challenging demands. FGMs are typically composed of three main categories: Ceramic/Metal, Ceramic/Ceramic, and Metal/Metal combinations. Different FGM variants are illustrated in Figure 2.9, and their gradients, which can be characterized by fraction, shape, orientation, and other factors, are shown in Figure 2.10. This tailored gradation of properties allows FGMs to excel in applications requiring high thermal, mechanical, or structural performance under severe conditions.

2.3.3.1 Effect of Centrifugal Force on Stir-Cast Composite to Develop FGMs

In the 1980s, FGM-MMCs were processed using SCP and centrifugal casting (CC) methods (Figures 2.11a and 2.11b). By the early 1990s, these techniques were widely used to develop

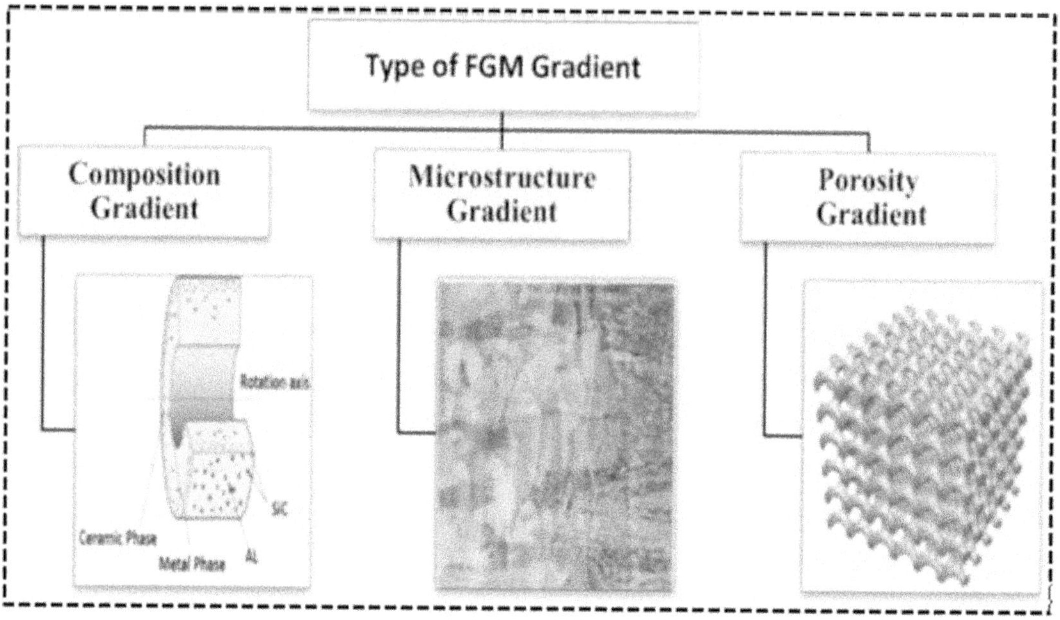

FIGURE 2.9 FGM variants.

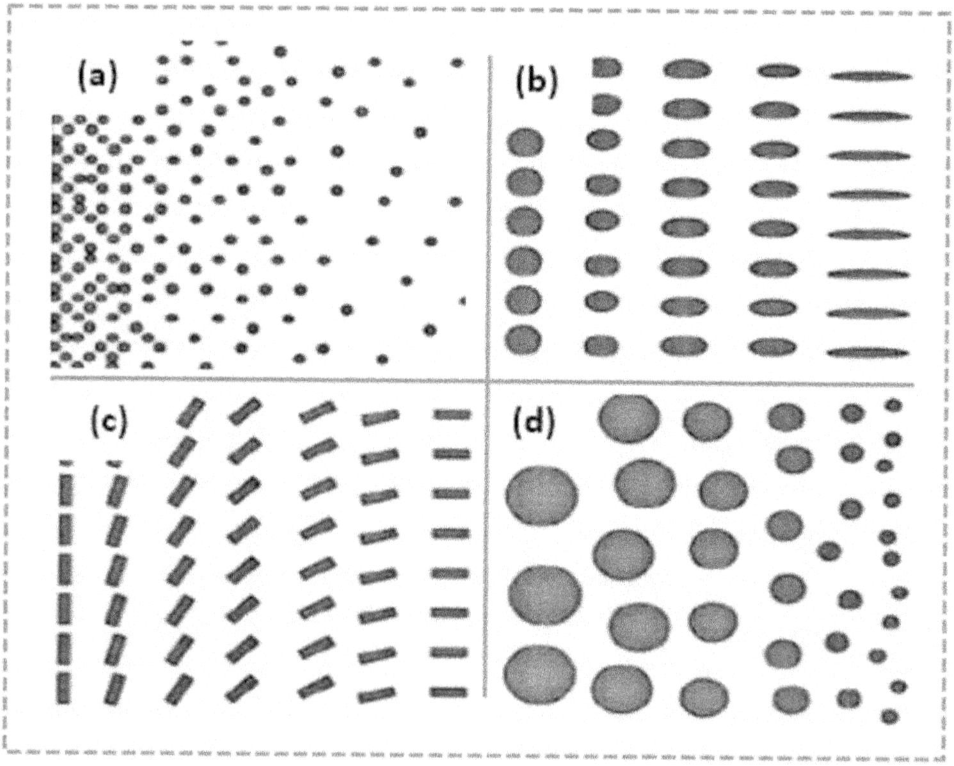

FIGURE 2.10 FGMs gradient of **a)** fraction, **b)** shape, **c)** orientation, **d)** size.

FGMs. In this process, molten composite with reinforcements added via SCP is poured into a rotating centrifugal mould (Figure 2.11c). The centrifugal force, influenced by particle density differences, propels denser metal toward the mould's outer region while lighter particles move toward the inner core. This mechanism also forces impurities like slag and gas bubbles to the inner zone, making them easier to remove by machining. The rotation speed and metal pouring rate depend on the alloy, size, and mould shape. The centrifugal force enhances uniformity and precision during casting. A specimen of a centrifugal cast aluminium composite with LM30 alloy and 5% graphite particles (Figure 2.11.d) demonstrated variations in microstructure and

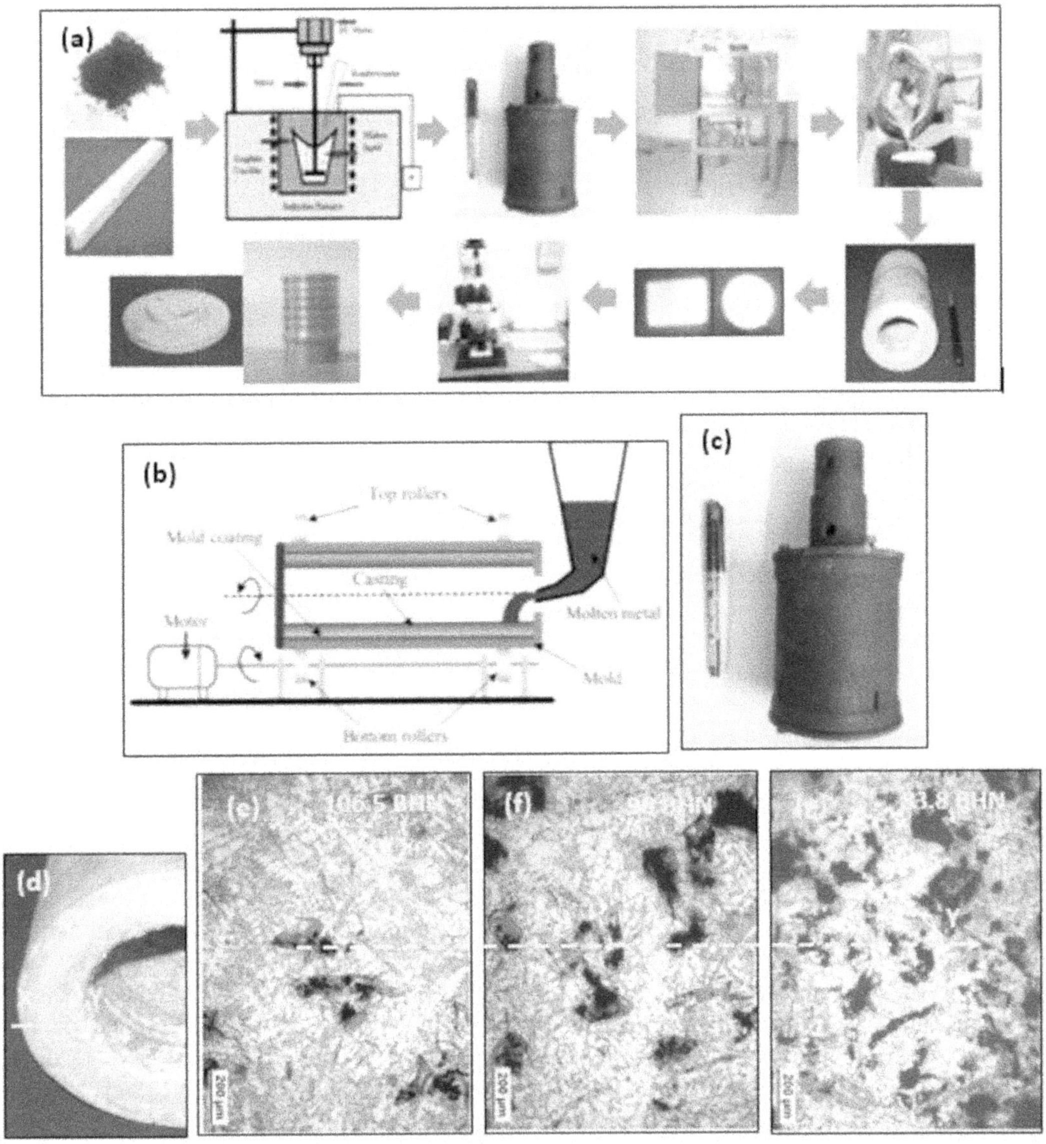

FIGURE 2.11 **a)** Development of Al-FGMs through centrifugal casting, **b)** horizontal centrifugal casting process, **c)** centrifugal cast mould, **d)** cast prepared, **e–g)** microstructures: Proportion of graphite reinforcement changes due to centrifugal force after SCP.

hardness across different radial distances. Microstructures at 6 mm (hardness 106.5 BHN), 8 mm (98 BHN), and 10 mm (83.8 BHN) from the outer to the inner periphery are shown in Figures 2.11 e–g. The incorporation of graphite into the LM30 alloy was achieved through the utilization of the SCP apparatus. Graphite facilitated the purification of alpha aluminium and silicon particles. Graphite (density 2.0 g/cm³) segregates toward the inner periphery due to its lower density compared to aluminium (2.7 g/cm³) and silicon (2.32 g/cm³). This is because of the centrifugal force. Maximum graphite (dark black colour in microstructure) and silicon (block grey colour in microstructure) concentrations were observed at 8 mm, attributed to centrifugal force. This segregation improves wear resistance at the inner periphery due to graphite's lubricating properties, showcasing centrifugal force's effectiveness in composite preparation [16].

2.4 EFFECT OF PROCESS PARAMETERS AND OPTIMIZATION ON SC

The performance and quality of MMCs developed through SCP significantly depend on the process parameters and the selection of additives. Proper parameter selection enhances the quality and consistency of the composite casts produced [17]. Key process parameters include stirring time, stirring speed, holding time, holding temperature, stirring temperature, particle density, stirrer blade angle, reinforcement addition rate, pouring temperature, and reinforcement addition temperature. Optimizing these parameters is essential to achieve uniform dispersion of reinforcements and improve the composite's mechanical and structural properties [18]. To achieve optimal results, parameters are fine-tuned using experimental design methods such as Taguchi techniques, response surface methodology (RSM), machining learning (ML), artificial neural networks (ANNs), etc. Such techniques allow for systematic analysis and determination of parameter combinations that yield the best mechanical and microstructural properties.

2.5 EFFECT OF SC ON MICROSTRUCTURE

Microscopic techniques, such as optical microscopy and field emission scanning electron microscopy (FE-SEM), are essential for analyzing the microstructure and morphology of stir-cast metal matrix composites (MMCs). These methods provide insights into grain size, particle distribution, and defects. The microstructure of the cast prepared by SCP plays a significant role. Key factors influencing the microstructure include: a) Matrix Material Microstructure: Grain structure and existing phases in the matrix metal (e.g., aluminium, magnesium) significantly affect the composite's properties [19]. b) Reinforcement Distribution: During SCP, uniform dispersion and strong bonding between the matrix and reinforcements, such as SiC or alumina, are crucial for achieving a desirable microstructure. c) Stirring parameters: Proper mechanical stirring of SCP breaks particle agglomerations and ensures homogeneity. Excessive agitation can damage particles or alter matrix properties. d) Cooling Rate: Controlled cooling influences solidification. Slow cooling leads to larger grains, while rapid cooling produces finer microstructures. e) Mould design: Mould geometry impacts cooling rates and grain orientation in the final product. f) inert atmosphere: It prevents oxidation during SCP, resulting in cleaner and more consistent microstructures. g) Post-casting treatments: Processes like solution treatment and ageing refine the microstructure and enhance mechanical properties. The steps for the sample preparation for microstructure analysis are shown in Figure 2.12a. Some of the microstructures of the AMCs processed through gravity SCP are given in Figure 2.12b–2.12f. Fine-tuning stirring speed, cooling rates, and mould design enable control over grain size, particle distribution, and structural uniformity. Understanding and optimizing the SCP process can help control and tailor the microstructure of metal matrix composites for specific applications.

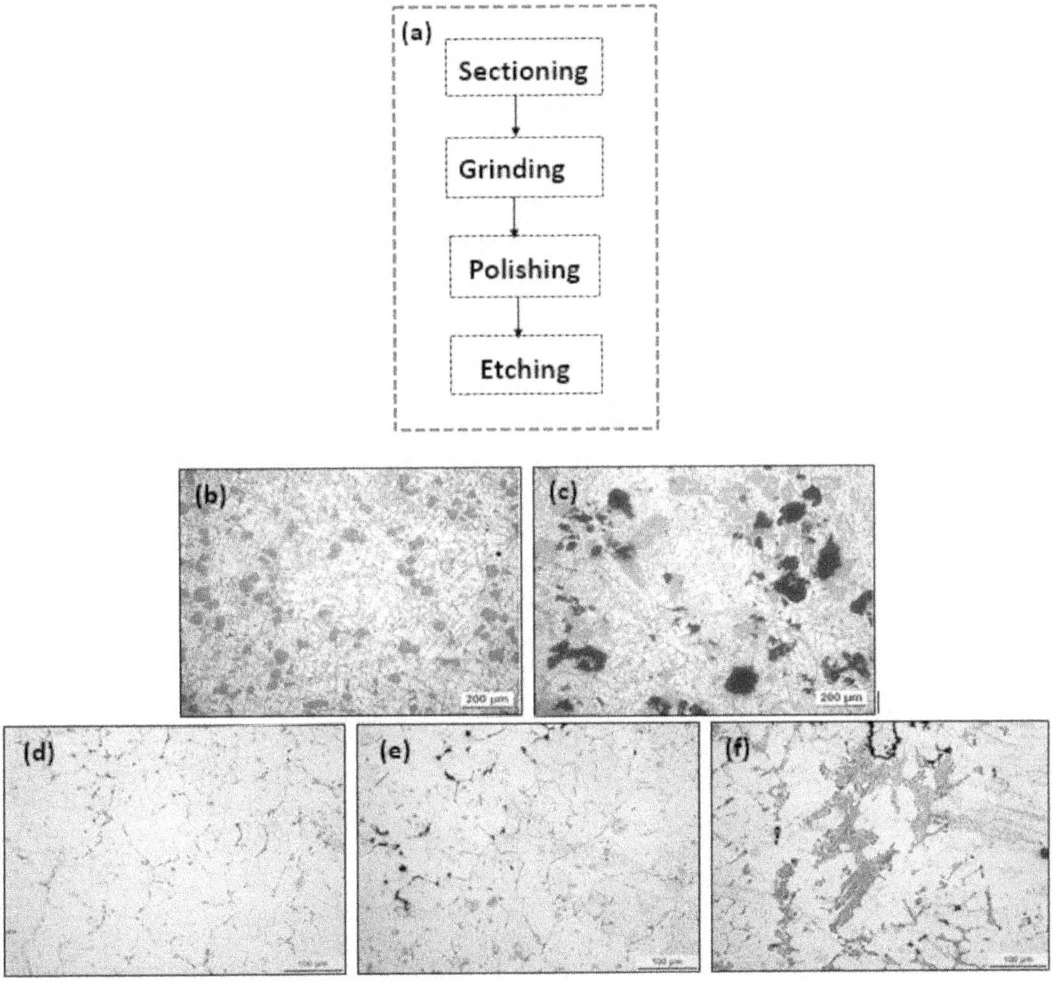

FIGURE 2.12 **a)** steps for specimen preparation, **b–c)** Microstructures of LM30 alloy + Graphite composite prepared by SCP, **d–f)** AA6061+CuO composite through the SCP.

2.6 EFFECT OF EQUAL CHANNEL ANGULAR PRESSING (ECAP) PROCESS ON STIR-CAST COMPOSITE

The SCP is instrumental in preparing metal-based composite samples for ECAP [20]. Figures 2.13a to 2.13g illustrate the production and microstructures of AA6061 alloy reinforced with CuO particles including the ECAP process effect. In ECAP, the sample undergoes severe plastic deformation by being pressed through an L-shaped channel die. This process generates submicron grain structures while maintaining the sample's cross-sectional area. Significant shear stress at the channel intersection induces substantial deformation, refining the grain structure and enhancing properties like strength, hardness, and creep resistance. The billet retains its shape and size after deformation, and the process is scalable for bulk production. Strain accumulation increases with repeated ECAP passes. For CuO-reinforced aluminium composite billets, molten metal is stirred at a constant speed to create a vortex, ensuring uniform distribution of preheated CuO particles. In ECAP, the transverse (X), flow (Y), and longitudinal (Z) planes represent different material orientations. To prevent

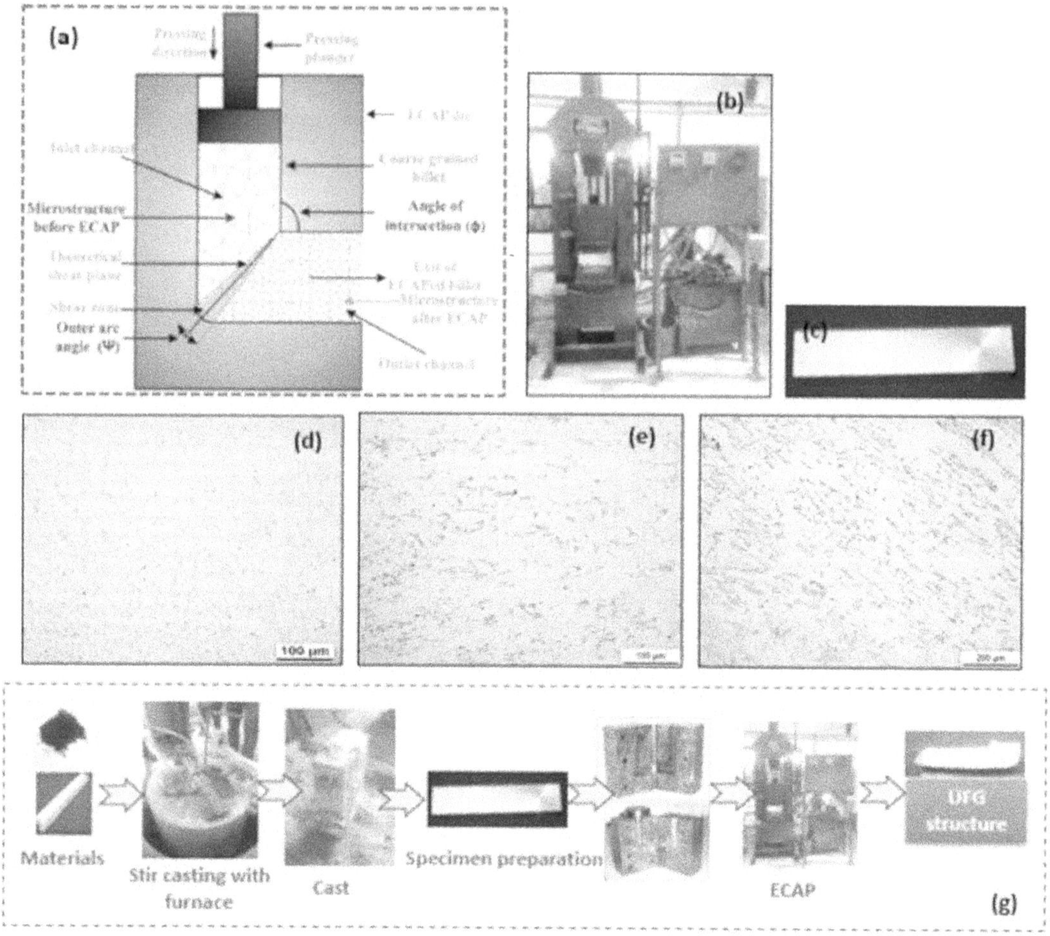

FIGURE 2.13 **a)** ECAP process, **b)** press of 15 ton, **c)** AMC billet for ECAP, **d–f)** microstructures with the reinforcement addition through SCP after ECAP process, **g)** ECAP process AMC sample prepared through SCP route.

fractures during ECAP, solution heat treatment at 530°C for 4 hours is essential, followed by age hardening at 175°C for 8 hours, which increases hardness by approximately 35 Hv. After one ECAP pass, grain size is refined to 25–40 μm, with multiple passes further improving grain refinement and mechanical properties.

2.7 INFLUENCE OF SC ON MECHANICAL PROPERTIES

The mechanical properties can be tailored by the proper SCP. Achieving the desired combination of properties requires careful material selection, optimization of processing parameters, and appropriate post-processing treatments. The mechanical properties of MMCs produced by SCP depend on factors such as the matrix material, reinforcing particles, processing conditions, and post-processing methods. The tensile strength is influenced by the strength of both the matrix material and the reinforcing particles. Effective dispersion and bonding of the particles within the matrix, achieved through proper SCP, enhance tensile strength. The hardness is typically improved by the addition of hard reinforcing particles. The uniform distribution of these particles, facilitated by stirring, leads to increased hardness. Like tensile strength, yield strength depends on the particle content and distribution. Effective stirring ensures even particle dispersion, improving yield strength. The compressive strength is determined by the characteristics and distribution of reinforcing particles. SCP promotes a uniform particle distribution, enhancing compressive strength. Figures 2.14.a and 2.14.b present the hardness values of the prepared MMCs, showcasing the improvement achieved through proper particle integration and distribution during SCP.

The flexural strength depends on the matrix and reinforcements. SCP ensures a homogeneous mixture, contributing to consistent flexural strength across the composite. SCP minimizes weak points through uniform particle distribution, improving overall impact resistance (toughness). Proper dispersion of reinforcing particles via SCP enhances the composite's ability to withstand cyclic loading (fatigue resistance). While reinforcing particles improve strength, they may reduce ductility. Achieving a balance during SCP is crucial for optimal strength and ductility. At elevated

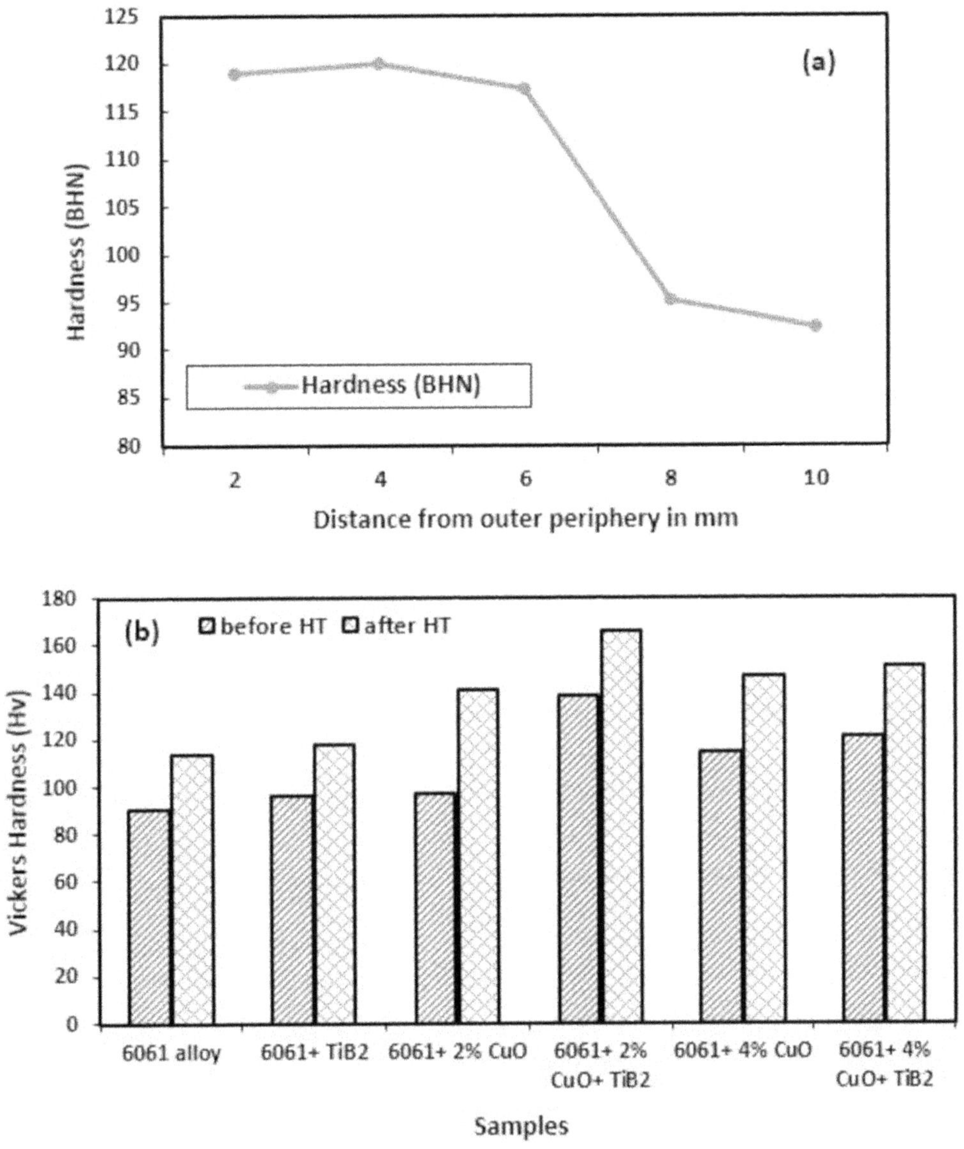

FIGURE 2.14 Hardness values measure. **a)** Brinell hardness value of LM30+Graphite composite, **b)** Vickers hardness values of AA6063 + CuO before and after heat treatment.

temperatures, creep resistance is influenced by the matrix and reinforcement selection. SCP ensures uniform reinforcement dispersion, improving resistance to deformation. Post-processing steps, such as heat treatment, can optimize the composite structure, further enhancing mechanical properties like strength and durability. SCP, combined with appropriate material selection and treatments, enables the tailoring of MMCs for specific performance requirements.

2.8 DEGASSING FOR SC

Degassing is a critical step in metal casting, including the SCP, to remove or reduce dissolved gases, particularly hydrogen, from molten metals. Excessive gas content can cause porosity and defects in the final composite. The choice of degassing method depends on the type of metal matrix, desired properties, and available equipment. Proper control of process parameters and adherence to safety protocols are essential during degassing. Table 2.1 outlines various degassing methods applicable to SCP. Figure 2.15a shows a sample without degassing, exhibiting porosity, while Figure 2.15b illustrates a degassed sample with reduced defects. Figure 2.15c displays PCl_5, a commonly used degassing agent in MMC production. Implementing effective degassing ensures improved composite quality, reduces porosity, and enhances mechanical performance.

2.9 CHALLENGES AND OPPORTUNITIES FOR SC

The SCP offers opportunities for tailoring material properties, cost-effectiveness, versatility, improved performance, and reduced environmental impact, despite challenges such as particle distribution, interface bonding, porosity, and tool wear. By optimizing process parameters, stir casting allows for the incorporation of a wide range of reinforcing materials, making it suitable for various applications [21, 22]. Compared to other methods like powder metallurgy, stir casting is cost-effective and scalable for mass production. It can produce composites with enhanced strength, stiffness, and wear resistance while reducing material waste through the use of recyclable materials. Table 2.2 outlines challenges and their mitigation strategies.

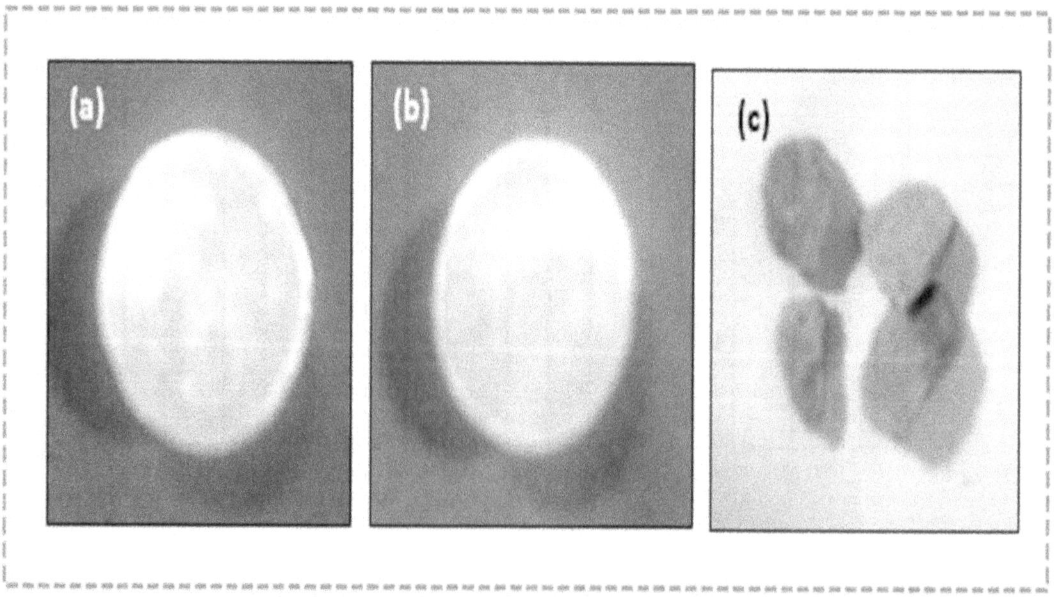

FIGURE 2.15 **a)** sample without degassing, **b)** sample with degassing, **c)** PCl_5 tablets for degassing.

TABLE 2.1
Different Degassing Methods during the Stir-Casting Process

Methods	Process
a) Vacuum degassing	Vacuum degassing is a widely used method to remove dissolved gases from molten metal. In SCP, a vacuum chamber creates a low-pressure environment, allowing dissolved gases, particularly hydrogen, to escape from the molten metal. This technique effectively reduces hydrogen content, improving the quality of the final composite.
b) Fluxing agent	Fluxing agents are sometimes used in the SCP to react with dissolved gases in molten metal, forming compounds that rise to the surface for easy removal. This method is particularly effective for aluminium alloys, enhancing the quality of the composite.
c) Inert gas atmosphere	Using an inert gas atmosphere, like argon or nitrogen, during SCP helps minimize exposure to atmospheric gases, reducing the risk of gas absorption during melting and stirring.
d) Stirring process	Effective stirring in SCP helps homogenize the metal matrix and aids in releasing dissolved gases by breaking up gas bubbles and promoting their rise to the surface.
e) Covering the molten metal	Using covers or lids on the crucible reduces the molten metal's exposure to the atmosphere, minimizing gas absorption, especially with reactive metals.
f) Proper alloying technique	Proper alloying techniques can control gas content by using deoxidizers to reduce oxygen levels in the molten metal, thereby lowering gas levels.
g) Optimizing melting conditions	Controlling melting conditions, such as temperature and heating rate, ensures a thorough melt before stirring, aiding in effective gas removal.

TABLE 2.2
Challenges during the SC Process and Mitigation

Parameter	Challenge	Mitigation
a) Particle distribution	Achieving a uniform distribution of reinforcing particles in the metal matrix can be challenging. Uneven distribution may result in inconsistent mechanical properties [23].	Optimal process parameters, such as stirring speed and time, must be precisely controlled to ensure homogeneity.
b) Particle-matrix interface	Achieving strong bonding between the matrix and reinforcing particles is essential for optimal mechanical properties. Poor bonding can lead to inadequate load transfer between the matrix and reinforcement.	Surface treatment of reinforcing particles or the use of wetting agents can enhance interface adhesion.
c) Porosity	The stirring process can introduce air or gas bubbles, leading to porosity in the final composite.	Proper degassing techniques, like vacuum or inert gas treatment, can minimize porosity.
d) Stirring tool wear	The high temperatures and abrasive nature of reinforcing particles can cause wear on the stirring tool, impacting the final product quality.	Using wear-resistant stirring tools or regularly replacing them can mitigate this issue.
e) Temperature	High temperatures can cause cracks in specimens due to rapid solidification.	Optimizing the process and preheating temperature helps prevent cracks during solidification.
f) Oxidation	Porosity may develop in casting due to oxidation, which can negatively impact the quality of the cast.	The risk of oxidation can be minimized by using inert gases, particularly argon.

2.10 APPLICATIONS OF THE SC PROCESS

The SCP is a versatile process used across various industries to produce MMCs with tailored properties. In the automotive industry, stir-cast MMCs are used for components like brake discs, pistons, and cylinder liners, benefiting from enhanced strength and wear resistance. Aerospace applications utilize stir-casting for lightweight yet strong components, such as aircraft and spacecraft parts. Stir-cast metal composites also offer improved thermal conductivity, making them ideal for electronic packaging and heat sinks. In the sports industry, SCP is used for durable, lightweight components like bicycle frames and golf club heads. Marine and defence industries benefit from stir-cast MMCs for corrosion-resistant and high-strength parts, including propellers and armour materials. Additionally, stir-cast composites are used in renewable energy (wind turbines), medical devices (implants, prosthetics), consumer goods (appliances, tools), and construction materials (reinforced concrete), providing a balance of mechanical properties. The flexibility of SCP in tailoring material properties makes it valuable for diverse applications, improving performance, and efficiency.

2.11 LIMITATIONS AND DISADVANTAGES OF THE SC PROCESS

While stir casting offers many advantages, it also has limitations that need to be addressed for optimal process and application selection. A key challenge is the flotation of low-density hollow particles, particularly at low particle volume fractions. Additionally, achieving good wetting of particles with the molten metal and the risk of particle fracture at high volume fractions are significant drawbacks [24]. Despite these challenges, ongoing research and process optimization, along with the use of advanced materials and techniques, continue to enhance the effectiveness and broaden the applications of stir casting across various industries. The limitations of the SCP are summarized in Table 2.3.

TABLE 2.3
Limitations of SC Method

Factors	Limitations
a) Particle sedimentation	During stirring, the risk of particle settling or sedimentation exists, particularly when the density of the reinforcing particles differs significantly from the matrix metal. This can lead to uneven distribution, negatively impacting the mechanical properties of the final composite.
b) Particle agglomeration	Agglomeration of reinforcing particles can occur, causing non-uniform distribution in the metal matrix. This leads to localized variations in material properties and reduces the reinforcement's effectiveness.
c) Reinforcement loading	The volume fraction of reinforcing particles is limited, as high reinforcement loading can cause uneven distribution, increased viscosity, and processing challenges.
d) Range of matrix alloys	SCP is best suited for specific matrix alloys, with its applicability limited by compatibility issues and melting temperature differences between the matrix and reinforcement materials.
e) Surface oxidation	Stirring exposes the molten metal to the atmosphere, potentially causing oxidation of the matrix and reinforcing particles, which can weaken the composite's strength and toughness.
f) Tool wear and contamination	The abrasive nature of reinforcing particles causes wear on the stirring tool, potentially contaminating the composite and requiring frequent tool replacement to maintain product quality.
g) Control over fibre orientation	Controlling the orientation of reinforcing fibres in fiber-reinforced composites can be challenging, potentially affecting the anisotropic properties and directional performance of the final composite.
h) Processing high melting point materials	SCP is primarily suited for materials with lower melting points. Processing high melting point materials often requires specialized equipment and techniques, adding complexity to the process.
i) Complex shapes products	SCP is better suited for simpler shapes and geometries, as casting intricate components often requires additional processing steps.
j) Energy consumption	The stirring process requires substantial energy to maintain the elevated temperatures needed for molten metal, which can affect the cost-effectiveness of the method.

2.12 ADVANTAGES OF THE SC PROCESS

The SCP process offers versatility by accommodating a wide range of matrix alloys and reinforcing materials, including metals and ceramics, enabling tailored material properties for specific applications. Compared to methods like powder metallurgy, SCP is cost-effective, requiring simpler equipment suitable for both small- and large-scale production. The process involves basic equipment—a crucible, stirring mechanism, and mould—allowing ease of setup and operation. By incorporating reinforcing particles or fibres, stir casting enhances mechanical properties such as strength, hardness, and wear resistance, with the flexibility to customize these properties through material selection and process control. Proper stirring ensures uniform particle distribution, contributing to consistent composite properties. The SCP supports the design of composites with varying reinforcement volumes, optimizing the balance of properties. Its faster processing time compared to other methods makes it efficient for quick-turnaround applications, and its scalability suits a variety of production scales and industries. The process also enables the development of composites with tailored thermal conductivity for applications requiring efficient heat management. Its compatibility with recyclable materials supports sustainability by reducing waste and environmental impact. Although better suited for simpler geometries, stir casting remains a valuable method for producing components across diverse industries.

2.13 COMPARISON OF SC WITH INFILTRATION AND SQUEEZE CASTING

The SCP, infiltration, and squeeze casting are key methods for fabricating MMCs, each with unique advantages and limitations. Each method offers specific benefits depending on the application requirements and budget constraints. *The SC process:* This involves mechanically stirring reinforcement particles into molten metal, which is then poured into a mould for solidification. SC is cost-effective, straightforward, and suitable for large-scale production, capable of forming complex shapes. However, it may result in poor reinforcement distribution and inhomogeneities and is limited to low-to-moderate reinforcement volume fractions. There is also a risk of particle segregation due to floating. *Infiltration:* Molten matrix metal infiltrates porous reinforcement preforms through capillary action, pressure, or vacuum. This method provides excellent reinforcement distribution, allows for high reinforcement volume fractions, and produces composites with superior mechanical properties. However, it is more complex and expensive than the SC process, requires precise control of process parameters, and is limited to specific preform shapes. *Squeeze casting:* In this process, molten metal is injected into a mould containing the reinforcement preform under high pressure, which is maintained during solidification. It produces low-porosity, high-density components with good wettability and strong matrix-reinforcement bonding. It is ideal for high-strength structural components and precision automotive and aerospace parts. However, the method involves high initial setup and operational costs, complex process control, and size limitations for the parts.

2.14 SAFETY CONSIDERATION DURING CASTING

The SC process involves high temperatures and molten metals, presenting potential hazards that require strict safety measures. Personnel should wear appropriate PPEs, including heat-resistant gloves, safety glasses, flame-resistant clothing, and closed-toe shoes to protect against molten metal splashes and hot surfaces. Adequate ventilation is necessary to control fumes and gases, particularly when melting metals that release hazardous byproducts. Provide thorough training to the personnel on hazards, emergency procedures, and proper equipment use. Establish clear emergency protocols, including first aid procedures and fire safety measures. Fire extinguishers, especially for metal fires, and automatic suppression systems should be readily available. Maintain updated Material Safety Data Sheets (MSDS) for all materials. Implement machine guards, lock-out/tag-out procedures, and safe handling of hot materials. When using an inert gas atmosphere to prevent oxidation during the process, be aware of the specific hazards associated with the gas. Ensure proper handling, ventilation, and monitoring of inert gases to minimize asphyxiation risks. Regular equipment inspections are essential to prevent safety hazards, and materials should be stored safely and organized. Ensure well-equipped first aid facilities and clear communication channels among staff. Prioritizing these safety practices will help create a safer stir-casting environment and reduce accident risks.

2.15 QUALITY CONTROL IN SC PROCESS AND MMCs

Quality control in stir casting and MMCs is crucial to ensure the final product meets specifications. It begins with selecting the right materials for the matrix and reinforcement, ensuring they meet mechanical, thermal, and chemical requirements. Monitoring process parameters, such as temperature, stirring speed, and pressure, is essential to avoid defects. Raw materials should undergo thorough inspection for composition and purity. Achieving a uniform distribution of reinforcement particles requires optimized stirring and mixing. Microstructural analysis (e.g., microscopy and XRD) evaluates dispersion, grain size, and defects. Mechanical testing assesses the performance of the casts and properties like tensile strength and hardness, while dimensional inspection ensures precision using tools like coordinate measuring machines (CMM). Surface quality is checked through visual inspection and roughness measurements. NDT techniques, such as dye penetrant testing, and ultrasonic or radiographic testing, ensure internal integrity. A quality management system, aligned with standards like ISO 9001, supports consistent production and continuous improvement. Regular audits and continuous improvement initiatives further enhance the effectiveness of the quality control process.

2.16 CONCLUSIONS

In conclusion, the stir casting technique stands out as a versatile and promising method for the development of MMCs. It offers flexibility in terms of material selection, allowing for a wide range of matrix and reinforcement combinations to be utilized. This flexibility enables the tailoring of composite properties to meet specific performance requirements across various industries and applications. This improvement in performance makes stir-cast MMCs attractive for use in demanding engineering applications where high strength, stiffness, and durability are essential. The relatively simple and cost-effective nature of the SCP makes it accessible to a wide range of manufacturers, both large-scale and small-scale. This accessibility facilitates the production of MMCs in diverse quantities and shapes, catering to different market needs. As research and development in this field continue to evolve, further refinements in the SCP are expected, leading to even greater advancements in MMCs and their widespread adoption in diverse industrial sectors.

REFERENCES

[1] Ramamoorthi, R., Justin, M.H.J., Sundaramoorthy, R., Dixon, J.J.J., Kalidas, K. and Manickaraj, K., 2021. Influence of stir casting route process parameters in the fabrication of aluminium matrix composites—a review. *Materials Today: Proceedings*, *45*(7), pp. 6660–6664.

[2] Mohit, K.S. and Raj, K.S., 2018. Fabrication of aluminum matrix composites by stir casting technique and stirring process parameters optimization. In *Advanced Casting Technologies*, Intechopen, pp. 111–126.

[3] Kareem, A., Qudeiri, J.A., Abdudeen, A., Ahammed, T. and Ziout, A., 2021. A review on AA6061 metal matrix composites produced by stir casting. *Materials*, *14*(1), pp. 1–22, 175.
[4] Arsha, A.G., Visakh, M., Ajay Raag, L., Akhil, M.G. and Rajan, T.P.D., 2022. Metallic nanocomposites for automotive applications. In *Nanotechnology in the Automotive Industry*, Elsevier, pp. 163–198.
[5] Senthil, M.S., Jegan, V. and Velmurugan, M., 2018. Mechanical properties of SiC, Al_2O_3 reinforced aluminium 6061-T6 hybrid matrix composite. *Journal of the Institution of Engineers (India) Series D*, *99*, pp. 71–77.
[6] Subramanian, S.M., Vijayan, J. and Muthaiah, V., 2017. Tribological, wear behaviour and hardness measurement of SiC, Al_2O_3 reinforced Al. matrix hybrid composite. *Journal of the Institution of Engineers (India) Series D*, *98*, pp. 291–296.
[7] Senthil, M.S., Velmurugan, M. and Jegan, V., 2016. Characterisation and optimisation of Al6061-Flyash-SiC composite in fabrication process using Taguchi method. *Indian Foundry Journal*, *62*(8), pp. 34–38.
[8] Senthil, M.S., Vigneshkumar, M., John Iruthaya Raj, M., Sivamani, S., Vinayaka, N., Srinivasan, M., Vijayan, V. and Addisalem, M., 2023. Stir casting processing, mechanical, and wear behavior of AA2024 +10 wt. % Flyash +5 wt. % graphite hybrid composite. *Advances in Materials Science and Engineering*, *9089620*, pp. 1–14.
[9] Coyal, A., Yuvaraj, N., Butola, R. and Tyagi, L., 2020. An experimental analysis of tensile, hardness, and wear properties of aluminium metal matrix composite through the stir casting process. *SN Applied Science*, *2*, p. 892.
[10] Ibrahim, T.K., Yawas, D.S., Dan-asabe, B. and Adebisi, A.A., 2023. Taguchi optimization and modeling of stir casting process parameters on the percentage elongation of aluminium, pumice, and carbonated coal composite. *Scientific Reports*, *13*, p. 2915.
[11] Sakthi, V.B., Senthil, M.S. and Nagaraj, P., 2015. Fabrication and characterization of Al-SiCp-fly ash composite using stir casting process. *Concurrent Advances in Mechanical Engineering*, *1*(1), pp. 19–29.
[12] Raj, T.K.R., Ramesh, S., Satish, B.B. and Pradeep, K.K., 2023. Sustainable development on production and characterization of metal matrix composites using stir casting. *Journal of Physics Communications*, *7*, p. 065002.
[13] Senthil, M.S., 2018. Processing and characterization of LM30 alloy + graphite reinforced composite through gravity and centrifugal casting. *Annales de Chimie—Science des Matériaux*, *4*, pp. 555–564.
[14] Arora, G.S., Saxena, K.K., Mohammed, K.A., Prakash, C. and Dixit, S., 2022. Manufacturing techniques for Mg-based metal matrix composite with different reinforcements. *Crystals*, *12*(7), p. 945.
[15] Bhaskar, C.K., Nitin, J., Lavish, K., Anjul, T., Vaibhav, J. and Ujjwal, G., 2022. Stir casting technology for magnesium-based metal matrix composites for bio-implants—a review. *Materials Today: Proceedings*, *62*(6), pp. 4519–4525.
[16] Singh, H., Kumar, D. and Singh, H., 2021. Development of magnesium-based hybrid metal matrix composite through in situ micro, nano reinforcements. *Journal of Composite Materials*, *55*(1), pp. 109–123.
[17] Sourav, D., Sanjay, G. and Mangal, R., 2020. Recent developments in magnesium metal–matrix composites for biomedical applications: A review. *ACS Biomaterials Science & Engineering*, *6*(9), pp. 4748–4773.
[18] Abbas, A., Rajagopal, V. and Huang, S.J., 2022. *Magnesium Metal Matrix Composites and Their Applications*. IntechOpen.
[19] Senthil, M.S. and Rajan, T.P.D., 2021. Characterization of graphite-reinforced LM30-aluminium matrix composite processed through gravity and vertical centrifugal casting processes. *Journal of the Institution of Engineers (India): Series D*, *102*, pp. 19–26.
[20] Senthil, M.S. and Maheswarn, K., 2018. Development of ultra fine-grained structure (UFG) on AA6061 and reinforced with CuO composite through equal channel angular pressing (ECAP) process. *Revue des Composites et des Matériaux Avancés*, *28*(4), pp. 481–494.
[21] Ananth, G., Muthu Krishnan, T., Thirugnanam, S. and Tariku Olkeba, T., 2023. Optimization on stir casting process parameters of Al7050/nano-B_4C metal matrix composites. *Journal of Nanomaterials*, *2023*, pp. 1–7, 3615093.
[22] Arunachalam, R., Pradeep, K.K. and Rajaraman, M., 2019. A review on the production of metal matrix composites through stir casting—furnace design, properties, challenges, and research opportunities. *Journal of Manufacturing Processes*, *42*, pp. 213–245.

[23] Ali, M., 2020. Review of stir casting technique and technical challenges for ceramic reinforcement particulate and aluminium matrix composites. *Journal of Silicate Based and Composite Materials*, *72*(6), pp. 198–204.

[24] Vyasaraj, M., Gururaj, P., Manoj, G. and Mrityunjay, D., 2021. Metal matrix syntactic composites. In *Encyclopedia of Materials: Composites*, D. Brabazon, ed., Elsevier, pp. 109–120.

3 Exploring Liquid Metallurgy Processes for Advanced Metal Matrix Composite

Abdur Rahman Mohammed Habeebullah, Sirajudeen Nizeeruddin, and Sathickbasha Katharbasha

3.1 INTRODUCTION

Advanced metal matrix composites (MMCs) signify a major leap in materials engineering, combining metal matrices with reinforcing phases to create materials with superior mechanical properties. These composites are tailored for various advanced applications.

MMCs typically consist of a metal matrix such as aluminium, magnesium, titanium, or copper, reinforced with materials like ceramics (e.g., silicon carbide, alumina) or carbon-based substances (e.g., carbon fibres, graphene). The selection of the matrix and reinforcement is determined by the required properties and the specific application. For instance, aluminium-based MMCs reinforced with silicon carbide SiC particles have been widely used in the aerospace and automotive industries due to their lightweight nature and excellent strength-to-weight ratios.

The manufacturing processes for MMCs are diverse, including powder metallurgy, casting, and advanced techniques like stir casting, squeeze casting, and in-situ fabrication. Each method influences the distribution and bonding of the reinforcement material within the matrix, thereby impacting the composite's overall performance. For instance, stir casting entails incorporating the reinforcement into the molten metal, which is then solidified, ensuring uniform dispersion and strong interfacial bonding.

Research in MMCs focuses on optimizing the bonding inside to prevent issues like agglomeration and wettability, which can compromise the material's performance. Advances in nanotechnology and surface treatment techniques have significantly improved the compatibility and efficiency of these composites. Hybrid composites possess better properties than traditional materials like steel, making them highly sought after, especially in industries such as automotive manufacturing where weight reduction is paramount. This drive toward weight reduction has spurred the adoption of metal matrix composites (MMCs) to replace conventional components [1–5].

Combining the strengths of various materials, hybrid composites offer superior properties compared to their individual components, making them ideal for applications in aerospace. Notably, the specific strength of hybrid composites surpasses traditional materials like steel, making them highly sought after, especially in industries such as automotive manufacturing where weight reduction is paramount. This drive toward weight reduction has spurred the adoption of MMCs to replace conventional components [1].

Innovative casting technologies have enabled mass production of various MMCs, utilizing a range of secondary particle materials. These hybrid composite materials typically fall into two categories: those comprising reinforcement particles and matrix materials. These include high resistance to temperature and oxidation, alongside superior electrical and thermal conductivity, all contributing to enhanced mechanical performance [2].

DOI: 10.1201/9781003499718-3

Within the range of materials used for the matrix in MMCs, including aluminium (Al), copper (Cu), magnesium (Mg), iron (Fe), and titanium (Ti), Al and Mg take centre stage due to their abundance and suitability for MMC production. Aluminium alloys, in particular, are extensively utilized owing to their lightweight nature, cost-effectiveness, and ease of processing through various methods, offering superior strength-to-mass ratios and corrosion resistance.

Reinforcement particles can take various forms including fibres, particulates, interpenetrating structures, or layered arrangements. Particulate reinforcements are notable for their availability, affordability, and ability to be homogeneously distributed within the metal matrix, contributing to the overall performance of the composite material. Aluminium matrix composites (AMCs) commonly incorporate reinforcements such as SiC, Al_2O_3, B_4C, or WC, presenting a cost-effective approach to achieving improved mechanical properties [3–5].

Primarily, achieving effective wettability—the essential bond between the solid reinforcement and the liquid matrix—poses a significant challenge. Additionally, ensuring a uniform distribution of reinforcement material, reducing porosity, and enhancing the mechanical strength of MMCs are complex tasks that require precise control.

3.2 THE BASIC PRINCIPLES AND FUNDAMENTALS OF THE STIR CASTING PROCESS

The stir casting process, also known as liquid-state processing or rheocasting, is a widely utilized technique for the production of MMCs. Stir casting achieves its goal by dispersing reinforcing materials throughout a molten metal matrix using a mechanical stirring mechanism. Here are the basic principles and fundamentals of the stir casting process:

3.2.1 Material Preparation

The primary matrix material, often an aluminium alloy, is melted to a liquid state in a furnace. The choice of matrix material depends on the desired properties of the final composite. A typical stir-casting experimental setup is shown in Figure 3.1.

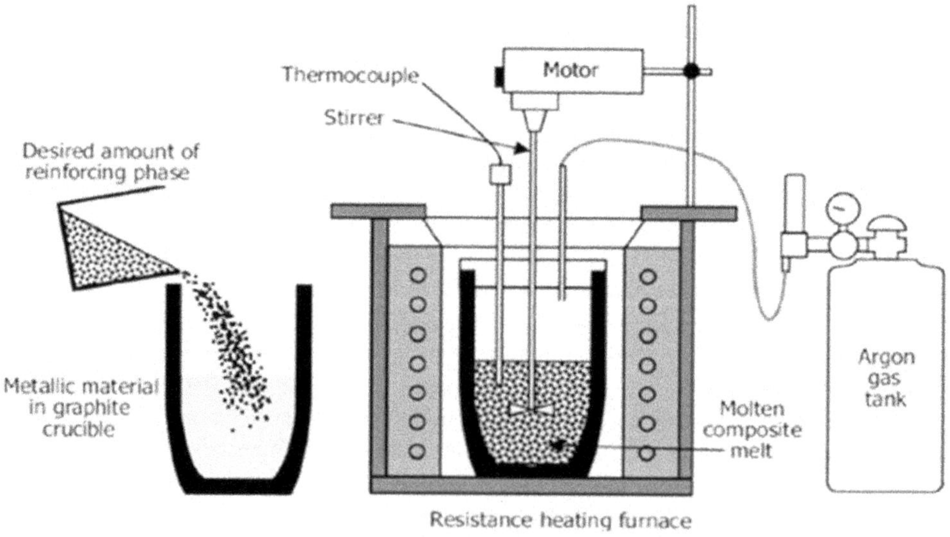

FIGURE 3.1 Typical stir-casting experimental setup [6].

3.2.2 Stir Casting Setup

A typical stir-casting experimental setup is shown in Figure 3.1. A mechanical stirrer, commonly a rotating impeller or blade, is introduced into the molten metal matrix. The stirring process plays a critical role in achieving a homogenous distribution of reinforcement particles within the molten metal matrix. To optimize this process for the desired composite properties, various parameters require meticulous control, including stirring speed, duration, temperature, and the size and shape of the reinforcing particles themselves [6].

3.2.3 Wettability and Bonding

Achieving good wettability is crucial for successful stir casting. In stir casting, wettability is paramount. It describes how effectively the molten metal (matrix) spreads and adheres to the reinforcement particles, creating a strong interfacial bond between the solid and liquid phases. To enhance this crucial interaction, surfactants or wetting agents can be introduced, promoting a robust connection between the matrix and the reinforcements [7].

3.2.4 Uniform Distribution

The stirring action prevents the settling of reinforcement particles, ensuring a homogeneous distribution throughout the matrix. This uniform dispersion contributes to enhanced mechanical properties in the final composite [8].

3.2.5 Cooling and Solidification

Once the stirring is complete, the composite is allowed to cool and solidify. Controlled cooling rates are employed to minimize porosity and achieve the desired microstructure. Understanding the solidification behaviour is crucial to control the formation of intermetallic phases and optimize the mechanical properties of the composite [9].

3.2.6 Post-Processing

Depending on the material combination, heat treatment processes may be employed to further refine the microstructure and enhance mechanical properties. After solidification, the composite may undergo machining and finishing processes to meet specific dimensional and surface quality requirements.

The stir casting process offers advantages such as simplicity, cost-effectiveness, and the ability to produce components with complex shapes. However, challenges include achieving optimal wettability, preventing particle settling, and minimizing porosity, which require careful control of process parameters. Continuous refinements in stir casting techniques are instrumental in the ongoing development of high-performance metal matrix composites, broadening their applicability across various industrial sectors [10].

3.3 OVERVIEW OF MATRIX MATERIALS COMMONLY USED IN MMCs AND THEIR PROPERTIES

MMCs represent a class of advanced materials where a metal matrix acts as the host for reinforcing elements like ceramic particles, fibres, or whiskers. These composites are engineered to combine the desirable properties of metals, such as ductility and toughness, with the high strength and modulus of ceramics. Below is an overview of common matrix materials used in MMCs and their properties.

3.3.1 Aluminium Alloys

Aluminium alloys are the most popular matrix materials in MMCs due to their low density, excellent corrosion resistance, good mechanical properties, and high thermal and electrical conductivity. The application ratio (wt%) of various structural materials in large commercial aircraft is depicted in Figure 3.2. The incorporation of reinforcements like silicon carbide (SiC) or aluminium oxide (Al_2O_3) into aluminium alloys enhances their stiffness, wear resistance, and high-temperature strength, making them suitable for automotive, aerospace, and sports equipment applications [11].

3.3.2 Magnesium Alloys

Magnesium alloys are known for their low density, which is even lower than that of aluminium alloys, making them ideal for applications where weight reduction is critical. Magnesium-based MMCs exhibit improved stiffness, strength, and high-temperature performance. However, their susceptibility to corrosion and lower creep resistance limit their application environments. They are often used in the automotive and aerospace sectors for components where weight saving is a priority [12].

3.3.3 Titanium Alloys

Titanium alloys are selected due to their impressive combination of high strength-to-weight ratio, exceptional corrosion resistance, and excellent performance at elevated temperatures. Reinforcing titanium with ceramics like SiC or boron fibres significantly enhances its mechanical properties, making titanium-based MMCs ideal for aerospace, military, and biomedical applications. Despite their advantages, the high cost of titanium alloys often limits their use in specialized, high-performance areas [13].

3.3.4 Copper Alloys

Copper alloys are selected for MMCs primarily due to their excellent thermal and electrical conductivity. The addition of reinforcements such as alumina or carbon fibres improves their hardness,

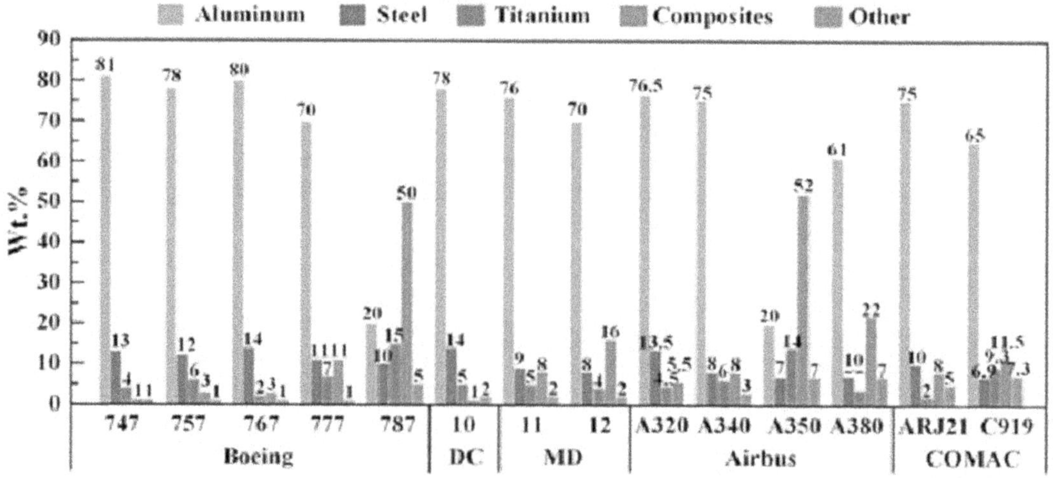

FIGURE 3.2 Structural materials in aircraft [11].

wear resistance, and mechanical strength, while somewhat reducing their thermal and electrical conductivities. Copper-based MMCs are commonly used in electrical applications, heat sinks, and wear-resistant components [14].

3.3.5 Nickel Alloys

Nickel alloys stand out for their superior ability to handle high temperatures, resist corrosion, and perform well in demanding environments. By incorporating ceramic reinforcements, nickel-based MMCs can achieve superior wear resistance and mechanical strength, making them suitable for high-temperature applications in the aerospace, energy, and chemical processing industries [15].

3.3.6 Properties and Implications

The incorporation of reinforcements typically translates to an enhancement in the yield strength and tensile strength of the base metal alloy. The modulus of elasticity is significantly enhanced in MMCs, providing greater rigidity than the base metal. While metals like aluminium and copper have high thermal conductivity, the introduction of ceramic particles can reduce this property. However, specific reinforcements are designed to maintain or enhance thermal performance. MMCs exhibit superior wear resistance compared to their matrix materials alone, making them ideal for applications involving abrasive environments [16].

3.4 REINFORCEMENT MATERIALS AND THEIR IMPACT ON MMC PROPERTIES

Reinforcement materials play a crucial role in MMCs, influencing their mechanical, thermal, and other properties. Various reinforcement materials are employed to enhance the overall performance of the composite. Here, we discuss several common reinforcement materials, such as silicon carbide (SiC), aluminium oxide (Al_2O_3), boron carbide (B_4C), and their impact on MMC properties.

3.4.1 Silicon Carbide (SiC)

It is extremely hard and wear-resistant, with high thermal conductivity and high strength and stiffness. MMC improves the hardness, wear resistance, and thermal conductivity of the composite and enhances the high-temperature strength and stability. It is suitable for applications requiring superior abrasion resistance, such as brake discs and engine components in the automotive and aerospace industries [17].

3.4.2 Aluminium Oxide (Al_2O_3)

It has high hardness, good thermal stability, and chemical inertness. In MMC it enhances hardness and wear resistance and provides good thermal stability. It is commonly used in applications where high-temperature stability and wear resistance are critical, such as cutting tools and abrasive components [18, 19].

3.4.3 Boron Carbide (B_4C)

It is extremely hard and lightweight, has high compressive strength and excellent neutron absorption. In MMC it enhances hardness and compressive strength. It increases wear resistance. It is suitable for applications requiring lightweight yet strong components, such as body armour and structural materials in aerospace.

3.4.4 Tungsten Carbide (WC)

It is exceptionally hard, with a high melting point and good thermal conductivity. In MMC it improves hardness and wear resistance and enhances thermal conductivity. It is commonly used in tooling and cutting applications for increased durability.

3.4.5 Carbon Fibers

It has a high strength-to-weight ratio, excellent stiffness, and low thermal expansion. In MMC it significantly improves tensile strength and stiffness. It reduces weight without sacrificing strength. Used in automotive and aerospace industries.

3.4.6 Graphene and Carbon Nanotubes

It has exceptional mechanical strength and high thermal and electrical conductivity. In MMC it enhances mechanical properties, including strength and modulus. It improves thermal conductivity. Utilized in advanced applications where lightweight and high-strength properties are critical, such as in nanocomposites for structural components.

3.4.7 Impact on MMC Properties in General

Reinforcement materials typically increase the strength of the MMC, providing improved load-bearing capabilities. Introducing rigid reinforcing materials leads to a significant increase in the modulus of elasticity, resulting in a composite with superior stiffness. Hard and abrasion-resistant reinforcements improve the wear resistance of MMCs, making them suitable for applications in harsh environments [19]. Certain reinforcements enhance the thermal conductivity of MMCs, making them effective in applications requiring good heat dissipation.

3.5 RECENT PROGRESS IN STIR CASTING TECHNIQUES

Recent advancements in stir-casting technology and methodology have significantly improved the production process and properties of MMCs. Current advancements in MMC technology strive to address longstanding challenges, including achieving uniform dispersion of reinforcement particles, minimizing porosity, and enhancing wettability between the matrix and reinforcements. Figure 3.3 depicts the two primary mechanisms [20]. Here are some of the notable recent advancements in stir-casting technology and methodology.

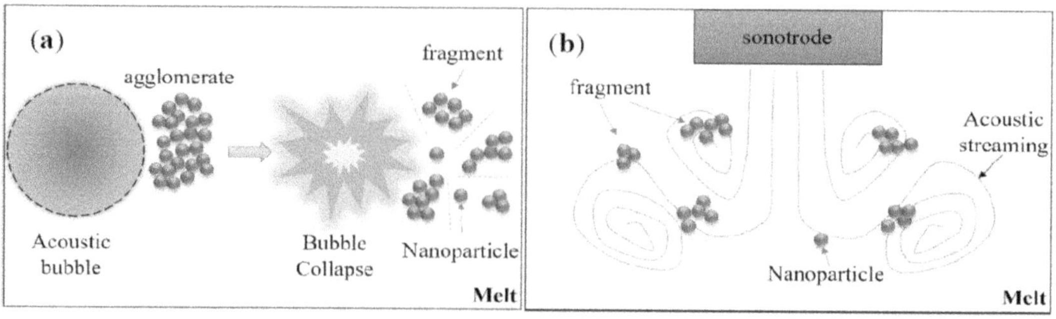

FIGURE 3.3 Illustrated here: The two main principles behind ultrasonic nanoparticle deagglomeration: (a) Cavitation bubble collapse and (b) acoustic streaming [20].

3.5.1 Controlled Atmosphere Stir Casting

To mitigate the formation of oxides and other defects during casting, advancements have included the development of controlled atmosphere stir casting. This technique involves performing the stir casting process under a controlled atmosphere, such as inert gas shielding, to reduce oxidation and improve the quality of the MMCs [21].

3.5.2 Ultrasonic-Assisted Stir Casting

Ultrasonic-assisted stir casting introduces ultrasonic vibrations into the molten metal during the mixing process. Figure 3.4 shows the experimental setup for ultrasonic processing of nanocomposites with UST done outside, and the mechanical properties are given in Figure 3.5. Ultrasonic-assisted stir casting introduces ultrasonic vibrations into the molten metal during the mixing process. This method improves the wettability of the reinforcement particles within the matrix, leading to a more uniform distribution. It also helps in breaking up agglomerates of reinforcement particles and reduces porosity, enhancing the mechanical properties of the composites [22]. The experimental setup of the ultrasonic assisted casting method is shown in Figure 3.6.

3.5.3 Semi-Solid Stir Casting

Semi-solid stir casting involves processing the metal matrix in a semi-solid state, where the metal is partially solid and partially liquid. The ultrasonic-assisted casting method (UACM) provides

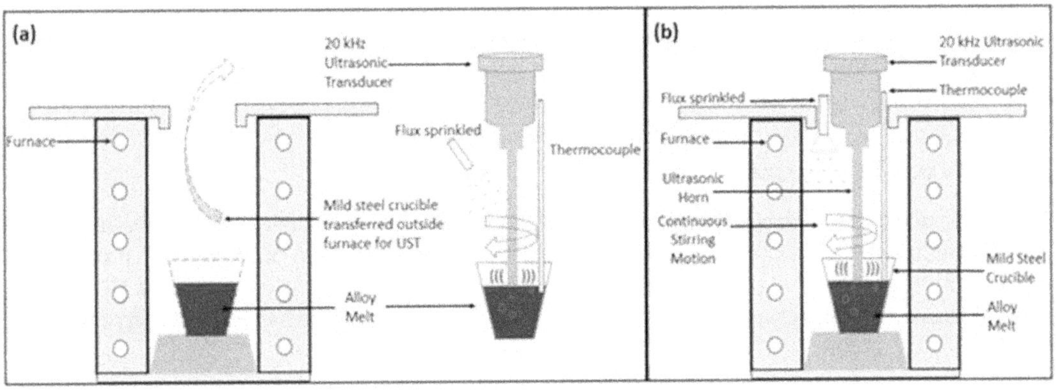

FIGURE 3.4 UST done in setup a) outside b) inside.

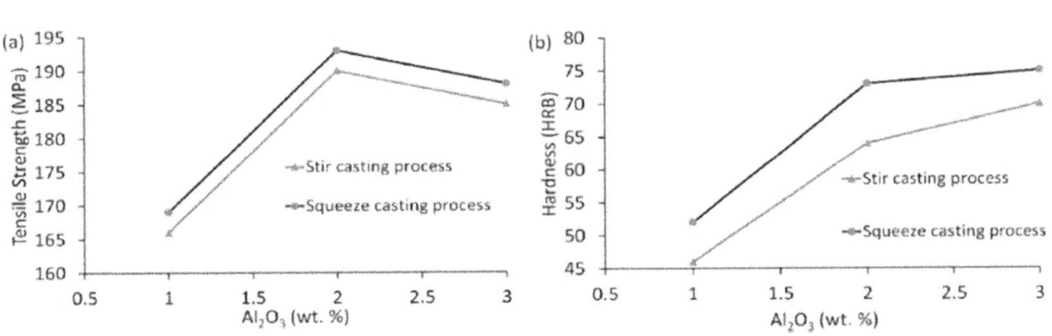

FIGURE 3.5 Comparison of methods (a) tensile strength and (b) hardness [22].

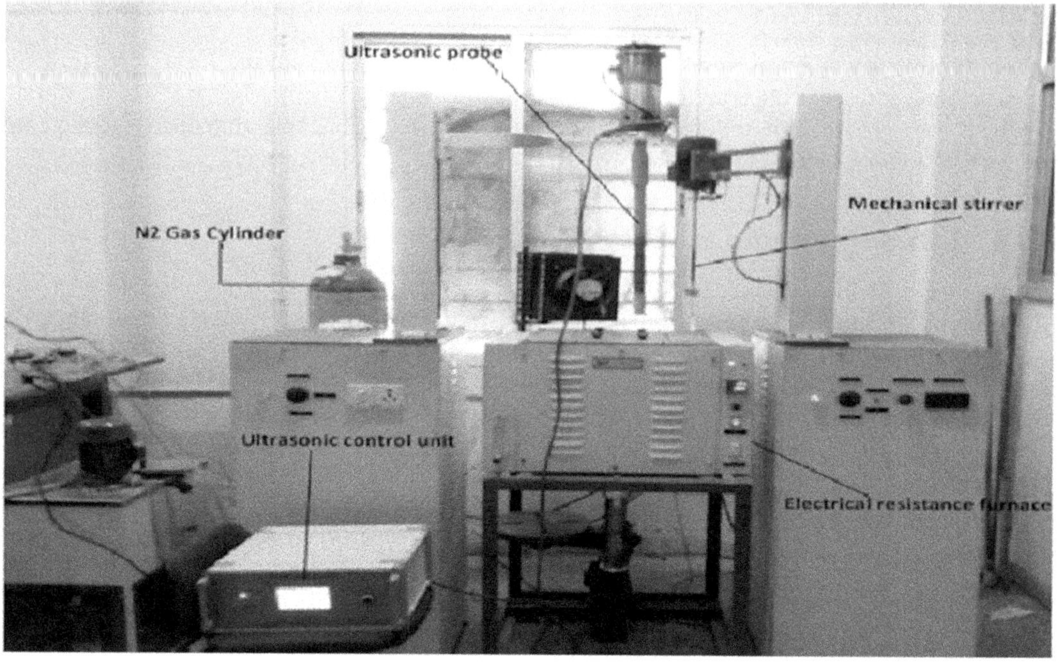

FIGURE 3.6 Ultrasonic assisted casting method [22].

greater control over the material's microstructure. This translates to a finer and more even distribution of reinforcing particles throughout the composite, along with minimized porosity. The combined effect is a considerable betterment in the mechanical properties of the final material [23].

3.5.4 Nano-Reinforcement Integration

The integration of nano-sized reinforcement particles has been a significant advancement in stir casting. Nano-reinforcements such as nano-SiC, nano-Al_2O_3, and graphene offer superior properties at smaller additions compared to their micro-sized counterparts. This leads to composites with enhanced strength, thermal stability, and wear resistance without significantly increasing the density [24].

3.5.5 Vacuum Stir Casting

Vacuum stir casting is an advancement aimed at reducing the incorporation of gases and minimizing porosity within the MMC. By performing the stir casting process under a vacuum, it's possible to significantly reduce the formation of gas porosity, leading to denser and stronger composites [25].

3.5.6 Intelligent Process Control Systems

The implementation of intelligent process control systems in stir-casting processes utilizes sensors and real-time monitoring to adjust parameters such as stirring speed, temperature, and time for optimal results. This automation and control lead to more consistent and high-quality MMCs by minimizing human error and variability in the manufacturing process.

3.5.7 Hybrid Reinforcement Techniques

Hybrid reinforcement involves using a combination of two or more types of reinforcement materials to achieve synergistic effects in the composite properties. This approach allows for the tailoring of

3.5.1 Controlled Atmosphere Stir Casting

To mitigate the formation of oxides and other defects during casting, advancements have included the development of controlled atmosphere stir casting. This technique involves performing the stir casting process under a controlled atmosphere, such as inert gas shielding, to reduce oxidation and improve the quality of the MMCs [21].

3.5.2 Ultrasonic-Assisted Stir Casting

Ultrasonic-assisted stir casting introduces ultrasonic vibrations into the molten metal during the mixing process. Figure 3.4 shows the experimental setup for ultrasonic processing of nanocomposites with UST done outside, and the mechanical properties are given in Figure 3.5. Ultrasonic-assisted stir casting introduces ultrasonic vibrations into the molten metal during the mixing process. This method improves the wettability of the reinforcement particles within the matrix, leading to a more uniform distribution. It also helps in breaking up agglomerates of reinforcement particles and reduces porosity, enhancing the mechanical properties of the composites [22]. The experimental setup of the ultrasonic assisted casting method is shown in Figure 3.6.

3.5.3 Semi-Solid Stir Casting

Semi-solid stir casting involves processing the metal matrix in a semi-solid state, where the metal is partially solid and partially liquid. The ultrasonic-assisted casting method (UACM) provides

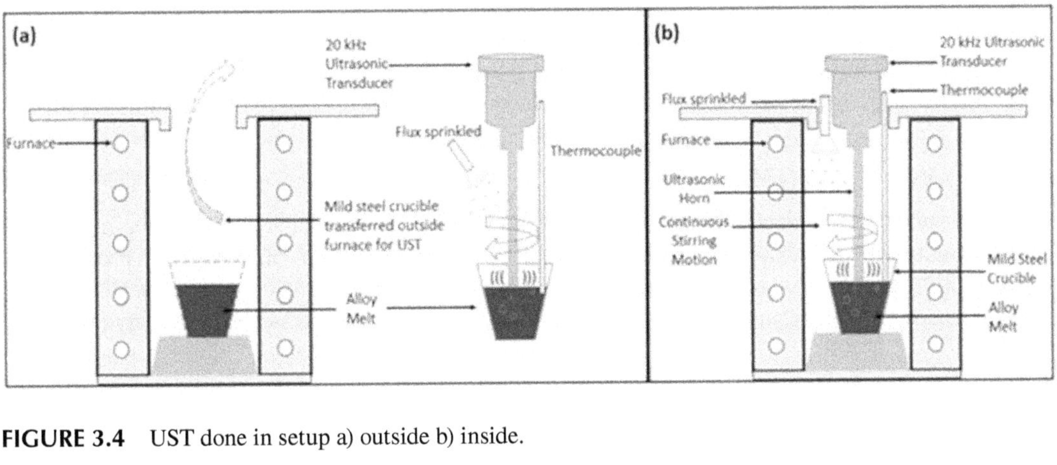

FIGURE 3.4 UST done in setup a) outside b) inside.

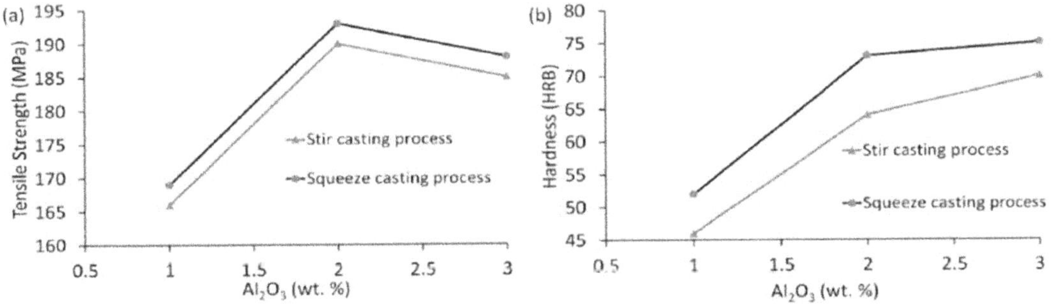

FIGURE 3.5 Comparison of methods (a) tensile strength and (b) hardness [22].

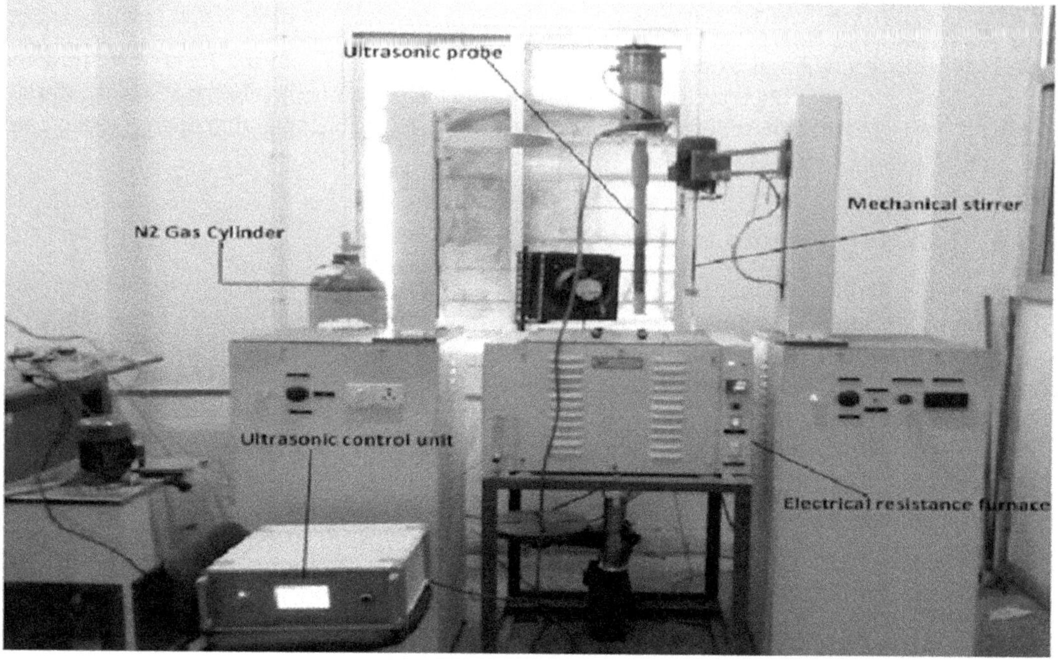

FIGURE 3.6 Ultrasonic assisted casting method [22].

greater control over the material's microstructure. This translates to a finer and more even distribution of reinforcing particles throughout the composite, along with minimized porosity. The combined effect is a considerable betterment in the mechanical properties of the final material [23].

3.5.4 Nano-Reinforcement Integration

The integration of nano-sized reinforcement particles has been a significant advancement in stir casting. Nano-reinforcements such as nano-SiC, nano-Al_2O_3, and graphene offer superior properties at smaller additions compared to their micro-sized counterparts. This leads to composites with enhanced strength, thermal stability, and wear resistance without significantly increasing the density [24].

3.5.5 Vacuum Stir Casting

Vacuum stir casting is an advancement aimed at reducing the incorporation of gases and minimizing porosity within the MMC. By performing the stir casting process under a vacuum, it's possible to significantly reduce the formation of gas porosity, leading to denser and stronger composites [25].

3.5.6 Intelligent Process Control Systems

The implementation of intelligent process control systems in stir-casting processes utilizes sensors and real-time monitoring to adjust parameters such as stirring speed, temperature, and time for optimal results. This automation and control lead to more consistent and high-quality MMCs by minimizing human error and variability in the manufacturing process.

3.5.7 Hybrid Reinforcement Techniques

Hybrid reinforcement involves using a combination of two or more types of reinforcement materials to achieve synergistic effects in the composite properties. This approach allows for the tailoring of

MMCs to specific application requirements, offering a balance between mechanical properties and cost-effectiveness [26].

3.5.8 Multi-Pass Stir Casting

Multi-pass stir casting employs a repetitive process of stirring and solidification cycles. This technique aims to achieve a more homogenous distribution and stronger interfacial bonding between the reinforcement particles and the surrounding metal matrix. This method can lead to composites with enhanced mechanical properties and reduced defects [27].

3.6 LATEST STUDIES IN PRODUCTION OF MMC USING CASTING TECHNIQUES

This research investigated the influence of multi-mode electromagnetic braking (EMBr) devices on molten steel flow behaviour in a continuous casting process. The EMBr device was modelled and its magnetic field strength was analyzed using MAXWELL software. FLUENT software with a magnetohydrodynamics (MHD) module was then used to simulate the molten steel/slag flow dynamics under the influence of the EMBr's electromagnetic field. Figure 3.7 depicts the velocity distribution across the surface of the funnel mould [28]. The close agreement between the numerical simulations and the industrial nail board experiment data validates the accuracy of the simulations [28].

A combined approach of theoretical simulations and experimental validation was used to determine the optimal pouring time interval for the dual-liquid casting process used in manufacturing LAS/HCCI bimetallic hammerheads (low alloy steel bonded with high chromium cast iron) [29]. Other research explores methods for assessing mould flux performance in continuous steel casting, beginning with standard analyses of particle size, phase identification, and transformations at high temperatures, alongside viscosity and break temperature evaluations. The study underscores the

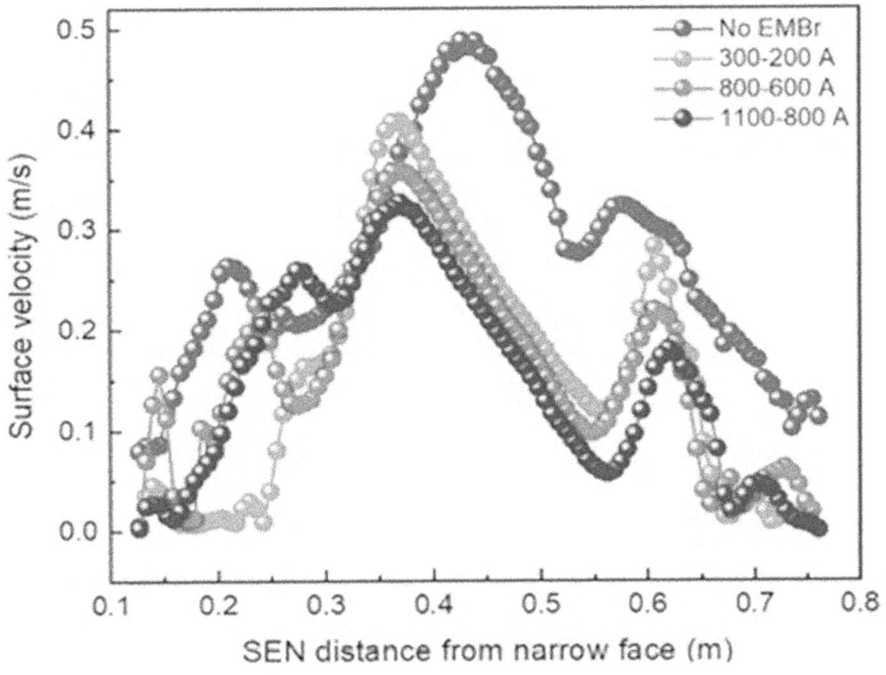

FIGURE 3.7 Funnel mould surface velocity [28].

combined use of a mould simulator and traditional techniques for a comprehensive assessment of mould flux performance in steel casting processes [30].

Continuous casting (CC) is a crucial steel production process known for its high efficiency and close-to-net shape outcomes. Enhancing the quality of CC products is a key research objective, focusing on the stability of the casting process. Submerged entry nozzle (SEN) clogging leads to issues like uneven solidification, mould flow asymmetry, meniscus fluctuations, and potential slag entrapment [31].

This paper introduces a novel theoretical framework designed to model the material properties of steel during the continuous casting process. It leverages the unique behaviours stemming from non-Newtonian dynamics, highlighting how these systems naturally align with fractional order transfer function models due to their distributed parameter characteristics. Additionally, the paper provides an analysis of how the implemented closed-loop control system impacts the final product quality, offering valuable insights into the end-to-end production line performance [32].

This study investigates the complex multi-phase interactions that occur during continuous casting (CC) with the influence of electromagnetic braking (EMBr). The findings suggest a calmer meniscus under EMBr for both conductive and semi-conductive mould domains, especially at a deeper immersion depth. Conversely, an insulated mould with shallow immersion depth led to self-induced MHD vortices, causing meniscus waving and potential entrapment of air bubbles. Additionally, the study employed power spectral density (PSD) analysis to further characterize the flow energy characteristics [33].

This research investigates the effect of ultrasonic vibrations applied during solidification on the aging behaviour of an AlSi7Mg alloy. The ultrasonic treatment, delivered through equipment attached to the mould walls, was implemented in two ways. Figure 3.8 presents backscatter electron (BSE) mode scanning electron microscope (SEM) micrographs of the solution-treated microstructures for samples A, B, and C. Additionally, highlighted intermetallic phases are shown in region (d) for each sample [34].

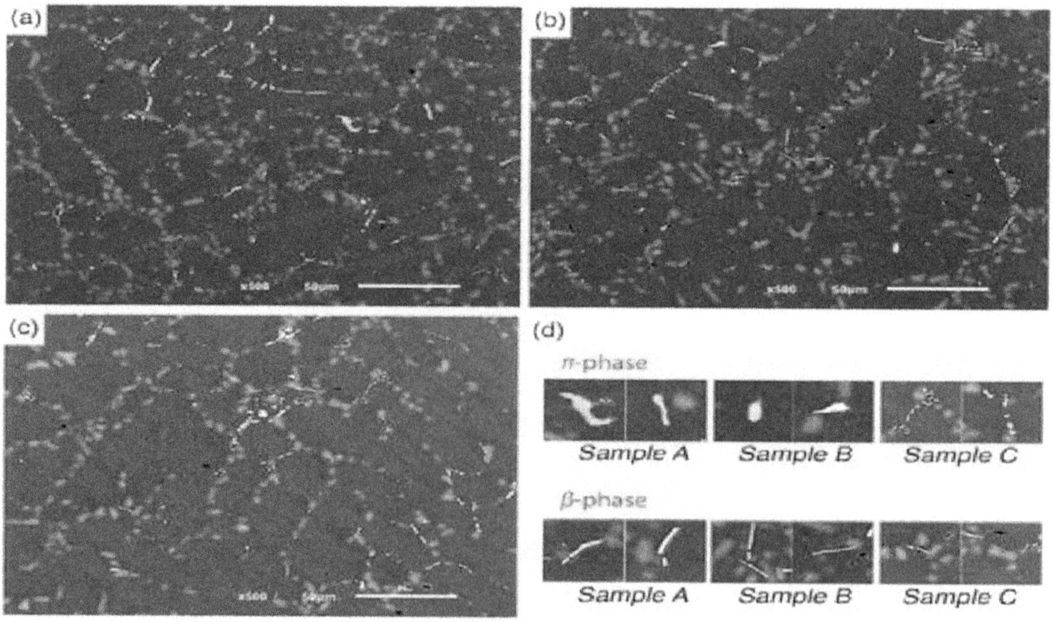

FIGURE 3.8 SEM micrographs of various samples [34].

This review discusses the evolution and advancements in gel casting methods for creating porous ceramics, focusing on the shift from using toxic acrylamide (AM) systems to more environmentally friendly alternatives involving natural polymers and minerals. The review underscores gel casting's role in advancing the fabrication of advanced, sustainable porous ceramics, serving as a key reference for future research in the field [35].

S. Kim and colleagues developed an innovative MMC that includes a durable TiN network, designed to endure the extreme temperatures of molten steel at 1540°C—well above the melting point of its AISI 316 L stainless steel base (1,430°C). This MMC is distinguished by its spider web-like TiN reinforcement, which significantly boosts its mechanical strength and toughness, ensuring it retains its form even when submerged in molten steel. The process of sintering, either during immersion or via heat treatment, enhances the TiN network's structure, with the formation of oxides contributing to better connectivity, strength, and toughness [36].

This research explores the properties of a novel ternary-reinforced metal matrix composite (n-HMMC) based on the AA7075 aluminium alloy. It significantly enhanced ultimate tensile strength by 24.4% and hardness by 32.8% over the base alloy but at the cost of ductility and impact strength due to microstructural inhomogeneities and agglomerates. Figure 3.9 presents the average ultimate tensile strength and percentage elongation values for the investigated materials [37].

This research explores a recent advancement in metal matrix composite (MMC) development: The utilization of in-situ generated core-shell structured particles as reinforcement. This breakthrough introduces a balanced paradigm in MMC fabrication, merging strength with

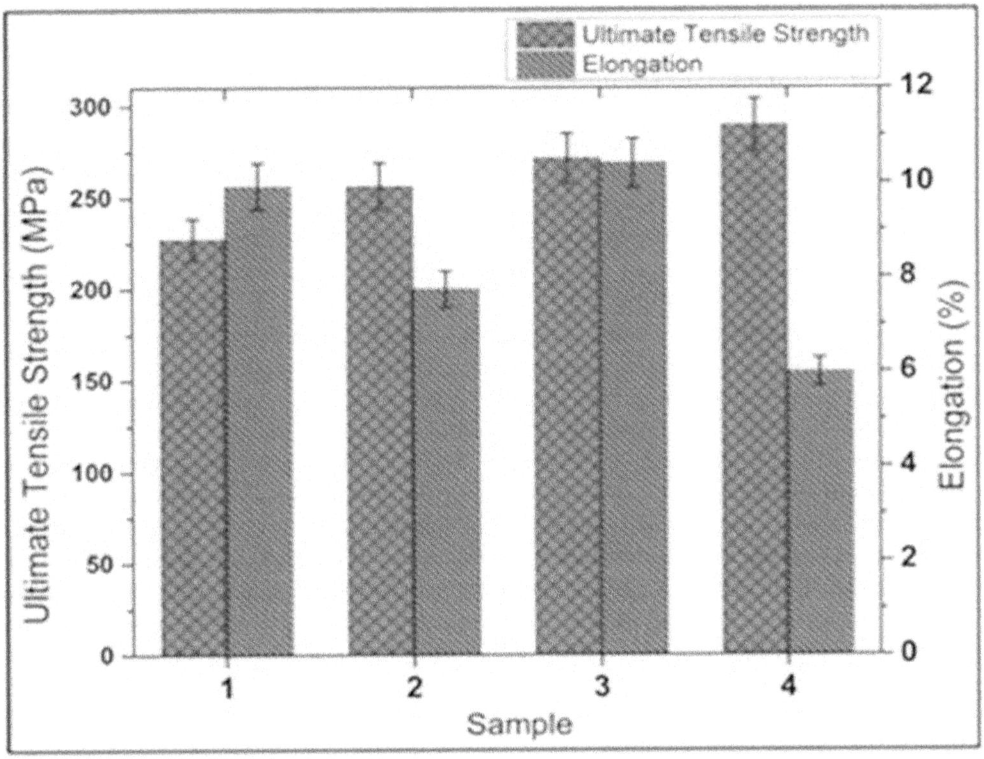

FIGURE 3.9 Presents the average ultimate tensile strength and percentage elongation values for the investigated materials [37].

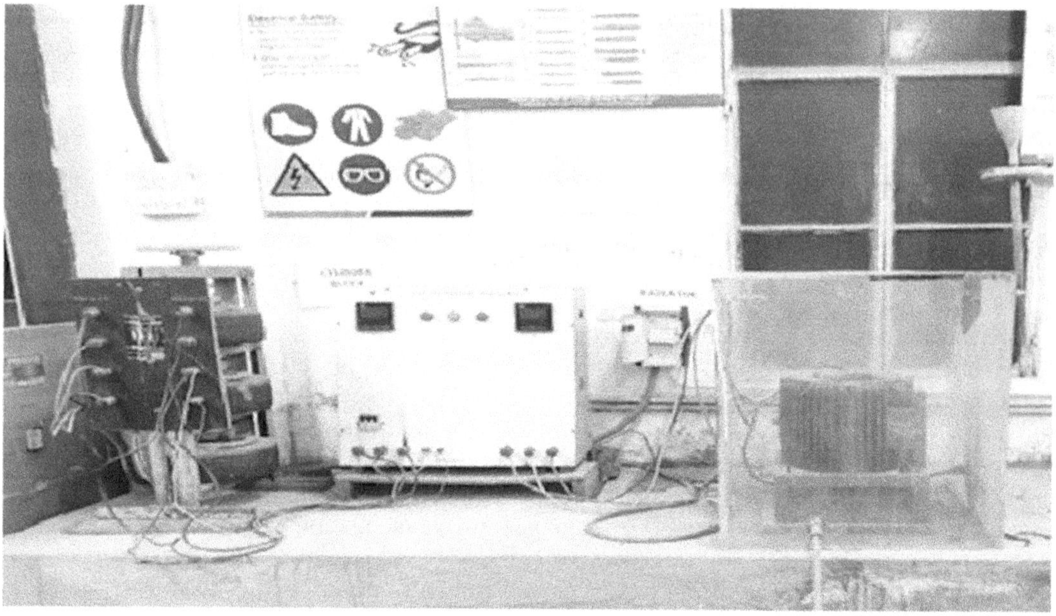

FIGURE 3.10 Setup of EMSC process [40].

flexibility. Despite the promising outlook, the field demands further investigation to refine these materials for optimization and real-world applications, marking a promising direction in materials science [38].

This research investigates enhancing low carbon cast steel's performance by integrating WC-metal matrix composites (WC-MMCs) directly within the cast, aiming to rival the properties of hard alloy steels. Mechanical assessments reveal significant improvements—hardness increases by threefold and wear resistance improves by 39% compared to the untreated base metal. This local reinforcement technique thus presents a promising method to significantly enhance the material properties of low-carbon cast steel [39].

Utilizing electromagnetic stir casting (EMSC), this study examines the production of a uniform MMC with AA6063-T6 alloy reinforced by 2wt% SiC and 5wt% waste bamboo powder (WBP). FESEM images demonstrate strong bonding with minimal porosity. XRD analysis indicates the absence of contaminants, with Al3Ca exhibiting the highest peak. The set-up used for this research is shown in Figure 3.10. Reinforcement incorporation led to significant improvements in mechanical properties [40].

This research focuses on aluminium-based metal matrix composites (AMCs). Incorporating SiC as a reinforcement enhances these attributes but at the cost of ductility. This research investigates SiC-reinforced aluminium MMCs fabricated using stir casting. The microstructures of the composites were characterized with SEM and EDX. Mechanical properties were also evaluated. Machining tests were conducted to determine the impact of varying cutting speeds and feed rates on the composites' machinability and surface quality. It was found that higher cutting speeds and lower feed rates facilitated easier machining and yielded better surface finishes, identifying optimal conditions for achieving the finest surface quality.

TABLE 3.1
Recent Works on Liquid Metallurgy

Manufacturing process	Tests conducted	Results	References
Stir casting with PG in 7075 aluminium alloy	BHN hardness, tensile strength, wear resistance	Significant improvements in BHN hardness, tensile strength, and wear resistance, especially with 10% PG content	Waleed T. Rashid
Stir casting of AA8011 with nano B_4C	Tribological behaviours, SEM analysis	Improved tribological performance with enhanced mechanical characteristics from nano B_4C reinforcement	N. Vinayaka et al.
Core–shell particle reinforced MMCs	Not specified	Enhanced strength without sacrificing ductility, overcoming brittleness. A new direction in materials science	D. Dudina and K. Georgarakis
WC-MMCs in low carbon cast steel	Microstructure analysis, mechanical testing	Threefold increase in hardness, 39% reduction in wear rate compared to base metal	Aida B. Moreira et al.
Nano B_4C reinforced aluminium alloy AA8011	Tribological behaviours, SEM analysis	Improved tribological performance with enhanced mechanical characteristics from nano B_4C reinforcement	N. Vinayaka et al.
Iron-based MMCs with graphite, alumina, zirconia nanoparticles	Surface topography analysis	Type and quantity of additives significantly influence surface geometry, aiding in selecting optimal combinations	R. Maruda et al.
EMBr effect on flow in slab continuous casting mould	Numerical simulations, industrial experiments	EMBr alters molten steel's flow patterns, reduces liquid-level fluctuation, mitigates risk of slag entrapment	Limin Zhang et al.
Dual-liquid casting of LAS/HCCI bimetallic hammerheads	Theoretical simulation, experimental validation	Optimized pouring time interval (40 seconds) results in superior interfacial width and bonding strength	Z. Xing et al.
How reinforcements, manufacturing, and microstructure shape wear resistance	Wear analysis, characterization	The manufacturing process significantly impacts the microstructure of aluminium matrix composites (Al-MMCs). Studies have shown that in-situ fabrication techniques generally yield Al-MMCs with superior microstructures and tribological performance compared to those produced through ex-situ methods	P Hariharasakthisudhan et al.

(Continued)

TABLE 3.1 (*Continued*)
Recent Works on Liquid Metallurgy

Manufacturing process	Tests conducted	Results	References
Tribological performance of a multiscale hybrid composite (AA6061/Al2O3/Si3N4/Graphite) sliding against EN31 steel	Ultrasonic treatment, wear study, structural characterization	This study demonstrates that incorporating dual-sized ceramic reinforcements enhances the material's wear resistance, particularly at extended sliding distances. Furthermore, pre-mixing the reinforcements with aluminium metal powder plays a crucial role. This pre-mixing process improves the wettability between the matrix and the reinforcements, ultimately leading to reduced wear loss	P Hariharasakthisudhan et al.

3.7 SUMMARY

The escalating demand for lightweight materials in diverse industries has propelled the exploration of advanced composites, with aluminium metal matrix composites (AMMCs) emerging as promising contenders for heightened material performance. This book chapter thoroughly explores the significance of AMMCs in structural and functional applications across various sectors, including marine, defence, automotive, aerospace, and high-temperature environments. This review article highlights stir casting as a widely adopted technique for fabricating metal matrix composites (MMCs) due to its simplicity and cost-efficiency, particularly for large-scale production. The review delves into critical aspects of the stir casting process, including its impact on the mechanical properties of the resulting MMCs. Furthermore, it explores the influence of incorporating diverse reinforcement materials and the challenges associated with stir casting of MMCs. By thoroughly examining mechanical properties such as strength, stiffness, and durability, the chapter aims to illuminate the performance enhancements achievable through the incorporation of reinforcing materials. Additionally, the chapter underscores the impact of different reinforcement elements on the overall characteristics of AMMCs, emphasizing the significance of meticulous material selection. Furthermore, the review addresses challenges associated with the stir casting method, offering valuable insights into current limitations and potential areas for improvement in the production process. Notably, advanced casting processes have gained prominence in recent years, and their significance in shaping the future of AMMCs is acknowledged. This chapter serves a dual purpose. It acts as a comprehensive resource by consolidating existing knowledge on advanced composite materials. This knowledge base proves valuable for researchers, engineers, and professionals working in this field. Furthermore, the chapter functions as a roadmap for future research by identifying key areas for further exploration. By providing this comprehensive overview of the current state of the art, the chapter effectively paves the way for significant advancements in the development of advanced composite materials.

REFERENCES

[1] Heitkamp, T., Girnth, S., Kuschmitz, S., Waldt, N., Klawitter, G. and Vietor, T., 2023. Experimental and numerical investigation of the mechanical properties of 3D-printed hybrid and non-hybrid composites. *Polymers*, *15*(5), p. 1164.

[2] Menachery, N., Thomas, S., Deepanraj, B. and Senthilkumar, N., 2023. Processing of nanoreinforced aluminium hybrid metal matrix composites and the effect of post-heat treatment: A review. *Applied Nanoscience*, *13*(6), pp. 4075–4099.
[3] Rahman, M.A., Sirajudeen, N. and Karunanithi, R., 2019. Hardness and dry sliding wear behavior of SiC reinforced Al 7150 alloy composite synthesized under the influence of aging, varying compaction pressures, particle size and volume fraction of ceramic particulates by hot compaction technique. *Materials Research Express*, *6*(6), p. 066536.
[4] Sirajudeen, N. and Rahman, M.A., 2020. Influence of aging and mass fraction of Al_2O_3 on mechanical properties of Al6082/Al_2O_3 composite produced by stir casting. *Metal Science and Heat Treatment*, *62*, pp. 205–209.
[5] Haque, S., Srirangarajalu, N., Abdur Rahman, M. and Sirajudeen, N., 2019. Effect of manufacturing process parameters on the wear rate of Al 6061–Cu–SiCp alloy composites. *Emerging Materials Research*, *8*(2), pp. 217–224.
[6] Gupta, M. and Ling, S.N.M., 2011. *Magnesium, Magnesium Alloys, and Magnesium Composites*. John Wiley & Sons.
[7] Dwivedi, S.P., Chaudhary, V. and Sharma, S., 2023. Effect of the addition of waste glass powder along with TiC as reinforcement on microstructure, wettability, mechanical and tribological behavior of AZ91D magnesium based alloy. *International Journal of Metalcasting*, *18*, pp. 1387–1400.
[8] Guo, Z., Liang, Y., Cao, J., Sun, T., Liu, X., Xue, J. and Lin, J., 2023. Fabrication of Al-YOF/TiB2 composites by molten-salt-assisted stirring casting. *Materials Letters*, *330*, p. 133385.
[9] Wu, P., Zhang, Y., Hu, J., Song, S., Li, Y., Wang, H., Yuan, G., Wang, Z., Wei, S. and Liu, F., 2023. Generalized stability criterion for controlling solidification segregation upon twin-roll casting. *Journal of Materials Science & Technology*, *134*, pp. 163–177.
[10] Logesh, K., Hariharasakthisudhan, P., Moshi, A.A.M., Rajan, B.S. and Sathickbasha, K., 2019. Mechanical properties and microstructure of A356 alloy reinforced AlN/MWCNT/graphite/Al composites fabricated by stir casting. *Materials Research Express*, *7*(1), p. 015004.
[11] Li, S.S., Yue, X., Li, Q.Y., Peng, H.L., Dong, B.X., Liu, T.S., Yang, H.Y., Fan, J., Shu, S.L., Qiu, F. and Jiang, Q.C., 2023. Development and applications of aluminum alloys for aerospace industry. *Journal of Materials Research and Technology*, *27*, pp. 943–983.
[12] Liu, B., Yang, J., Zhang, X., Yang, Q., Zhang, J. and Li, X., 2023. Development and application of magnesium alloy parts for automotive OEMs: A review. *Journal of Magnesium and Alloys*, *11*(1), pp. 15–47.
[13] Li, G., Ma, F., Liu, P., Qi, S., Li, W., Zhang, K. and Chen, X., 2023. Review of micro-arc oxidation of titanium alloys: Mechanism, properties and applications. *Journal of Alloys and Compounds*, *948*, p. 169773.
[14] Sun, Y., Gong, W., Zhu, R., Li, Y. and Sun, S., 2023. Research progress and prospect of friction stir welding of copper and copper alloys. *Metallurgical Research & Technology*, *120*(3), p. 309.
[15] Rathi, N., Kumar, P. and Gupta, A., 2023. Non-conventional machining of nickel based superalloys: A review. *Materials Today: Proceedings*. In press.
[16] Ostolaza, M., Arrizubieta, J.I., Lamikiz, A., Plaza, S. and Ortega, N., 2023. Latest developments to manufacture metal matrix composites and functionally graded materials through AM: A state-of-the-art review. *Materials*, *16*(4), p. 1746.
[17] Sharma, S., Goyal, A., Bharadwaj, P., Oza, A.D. and Pandey, A., 2023. Application of metal matrix composite fabricated by reinforcement materials—a review. *Materials Today: Proceedings*. In press.
[18] Sirajudeen, N., Rahman, M.A., Haque, S., Karunanithi, R. and Patnaik, S., 2019. Influence of aging & varying weight fraction of Al_2O_3 particles on the mechanical behaviour & volumetric wear rate of Al 7075 alloy composite produced by liquid metallurgy route. *Materials Research Express*, *6*(8), p. 086597.
[19] Hariharasakthisudhan, P., Sathickbasha, K., Surya Rajan, B. and Jeyanthram, V., 2020. Steady-state wear behavior of multi-scale hybrid composite—AA6061/Al_2O_3 (μ m)/Si_3N_4 (nm)/graphite—EN31 steel tribo system. *Journal of Tribology*, *142*(1), p. 011702.
[20] Khandelwal, A., Mani, K., Srivastava, N., Gupta, R. and Chaudhari, G.P., 2017. Mechanical behavior of AZ31/Al_2O_3 magnesium alloy nanocomposites prepared using ultrasound assisted stir casting. *Composites Part B: Engineering*, *123*, pp. 64–73.
[21] Hariharasakthisudhan, P., Sudhan, H.H.T., Karthik, S., Sathickbasha, K. and Surya Rajan, B., 2022. Quasi-static compression behavior and microstructure changes in low-cost AA6061 composites. *Proceedings of the Institution of Mechanical Engineers, Part C: Journal of Mechanical Engineering Science*, *236*(9), pp. 4876–4884.
[22] Rana, R.S. and Purohit, R., 2015. Development and analysis of Al-matrix nano composites fabricated by ultrasonic assisted squeeze casting process. *Materials Today: Proceedings*, *2*(4–5), pp. 3697–3703.

[23] Zhang, Q., 2023. *Mechanical Properties of Semi-Solid Al Castings: Role of Stirring*, Doctoral dissertation, Jönköping University, School of Engineering.

[24] Sobhani, E., 2023. Improvement of vibrational characteristics of multipurpose structures (plate and shells) used in aerospace components by deploying graphene oxide powders (GOPs) in a matrix as a nano-reinforcement: A comprehensive study. *Engineering Analysis with Boundary Elements*, *146*, pp. 598–635.

[25] Baraniraj, A., Sathiyagnanam, A.P., Venkatesh, R. and De Poures, M.V., 2023. Vacuum stir cast developed aluminium alloy hybrid nanocomposite performance compared with gravity cast: Mechanical and tribological characteristics study. *International Journal of Metalcasting*, *18*, pp. 1273–1283.

[26] Abebe Emiru, A., Sinha, D.K., Kumar, A. and Yadav, A., 2023. Fabrication and characterization of hybrid aluminium (Al6061) metal matrix composite reinforced with SiC, B4C and MoS2 via stir casting. *International Journal of Metal Casting*, *17*(2), pp. 801–812.

[27] Jolokhani, A., Razaghian, A., Moharami, A. and Emamy, M., 2023. Microstructure and tribological properties of as-cast and multi-pass friction stir processed Mg-0.5 Zn-0.5 Zr/SiC composite fabricated by stir casting technique. *Journal of Materials Research and Technology*, *27*, pp. 7823–7838.

[28] Zhang, L., Xiao, P., Wang, Y., Zhang, C. and Zhu, L., 2023. Effect of EMBr on flow in slab continuous casting mold and industrial experiment of nail dipping measurement. *Metals*, *13*(1), p. 167.

[29] Xing, Z.G., He, L.X., Liang, S.X., Chang, L.B., Xiao, Z.X., Xing, W.L., Shen, H.B., Cao, J.J. and Liu, H.J., 2023. Process optimization of dual-liquid casting and interfacial strength–toughness of the produced LAS/HCCI bimetal. *Materials*, *16*(5), p. 2008.

[30] Kamaraj, A., Tripathy, S., Chalavadi, G., Sahoo, P.P. and Misra, S., 2022. Characterization and assessment of mold flux for continuous casting of liquid steel using an inverse mold simulator. *Steel Research International*, *93*(3), p. 2100121.

[31] Vakhrushev, A., Kharicha, A., Wu, M., Ludwig, A., Tang, Y., Hackl, G., Nitzl, G., Watzinger, J. and Bohacek, J., 2021. On modelling parasitic solidification due to heat loss at submerged entry nozzle region of continuous casting mold. *Metals*, *11*(9), p. 1375.

[32] Birs, I., Muresan, C., Copot, D. and Ionescu, C., 2023. Model identification and control of electromagnetic actuation in continuous casting process with improved quality. *IEEE/CAA Journal of Automatica Sinica*, *10*(1), pp. 203–215.

[33] Vakhrushev, A., Karimi-Sibaki, E., Bohacek, J., Wu, M., Ludwig, A., Tang, Y., Hackl, G., Nitzl, G., Watzinger, J. and Kharicha, A., 2023. Impact of submerged entry nozzle (SEN) immersion depth on meniscus flow in continuous casting mold under electromagnetic brake (EMBr). *Metals*, *13*(3), p. 444.

[34] Gomes, I.V., Grilo, J., Carneiro, V.H. and Puga, H., 2023. Impact of the ultrasonic-assisted casting of an AlSi7Mg alloy on T6 heat treatment. *Metals*, *13*(2), p. 255.

[35] Putri, S.E., Ahmad, A., Raya, I., Triandi, R., Tjahjanto, H.N., Taba, P. and Karim, H., 2023. A review of the development of the gel casting method for porous ceramic fabrication. *Rasayan Journal of Chemistry*, *16*(1), pp. 48–60.

[36] Kim, S.B., Kim, S.H. and Cho, J.W., 2021. High temperature endurable metal matrix composite reinforced with continuously networked TiN. *Journal of Alloys and Compounds*, *889*, p. 161633.

[37] Khan, A.H., Shah, S.A.A., Umar, F., Noor, U., Gul, R.M., Giasin, K. and Aamir, M., 2022. Investigating the microstructural and mechanical properties of novel ternary reinforced AA7075 hybrid metal matrix composite. *Materials*, *15*(15), p. 5303.

[38] Dudina, D.V. and Georgarakis, K., 2022. Core–shell particle reinforcements—a new trend in the design and development of metal matrix composites. *Materials*, *15*(7), p. 2629.

[39] Moreira, A.B., Ribeiro, L.M., Lacerda, P., Pinto, A.M. and Vieira, M.F., 2022. A study on a cast steel reinforced with WC–metal matrix composite. *Materials*, *15*(18), p. 6199.

[40] Sharma, O., Gupta, P. and Kumar, T., 2022, July. Fabrication and characterisation of AA6063-T6/SiC/waste bone powder using electromagnetic stir casting. *IOP Conference Series: Materials Science and Engineering*, *1248*(1), p. 012076.

4 Development of Metal Matrix Composites Through Solid-State Techniques

Saravanan Rathinasamy, Omkumar Meenakshisundaram, Sathish Thanikodi, and Jothi Arunachalam Solairaju

4.1 INTRODUCTION

4.1.1 Overview of MMCs

A metal matrix reinforced with a secondary material, usually a ceramic or another metal, is known as a metal matrix composite (MMC) class of materials. This combination creates a unique synergy that imparts enhanced thermal, electrical, and mechanical properties compared to traditional monolithic metals. The versatility and tolerability of MMCs have led to their extensive use in high-performance applications, like automotive, defence, electronics, and aerospace. By carefully choosing the type of metal matrix, the material for the reinforcement, and the volume fraction of the reinforcement, MMCs can be made to meet specific requirements [1]. Aluminium, titanium, magnesium, and copper are common metal matrices, and ceramics like silicon carbide, aluminium oxide, or carbon-based materials like graphene are frequently used as reinforcements. Because of their versatility, MMCs can provide a range of benefits, including enhanced thermal and electrical conductivity, high strength, low weight, and corrosion resistance. These benefits are especially useful in situations where conventional metals aren't a good fit [2].

4.1.2 Importance of Synthesis Techniques

The final properties and performance characteristics of MMCs are largely determined by the synthesis techniques employed during their production. For this reason, variables like thermal stability, durability, mechanical strength, and even manufacturing costs can be greatly impacted by the synthesis method chosen. Techniques like powder metallurgy and spark plasma sintering offer precise control over these microstructural aspects, allowing for the design of composites with specific mechanical properties. Uniform distribution of reinforcement is critical for ensuring consistent performance throughout the composite, and synthesis techniques that promote uniformity can lead to stronger, more reliable materials [3].

4.1.3 Role of Solid-State Methods

Solid-state methods are fundamental to the synthesis of metal matrix composites (MMCs), offering key advantages over traditional melting-based techniques. These methods, which do not require the complete melting of the metal matrix, provide enhanced control over the composite's microstructure, allowing for uniform distribution of reinforcement and finer grain sizes. This leads to consistent mechanical properties and minimizes issues such as segregation and porosity, which are

DOI: 10.1201/9781003499718-4

common in melting-based processes. Moreover, solid-state methods reduce the risk of chemical reactions between the metal matrix and the reinforcement, resulting in stronger and more stable interfacial bonding [4].

4.2 POWDER METALLURGY

4.2.1 Principle and Process Overview

The principle behind solid-state methods for synthesizing metal matrix composites (MMCs) involves creating a composite material without fully melting the metal matrix, which helps maintain greater control over the microstructure and avoids issues related to excessive heat, such as chemical reactions, oxidation, and porosity [5].

4.2.1.1 Friction Stir Processing

Friction stir processing (FSP) involves using a rotating tool to generate frictional heat and mechanical deformation within a solid metal matrix. As the tool moves through the material, the friction-generated heat softens the metal without melting it, creating a plastically deformed region. Reinforcement materials can be introduced during this process, allowing them to mix thoroughly with the softened metal, leading to a fine-grained microstructure and strong interfacial bonding.

4.2.2 Powder Selection and Preparation

In powder metallurgy in particular, the choice and preparation of powders are crucial processes in the solid-state synthesis of MMCs. The final composite's microstructure, mechanical attributes, and performance are directly impacted by the calibre and features of the powders used in these procedures.

4.2.2.1 Metal Matrix Selection

The desired qualities of the composite and its intended use determine which metal should be used for the matrix. Copper, titanium, magnesium, and aluminium are typical matrix metals. Because of its excellent mechanical qualities and low weight, aluminium is used extensively.

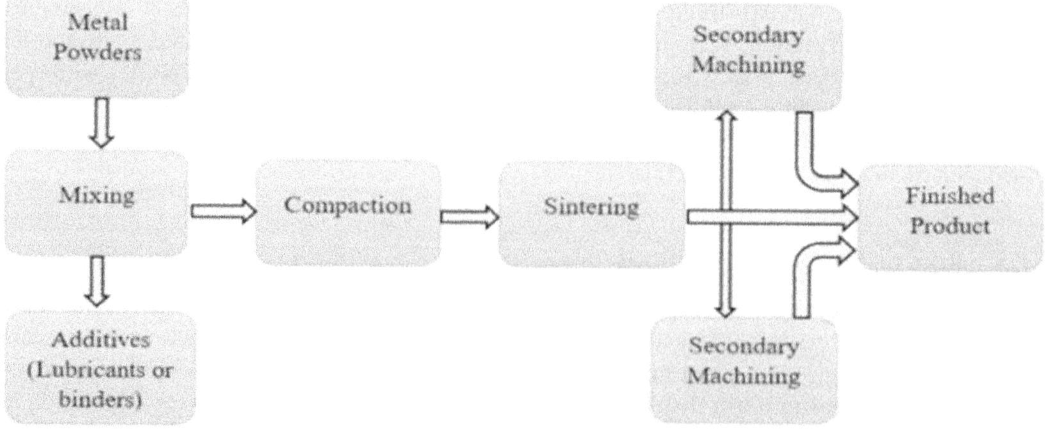

FIGURE 4.1 Powder metallurgy process.

4.2.2.2 Reinforcement Selection

Reinforcement materials typically include ceramics like silicon carbide, aluminium oxide, and boron carbide, as well as carbon-based materials like graphene and carbon nanotubes. The selection depends on the desired mechanical, thermal, and electrical properties of the MMC.

4.2.2.3 Powder Characteristics

Key characteristics to consider when selecting powders for MMC synthesis are particle size, shape, and distribution. Fine powders generally provide better compaction and sintering, leading to denser composites. The shape of the powder particles can influence flowability and compaction behaviour [6].

4.2.2.4 Powder Preparation

The preparation of powders involves several steps to ensure they are ready for compaction and sintering. Homogeneous mixing of metal matrix powders and reinforcement materials is crucial for uniformity.

4.2.3 Consolidation Technique

A crucial stage in the solid-state synthesis of metal matrix composites (MMCs) is consolidation. Through the use of pressure and heat, loose powders are transformed into a dense, solid structure in this process [7].

4.2.3.1 Hot Pressing

Hot pressing involves simultaneously applying heat and pressure to powder compacts. This technique is typically performed in a vacuum or inert atmosphere to prevent oxidation and contamination. The heat promotes diffusion among the metal particles, while the pressure facilitates compaction and reduces porosity.

4.2.3.2 Cold Pressing and Sintering

In cold pressing, powders are compacted at room temperature to create a "green" compact, which is then sintered at elevated temperatures to achieve densification. This two-step process is common in powder metallurgy, where the green compact is sintered in a furnace to enable diffusion among the metal particles. Figure 4.2 represents how the cold pressing and sintering are done.

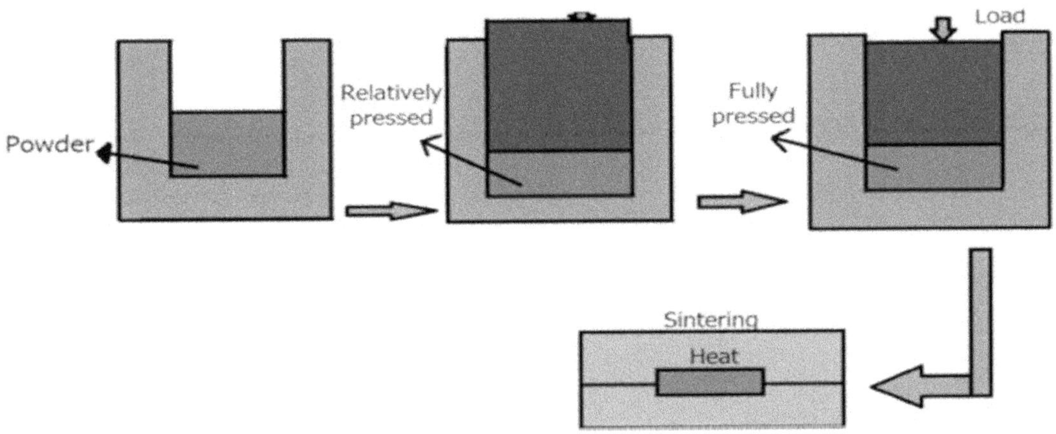

FIGURE 4.2 Cold pressing and sintering methods in MMC.

4.2.3.3 Spark Plasma Sintering

Spark plasma sintering (SPS) uses pulsed electric currents to rapidly heat and consolidate powder compacts. The electrical pulses generate localized heat, which, combined with applied pressure, leads to fast and efficient densification.

4.2.3.4 Isostatic Pressing

Isostatic pressing applies uniform pressure to a powder compact from all directions, promoting even compaction and minimizing stress concentrations [8].

4.2.4 Sintering and Densification Mechanisms

Sintering and densification are essential mechanisms in the solid-state synthesis of metal matrix composites (MMCs). These processes involve bonding particles at elevated temperatures, without fully melting the metal matrix, to achieve a dense and mechanically robust structure. The following sections describe the key principles and mechanisms of sintering and densification in the context of MMC production [9].

4.2.4.1 Sintering Fundamentals

Sintering is the process where particles, typically powders, are heated to a temperature below their melting point, allowing atoms to diffuse across particle boundaries and form solid bonds.

4.2.4.2 Diffusion Mechanisms

During sintering, diffusion is the primary mechanism through which particles bond and densification occurs.

4.2.4.3 Densification Techniques

Combines heat and pressure to accelerate densification, resulting in higher density and reduced porosity. Uses pulsed electric currents to create localized heating and pressure, leading to rapid densification and fine-grained microstructures. Applies uniform pressure and heat to achieve complete densification and eliminate internal voids [10].

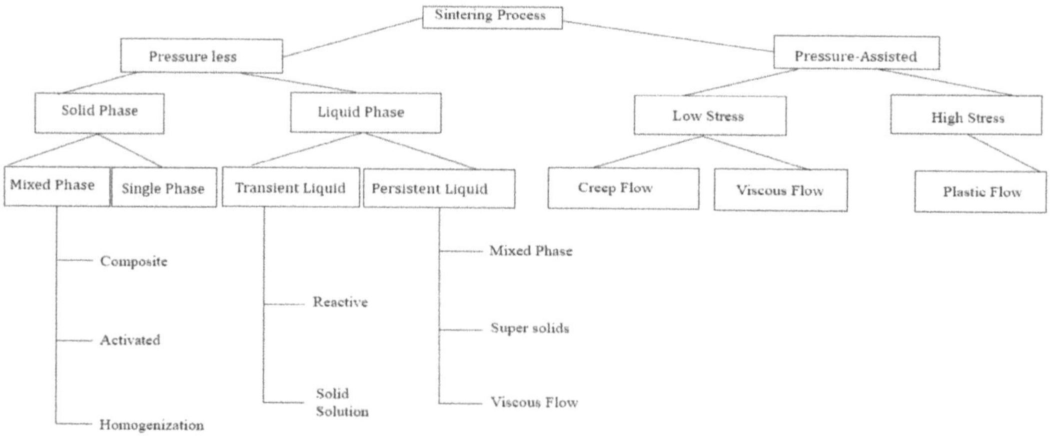

FIGURE 4.3 Sintering process.

4.2.5 Control of Microstructure and Properties

4.2.5.1 Powder Characteristics

The size, shape, and distribution of powder particles influence the final microstructure. Uniform, fine powders generally lead to denser composites with better mechanical properties.

4.2.5.2 Compaction and Consolidation Techniques

Methods like hot pressing, spark plasma sintering, and hot isostatic pressing affect grain size, porosity, and overall densification. Optimizing temperature and pressure helps achieve the desired structure and properties.

4.2.5.3 Sintering Temperature and Time

Controlling grain growth and densification requires a precise balance between sintering temperature and duration. Reduced mechanical strength and excessive grain growth can result from sintering at too high of a temperature or for too long of a time.

4.2.5.4 Interfacial Bonding

It is essential that the reinforcement material and metal matrix have a strong bond. Friction stir processing and diffusion bonding are examples of solid-state methods that reduce chemical reactions and strengthen interfacial layers.

4.2.5.5 Reinforcement Distribution and Orientation

The uniform distribution and proper orientation of reinforcement materials impact the composite's mechanical properties. Solid-state techniques allow for tailored reinforcement patterns, affecting properties like tensile strength and wear resistance.

4.2.5.6 Post-Processing Treatments

Heat treatments, mechanical working (like rolling or forging), and surface finishing can refine microstructure and enhance mechanical properties. These treatments can also relieve residual stress and improve wear and corrosion resistance [11].

4.3 MECHANICAL ALLOYING

4.3.1 Fundamentals of Mechanical Alloying

Mechanical alloying (MA) is a solid-state process for creating metal matrix composites (MMCs) and advanced alloys. It involves repeated mechanical deformation, fracturing, and cold welding of powder particles in a high-energy ball mill, resulting in thorough mixing and homogenization at a microscopic or atomic level.

4.3.2 High-Energy Ball Milling

High-energy ball milling is a mechanical process that uses repeated impact, friction, and shear forces to grind materials into fine powders or mix them to create alloys or composites [12].

4.3.2.1 Principles of High-Energy Ball Milling

High-energy ball milling involves intense mechanical action to reduce particle size, mix different materials, and initiate solid-state reactions.

4.3.3 Parameters Influencing the Process

The high-energy ball milling process, which is crucial for mechanical alloying and the synthesis of metal matrix composites (MMCs), is influenced by several parameters that determine the outcome. These parameters include the microstructure, particle size, distribution, and mechanical properties of the final product [13].

4.3.4 Mechanisms of Particle Refinement and Mixing

In high-energy ball milling, particle refinement and mixing are key mechanisms that lead to the creation of metal matrix composites (MMCs), advanced alloys, and other materials with unique properties.

4.3.4.1 Particle Refinement

Particle refinement refers to the reduction of powder particles to smaller sizes, ultimately contributing to a fine-grained microstructure in the final product.

4.3.4.2 Mixing and Homogenization

Mixing and homogenization are crucial for achieving a uniform distribution of different materials within the metal matrix. This is especially important in MMCs, where reinforcement particles must be evenly distributed to ensure consistent mechanical properties.

4.3.4.3 Role of Milling Media

The type and size of milling media play a significant role in particle refinement and mixing. Smaller milling media increase the surface area for impact and friction, promoting more efficient grinding and mixing. The material of the milling media, such as steel or tungsten carbide, affects the intensity of mechanical action and the risk of contamination.

4.3.4.4 Factors Influencing Particle Refinement and Mixing

Longer milling times typically lead to greater particle refinement and improved mixing, but excessive milling can cause unwanted grain growth and contamination. Higher milling speeds increase the energy imparted to the powders, promoting more intense impact and shear forces, thus enhancing refinement and mixing [14].

4.3.5 Applications in MMC Synthesis

4.3.5.1 Reinforced Metal Matrix Composites

High-energy ball milling allows for the integration of reinforcement materials into a metal matrix, producing composites with enhanced mechanical properties. Reinforcements can include ceramic particles and carbon-based materials.

4.3.5.2 Nanocomposites

This technique is effective in creating nanocomposites, where reinforcement particles are on the nanometre scale. High-energy ball milling can reduce particle size to the nanoscale, allowing for the production of MMCs with nanostructured reinforcements.

4.3.5.3 High-Temperature MMCs

High-energy ball milling is useful for producing MMCs designed for high-temperature applications. By integrating ceramic reinforcements into metal matrices, this process creates composites that can withstand extreme temperatures without losing mechanical strength. These high-temperature MMCs are valuable in aerospace, automotive, and energy industries, where materials must maintain their properties under harsh conditions.

4.3.5.4 High-Entropy Alloys with MMC Structures

High-entropy alloys (HEAs) are a class of materials composed of multiple principal elements, resulting in unique properties. High-energy ball milling is used to create MMCs with HEA matrices, leading to composites with exceptional strength, toughness, and resistance to wear and corrosion. This combination of HEAs and MMCs opens new possibilities for advanced engineering materials.

4.3.5.5 Functionally Graded Materials

High-energy ball milling can be used to create functionally graded materials (FGMs), where the composition and properties vary across the material's thickness. This allows for the design of MMCs with tailored gradients, enhancing performance in specific applications. FGMs are often used in wear-resistant coatings, thermal barriers, and other specialized applications.

4.3.5.6 MMCs for Additive Manufacturing

Additive manufacturing (AM) is becoming increasingly popular in various industries, and high-energy ball milling plays a role in creating MMC powders for AM processes. The thorough mixing and refinement achieved through ball milling ensure uniform powder characteristics, enabling consistent and reliable AM production. MMCs for AM offer a range of applications, from aerospace components to medical devices.

4.3.5.7 MMCs for Structural Applications

Composite materials are frequently employed in the automotive and aerospace sectors to decrease weight while simultaneously preserving or enhancing mechanical characteristics. MMCs containing ceramic or carbon-based reinforcements are especially valuable due to their exceptional strength-to-weight ratios [15].

4.4 IN-SITU FABRICATION TECHNIQUES

4.4.1 In-Situ Reaction Synthesis

In-situ reaction synthesis is a process used to create metal matrix composites (MMCs) by inducing chemical reactions within the metal matrix, leading to the formation of reinforcing phases during the synthesis. This method differs from traditional composite fabrication, where reinforcement materials are added externally [16].

4.4.1.1 Principles of In-Situ Reaction Synthesis

Precursors such as metal powders, alloys, or reactive compounds are mixed to create a homogeneous starting material. These precursors contain the necessary elements to form the desired reinforcement phases during synthesis. The mixture is subjected to high temperatures, triggering chemical reactions that lead to the formation of reinforcing phases, such as ceramics or intermetallics.

4.4.1.2 Challenges and Considerations

The synthesis process must be carefully controlled to avoid unwanted reactions and ensure the desired reinforcement structure forms. The formation of reinforcing phases during synthesis can create residual stresses, potentially affecting the composite's mechanical properties [17].

4.4.2 Reactive Milling

Reactive milling is a specialized form of high-energy ball milling where chemical reactions occur between different materials during the milling process. This approach combines the mechanical energy from ball milling with chemical reactivity, allowing for the synthesis of metal matrix composites (MMCs), intermetallics, ceramics, and other advanced materials. Reactive milling is particularly useful for creating new phases, refining microstructures, and enhancing material properties. Here's an overview of reactive milling, focusing on its principles, applications, and key considerations [18].

4.4.3 Spark Plasma Sintering (SPS)

SPS, also referred to as Field Assisted Sintering Technique (FAST), is a sophisticated sintering process employed to compact powders into solid, high-strength materials. It employs a combination of uniaxial pressure and electric current, creating rapid heating and promoting efficient sintering [19].

4.4.3.1 Spark Plasma Sintering

SPS operates by applying uniaxial pressure to powder compacts while simultaneously passing a pulsed electric current through them. This combination of pressure and electric current leads to rapid heating and enhanced mass transport, enabling efficient densification.

4.4.4 Hot Pressing

Hot pressing is a common technique used to consolidate powders and create dense metal matrix composites (MMCs) and other advanced materials. It combines high temperatures and uniaxial pressure to promote particle bonding, allowing for efficient densification and enhanced mechanical properties. Hot pressing has been widely adopted in industries where the production of dense, high-strength materials is critical, such as aerospace, automotive, and electronics [20].

4.4.5 Microwave Sintering

Microwave sintering is a unique sintering technique that uses microwave radiation to heat and consolidate powder materials into dense compacts. This method is gaining traction in the synthesis of metal matrix composites (MMCs), ceramics, and other advanced materials due to its ability to achieve rapid heating, uniform temperature distribution, and reduced sintering times compared to conventional sintering processes [21].

4.5 SOLID STATE DIFFUSION BONDING

4.5.1 Principles of Solid-State Bonding

Solid-state bonding refers to a group of processes used to join materials without melting them, relying instead on diffusion, deformation, and other mechanisms to create a strong bond. This method is extensively used in the production of metal matrix composites (MMCs), as well as in other situations where there is a need for strong and high-quality connections.

4.5.2 Diffusion Mechanisms

Diffusion is a primary mechanism in solid-state bonding. It involves the movement of atoms or molecules across the interface between two materials. The high temperatures and pressures typically used in solid-state bonding increase atomic mobility, promoting diffusion. As atoms move across the interface, they create metallurgical bonds, effectively joining the two materials at the atomic level [22].

4.5.2.1 Pressure and Deformation

Solid-state bonding often relies on the application of pressure to promote mechanical contact and increase diffusion. Pressure can cause plastic deformation at the interface, which enhances atom-to-atom contact and reduces voids or gaps between the materials. Deformation also disrupts surface oxides and contaminants, allowing for better bonding.

4.5.2.2 Heat and Temperature Control

Although solid-state bonding does not involve melting, heat is often used to accelerate diffusion and increase atomic mobility. Temperature control is crucial to ensure that the materials do not reach their melting points, as this would lead to a loss of structural integrity. The heat may also serve to reduce surface oxides and impurities, further facilitating bonding.

4.5.2.3 Surface Preparation and Cleaning

Solid-state bonding requires clean and prepared surfaces to ensure effective contact and diffusion. Surface preparation may involve mechanical abrasion, chemical cleaning, or other methods to remove contaminants and oxides. Proper surface preparation is crucial for achieving high-quality solid-state bonds.

4.5.2.4 Interfacial Bonding and Structure

The structure of the interface plays a significant role in solid-state bonding. A well-bonded interface has strong metallurgical connections, while weak or contaminated interfaces can lead to poor bonding and reduced mechanical properties. The goal of solid-state bonding is to create a seamless transition between the materials, ensuring high strength and durability.

4.5.2.5 Diffusion Mechanisms

Diffusion mechanisms are fundamental to understanding how atoms or molecules move through solid materials, contributing to processes like solid-state bonding, sintering, and various metallurgical phenomena. Diffusion is a key principle in metal matrix composites (MMCs), as it plays a role in the distribution of reinforcing phases, the consolidation of metal powders, and the formation of strong bonds.

4.5.2.6 Vacancy Diffusion

Vacancy diffusion is a mechanism where atoms move through a lattice by swapping positions with vacant lattice sites or "vacancies." Vacancies are naturally occurring defects in a crystal structure where an atom is missing. Vacancy diffusion is temperature dependent, as higher temperatures increase atomic mobility and vacancy concentration. It is a common mechanism in metals and is influenced by the lattice structure, impurity concentration, and temperature.

4.5.2.7 Interstitial Diffusion

Interstitial diffusion involves the movement of smaller atoms through the interstices or spaces between the larger atoms in a crystal lattice. This mechanism typically applies to elements like hydrogen, carbon, or nitrogen diffusing through metal lattices.

4.5.2.8 Grain Boundary Diffusion

Grain boundary diffusion occurs along the interfaces between individual grains in a polycrystalline material. These grain boundaries have a different structure from the grains themselves, often with higher vacancy concentrations and disordered atom arrangements.

4.5.2.9 Surface Diffusion

Surface diffusion involves the movement of atoms or molecules along the surface of a material. This mechanism plays a role in surface reactions, sintering, and the formation of thin films. Surface diffusion is particularly important in processes involving surface treatments, thin film deposition, and the early stages of sintering [23].

4.5.3 Process Parameters and Control

Process parameters and control are critical in the synthesis and manufacturing of metal matrix composites (MMCs). Proper control over these parameters ensures the desired microstructure, properties, and performance in the final composite. This involves adjusting various factors during the fabrication process to achieve specific outcomes [24]

4.5.4 Applications in MMCs Joining and Fabrication

MMCs are designed materials that incorporate a metal matrix with reinforcing substances such as ceramics, fibres, or particles to improve mechanical characteristics such as strength, toughness, wear resistance, and thermal stability. The process of joining and fabricating metal matrix composites (MMCs) requires the use of specialized techniques to achieve robust and dependable connections while preserving the distinctive characteristics of the composite material [25].

4.5.4.1 Welding and Joining

Joining MMCs with other MMCs or traditional metals poses unique challenges due to differences in thermal expansion, material composition, and melting points [26].

4.6 LASER AND ELECTRON BEAM PROCESSING

4.6.1 Laser Beam Melting (LBM)

LBM, alternatively referred to as selective laser melting (SLM), is a sophisticated additive manufacturing technique that fabricates metal components by liquefying and merging powdered metals using a high-energy laser beam. The utilization of computer-controlled systems in a step-by-step manner enables the production of intricate internal structures and complex geometries, making it particularly well-suited for manufacturing customized components.

4.6.2 Electron Beam Melting (EBM)

EBM is a method of additive manufacturing where a concentrated electron beam is employed to liquefy and bond metal powders together, forming solid components in a sequential layer-by-layer fashion. EBM, unlike laser beam melting, functions within a vacuum setting, thereby minimizing oxidation and contamination during the manufacturing process.

4.6.3 Selective Laser Melting (SLM)

Selective laser melting (SLM), also known as laser beam melting (LBM), is a popular additive manufacturing technique used to create complex metal parts by selectively melting and fusing metal powders. The process relies on a high-energy laser beam to build components layer by layer, guided by a computer-controlled system that follows a digital 3D model.

4.6.4 Applications in MMCs Manufacturing

Metal Matrix Composites (MMCs) are increasingly used across various industries due to their unique combination of metal and reinforcement properties. The applications of MMCs in manufacturing are diverse, leveraging their enhanced strength, stiffness, wear resistance, and thermal stability [27].

4.6.4.1 Aerospace and Defence

MMCs are used to manufacture structural components, such as wings, fuselages, and landing gear, offering weight savings while maintaining structural integrity.

4.6.4.2 Automotive and Transportation

MMCs are used in engine blocks, pistons, and cylinder heads, providing high strength and reduced weight for better fuel efficiency. MMCs with fibre reinforcements are used in suspension components, enhancing durability and reducing overall weight.

4.6.4.3 Electronics and Thermal Management

MMCs are used in electronics and thermal management applications due to their high thermal conductivity and mechanical strength. MMCs with aluminium or copper matrices and high-conductivity reinforcements are used to create heat sinks for electronic components, providing efficient heat dissipation.

4.6.4.4 Industrial Equipment and Machinery

MMCs with carbide reinforcements are used in cutting tools and machining equipment, providing increased wear resistance and longer tool life. MMCs are used in conveyor belts and rollers, benefiting from their high strength and durability in industrial environments. MMCs with corrosion-resistant reinforcements are used in pumps and valves, offering enhanced longevity and reduced maintenance [28].

4.7 CHALLENGES AND FUTURE DIRECTIONS

4.7.1 Process Optimization

Process optimization in manufacturing metal matrix composites (MMCs) involves improving production methods to ensure maximum efficiency, quality, and performance while minimizing costs, time, and material waste.

4.7.1.1 Identifying Key Parameters

The first step in process optimization is identifying critical parameters that influence the synthesis and performance of MMCs.

4.7.1.2 Implementing Statistical Process Control (SPC)

Statistical process control (SPC) involves using statistical methods to monitor and control manufacturing processes. By tracking key parameters and identifying trends, SPC helps maintain process stability and reduce variability. Strategies for SPC include:

4.7.1.3 Applying Design of Experiments (DOE)

Design of experiments (DOE) is a structured approach to experimentation that allows for efficient exploration of process variables and their interactions [29]. By systematically varying parameters, DOE helps optimize processes for MMCs. Key aspects of DOE include:

4.7.2 Scalability and Industrial Adoption

Scalability and industrial adoption are key considerations when transitioning Metal Matrix Composites (MMCs) from research and development to widespread use in manufacturing. This transition involves overcoming technical, logistical, and economic challenges to ensure that MMCs can be produced at scale with consistent quality and reasonable costs. Here's an overview of the factors influencing scalability and industrial adoption of MMCs:

4.7.2.1 Challenges in Scaling MMC Production

As production volume increases, maintaining consistent quality and microstructure becomes more challenging. Variability in raw materials, process parameters, and equipment can impact the final product. MMCs often require specialized materials and manufacturing processes, leading to higher costs compared to traditional metals. Reducing these costs is essential for broader industrial adoption.

4.7.2.2 Strategies for Achieving Scalability

Developing standardized processes and protocols helps ensure consistent quality and reduces variability. Standardization also facilitates automation and reduces reliance on skilled labour. Using advanced manufacturing technologies, such as additive manufacturing and automated processes, can improve scalability and reduce costs. These technologies offer flexibility and precision, enabling more efficient production. Identifying cost drivers and implementing cost-saving measures can make MMC production more economically viable.

4.7.2.3 Industrial Adoption of MMCs

Identifying applications where MMCs offer significant benefits over traditional materials can drive industrial adoption. This may include aerospace, automotive, electronics, and medical devices. Ensuring compliance with industry-specific regulations and standards is essential for industrial adoption. MMC manufacturers must meet safety, environmental, and quality requirements [30].

4.7.3 Integration with Additive Manufacturing

Integration with additive manufacturing (AM) in the production of Metal Matrix Composites (MMCs) is revolutionizing traditional manufacturing processes by enabling complex geometries, customization, and reduced waste. Techniques like selective laser melting (SLM), electron beam melting (EBM), directed energy deposition (DED), and binder jetting offer precise control over component design, allowing for intricate internal structures and tailored reinforcement distributions within MMCs [31].

4.7.4 Environmental and Health Considerations

Environmental and health considerations play a critical role in the production and use of metal matrix composites (MMCs). During manufacturing, processes such as powder metallurgy and additive manufacturing can generate fine particulates and emissions, potentially impacting air quality and worker health [32].

4.8 CASE STUDIES

4.8.1 Application of Solid-State Techniques in Specific MMC Systems

Solid-state techniques play a crucial role in the manufacturing of specific Metal Matrix Composites (MMCs), offering advantages in terms of process control, material properties, and environmental impact. Here are some notable applications of solid-state techniques in specific MMC systems:

4.8.1.1 Aluminium-Based MMCs

Solid-state methods like powder metallurgy and friction stir processing are widely used for aluminium-based MMCs, which are common in aerospace and automotive industries due to their lightweight and high strength-to-weight ratio.

4.8.1.2 Titanium-Based MMCs

Titanium-based MMCs, used in aerospace and medical applications, benefit from solid-state methods like hot isostatic pressing (HIP) and diffusion bonding. These techniques allow for strong interfacial bonding without compromising the integrity of the metal matrix, leading to high-performance components with enhanced fatigue resistance and biocompatibility.

4.8.1.3 Copper-Based MMCs

Copper-based MMCs, valued for their thermal and electrical conductivity, find applications in electronics and thermal management. Solid-state techniques like powder compaction and sintering are used to create components with high thermal conductivity and mechanical strength, ideal for heat sinks and electronic packaging.

4.8.2 Performance Evaluation and Comparative Analysis

Performance evaluation and comparative analysis are crucial components in the study of metal matrix composites (MMCs). These processes help determine the effectiveness of different manufacturing methods, the influence of various reinforcements, and the overall performance of MMCs in specific applications.

4.8.2.1 Metrics for Performance Evaluation

Evaluating these properties helps assess the durability and strength of MMCs. Metrics like thermal conductivity, heat capacity, and thermal expansion are critical for applications requiring heat management or exposure to extreme temperatures. Evaluating wear resistance and corrosion behaviour is essential for MMCs used in automotive, aerospace, and industrial environments [33].

4.8.2.2 Comparative Analysis of Manufacturing Methods

Comparing the cost and scalability of different methods, such as powder metallurgy, additive manufacturing, and solid-state processes, helps identify the most economical and scalable approach for MMC production. Analyzing the consistency of MMCs produced through various methods is crucial for ensuring reliable performance. Methods with robust quality control tend to yield more consistent products.

4.8.2.3 Performance Evaluation in Specific Applications

MMCs in aerospace must exhibit high strength-to-weight ratios, fatigue resistance, and thermal stability. Comparative analysis helps identify the best MMC compositions and manufacturing methods for these requirements. Comparative analysis can identify which MMCs and manufacturing methods offer the best balance of performance and cost [34].

4.9 CONCLUSION

MMCs are rapidly developing and dynamic area of study that finds applications in diverse industries such as aerospace, automotive, electronics, and medical devices. Solid-state techniques and additive manufacturing play critical roles in enabling the production of MMCs with tailored properties and complex geometries, contributing to their versatility and growing industrial adoption. Moreover, integrating MMCs with additive manufacturing opens new possibilities for design innovation, customization, and reduced waste. Environmental and health considerations must also be factored into the manufacturing processes to ensure sustainable practices and safe working conditions. Performance evaluation and comparative analysis are essential for selecting the best MMC compositions and manufacturing methods, guiding industries toward the most effective solutions for their specific needs. The continuous advancements in MMC technology hold the potential for additional enhancements in performance and cost-efficiency, facilitating wider acceptance and the development of creative applications. By working together and conducting ongoing research, MMCs are positioned to have a growing impact in addressing the needs of contemporary engineering and manufacturing. They provide lightweight, strong, and long-lasting solutions for a range of intricate problems.

REFERENCES

[1] Mussatto, A., Ahad, I.U., Mousavian, R.T., Delaure, Y. and Brabazon, D., 2021. Advanced production routes for metal matrix composites. *Engineering Reports*, *3*(5), p.e12330.

[2] Bahl, S., 2021. Fiber reinforced metal matrix composites-a review. *Materials Today: Proceedings*, *39*, pp. 317–323.

[3] Nieto, A., Agarwal, A., Lahiri, D., Bisht, A. and Bakshi, S.R., 2021. *Carbon Nanotubes: Reinforced Metal Matrix Composites*. CRC Press.

[4] Parikh, V.K., Badheka, V.J., Badgujar, A.D. and Ghetiya, N.D., 2021. Fabrication and processing of aluminum alloy metal matrix composites. *Materials and Manufacturing Processes*, *36*(14), pp. 1604–1617.

[5] Chauhan, A., Vates, U.K., Kanu, N.J., Gupta, E., Singh, G.K., Sharma, B.P. and Gorrepati, S.R., 2021. Fabrication and characterization of novel nitinol particulate reinforced aluminium alloy metal matrix composites (NiTip/AA6061 MMCs). *Materials Today: Proceedings*, *38*, pp. 3027–3034.

[6] Jayappa, K., Kumar, V. and Purushotham, G.G., 2021. Effect of reinforcements on mechanical properties of nickel alloy hybrid metal matrix composites processed by sand mold technique. *Applied Science and Engineering Progress*, *14*(1), pp. 44–51.

[7] Singh, B., Kumar, I., Saxena, K.K., Mohammed, K.A., Khan, M.I., Moussa, S.B. and Abdullaev, S.S., 2023. A future prospects and current scenario of aluminium metal matrix composites characteristics. *Alexandria Engineering Journal*, *76*, pp. 1–17.

[8] Raj, P.P., Sridhar, R., Pugazhenthi, R., Anbuchezhiyan, G. and Ganesh, M., 2023. Investigating the mechanical properties of tungsten carbide metal matrix composites with Al 6061. *Materials Today: Proceedings*. https://doi.org/10.1016/j.matpr.2023.03.605

[9] Chaudhary, B., Patel, M., Jain, N.K., Murugesan, J. and Patel, V., 2023. Friction stir powder additive manufacturing of Al 6061/FeCoNi and Al 6061/Ni metal matrix composites: Reinforcement distribution, microstructure, residual stresses, and mechanical properties. *Journal of Materials Processing Technology*, *319*, p. 118061.

[10] Dewangan, S., Bohra, P.M. and Sharma, P., 2024. Mechanical properties and microstructure analysis of Al-6061 metal matrix prepared with alumina and silicon carbide powders. *Journal of the Institution of Engineers (India): Series D*, pp. 1–11.

[11] Lal, C., Dhiman, B., Tejyan, S. and Singh, V., 2024. Aluminum 6061 alloy and metal matrix composites: A bibliometric analysis (1973–2023). *Materials Today: Proceedings*. https://doi.org/10.1016/j.matpr.2024.04.018

[12] Gafur, M.A., Abrar, R. and Soshi, S.S., 2023. Development and characterization of aluminium-based metal matrix composites. *Materials Sciences and Applications*, *14*(1), pp. 1–19.

[13] Abebe Emiru, A., Sinha, D.K., Kumar, A. and Yadav, A., 2023. Fabrication and characterization of hybrid aluminium (Al6061) metal matrix composite reinforced with SiC, B_4C and MoS_2 via stir casting. *International Journal of Metalcasting*, *17*(2), pp. 801–812.

[14] Hima Gireesh, C., Durga Prasad, K.G. and Ramji, K., 2018. Experimental investigation on mechanical properties of an Al6061 hybrid metal matrix composite. *Journal of Composites Science*, *2*(3), p. 49.

[15] Asif, M., Chandra, K. and Misra, P.S., 2011. Development of aluminium based hybrid metal matrix composites for heavy duty applications. *Journal of Minerals & Materials Characterization & Engineering*, *10*(14), pp. 1337–1344.

[16] Lee, T., Jeong, W., Chung, S., Kim, J.J., Lee, D., Kim, J. and Ryu, H.J., 2023. Role of B4C in the directed energy deposition of aluminum alloy 6061/B_4C composite using core-shell powder for crack annihilation, pore mitigation, and grain refinement. *Additive Manufacturing*, *72*, p. 103622.

[17] Bharathi, P. and Kumar, T.S., 2023. Mechanical characteristics and wear behaviour of Al/SiC and Al/SiC/B4C hybrid metal matrix composites fabricated through powder metallurgy route. *Silicon*, *15*(10), pp. 4259–4275.

[18] Arunachalam, J., Sathish, T., Makki, E. and Giri, J., 2024. Influence of machining process of MoS_2/B_4C/AZ31 Mg alloy composite and its tribological characteristics. *AIP Advances*, *14*(3).

[19] Senthilnathan, T. and Balachandar, K., 2023. Mechanical and microstructure evaluation of novel hybrid TiB_2/B_4C aluminium metal matrix composite. *Materials Research Express*, *10*(6), p. 066509.

[20] Göçer, A. and Karamis, M.B., 2024. Production and mechanical characterization of steel/Al-B_4C layered circular hybrid composite materials. *Arabian Journal for Science and Engineering*, pp. 1–19.

[21] Karabulut, Ş., Karakoç, H., Bilgin, M., Canpolat, H., Krolczyk, G.M. and Sarıkaya, M., 2023. A comparative study on mechanical and ballistic performance of functionally graded Al6061 composites reinforced with B_4C, SiC, and Al_2O_3. *Journal of Materials Research and Technology*, *23*, pp. 5050–5065.

[22] Cramer, C.L., Cakmak, E. and Unocic, K.A., 2023. Hardness measurements and interface behavior of SiC-B_4C-Si multiple phase particulate composites made with melt infiltration and additive manufacturing. *Journal of Composites Science*, *7*(4), p. 172.

[23] Tasci, U., Yilmaz, T.A. and Bostan, B., 2023. Investigation of microstructure, wear and transverse rupture strength of WE43/nano B_4C composites produced by powder metallurgy method. *Tribology International*, *180*, p. 108231.

[24] Sivalingam, S., Kumaresan, G., Ganesh, P. and Azhagan, M.T., 2023. Superplastic formability and cavitation analysis of AA7075 matrix composite reinforced with B_4C particles produced by stir casting. *International Journal of Metalcasting*, *17*(4), pp. 2917–2927.

[25] Ranjitha, P., Bavan, D.S., Ajaykumar, B.S., Raju, T.H. and Udayashankar, S., 2024. Investigation of mechanical properties of Al6061–SiC–B_4C composites produced by using stir casting method. *Journal of the Institution of Engineers (India): Series D*, pp. 1–13.

[26] Mydur, K., Kumar, M., Nagaral, M., Auradi, V., Bharath, V. and Sudarshan, T.A., 2023. Microstructure, physical, tensile and wear behaviour of B_4C particles reinforced Al7010 alloy composites. *Manufacturing Review*, *10*, p. 3.

[27] Angadi, S.B., Nagaral, M., Pilankar, V., Yarnalkar, P., Purohit, B. and Shewale, S., 2023. Assessment of mechanical behaviour and microstructure of micro boron carbide particles reinforced Al2011 alloy metal matrix composite for aerospace applications. *International Journal of Vehicle Structures & Systems*, *15*(2).

[28] Gaylan, Y., Avar, B., Panigrahi, M., Aygün, B. and Karabulut, A., 2023. Effect of the B_4C content on microstructure, microhardness, corrosion, and neutron shielding properties of Al–B_4C composites. *Ceramics International*, *49*(3), pp. 5479–5488.

[29] Kumar, C., Sarkar, S., Mukhopadhyay, G., Chakraborti, P.C., Sen, I. and Roy, S., 2023. Systematic study of the effect of K2TiF6 flux content on the microstructure and mechanical properties of Al–B_4C composites. *Materials Science and Engineering: A*, *871*, p. 144913.

[30] Ananth, G., Muthu Krishnan, T., Thirugnanam, S. and Tariku Olkeba, T., 2023. Optimization on stir casting process parameters of Al7050/nano-B_4C metal matrix composites. *Journal of Nanomaterials*, *2023*.

[31] Raksha, M.S., Nagaral, M., Boppana, S.B., Anjinappa, C., Khan, M.S., Wahab, M.O.A., Islam, S., Bhardwaj, V., Palavalasa, R.K., Khan, M.A. and Razak, A., 2023. Impact of boron carbide particles and weight percentage on the mechanical and wear characterization of Al2011 alloy metal composites. *ACS Omega*, *8*(26), pp. 23763–23771.

[32] Yunus, M. and Alfattani, R., 2023. Assessment of mechanical and tribological behavior of AA6061 reinforced with B_4C and Gr hybrid metal matrix composites. *Coatings*, *13*(9), p. 1653.

[33] Ramadoss, N., Pazhanivel, K., Ganeshkumar, A. and Arivanandhan, M., 2023. Microstructural, mechanical and corrosion behaviour of B_4C/BN-reinforced Al7075 matrix hybrid composites. *International Journal of Metalcasting*, *17*(1), pp. 499–514.

[34] Yi, J., Zhang, X., Liu, G., Rao, J.H. and Liu, H., 2023. Microstructure and dynamic microhardness of additively manufactured (TiB2+ TiC)/AlSi10Mg composites with AlSi10Mg and B_4C coated Ti powder. *Journal of Alloys and Compounds*, *939*, p. 168718.

5 Additive Manufacturing of Metal Matrix Composites

Ved Prakash Dubey, Shivam Sharma, Mateusz Kopec, and Zbigniew Kowalewski

5.1 INTRODUCTION

Titanium (Ti) and its alloys are extensively used in diverse engineering applications including automotive, aircraft, biomedical, military, and other structural applications, attributed to their distinctive combination of desirable properties. These include excellent strength-to-weight ratio, high corrosion resistance, and biocompatibility [1]. However, their inherent limitations, including relatively low wear resistance and limited mechanical performance at elevated temperatures, restrict their broader applications in certain demanding environments [2]. Titanium-based matrix composites (TMCs) offer a potential solution to overcome these limitations by strategically embedding ceramic or intermetallic reinforcements within a titanium alloy matrix [3, 4]. Metal matrix composites (MMCs), particularly those composed of titanium, magnesium, and cobalt-chromium, are increasingly used in biomedical applications. This comes from their distinctive ability to efficiently combine the desirable properties of metals, such as high strength and rigidity, with the beneficial characteristics of other reinforcements, such as biocompatibility or enhanced wear resistance [5].

In recent years, AM technologies, additionally referred to as 3D printing, have offered a revolutionary approach for fabricating complex-shaped, high-performance components. This technology is based on a layer-by-layer deposition of materials, enabling the manufacturing of complex geometries and near-net shape components. This procedure minimizes material waste and the need for additional post-processing after manufacturing. AM processes are particularly advantageous for TMCs, as they allow for the precise control of reinforcement distribution, leading to parts with desired microstructure and mechanical properties tailored to specific applications [6, 7].

This chapter provides a comprehensive analysis of the most significant latest advancements in AM of TMCs. It starts with a brief introduction of various AM processes utilized for the fabrication of TMCs. Subsequently, an in-depth examination of additive manufacturing techniques is discussed with a specific focus on key factors such as processing parameters and their effects on microstructures, mechanical and wear properties. Finally, the chapter concludes by addressing the challenges and providing a prospective analysis of the potential opportunities in the field of additively manufactured TMCs.

5.2 ADDITIVE MANUFACTURING PROCESSES FOR FABRICATION OF TMCs

Based on the machine types sharing similar processing, the AM process can be categorized into seven techniques, as referred to by the ASTM standards [8]. Table 5.1 provides an overview of the categories and methodologies that can be utilized for each process.

Additive manufacturing could also be categorized into "direct" and "indirect" techniques based on their consolidation mechanisms. The key differences between these two AM processes are listed in Table 5.2. Direct AM techniques achieve simultaneous shaping and densification through the complete melting of material powder followed by solidification. These techniques provide a high part density and need minimal post-processing, making them suitable for manufacturing high-performance

DOI: 10.1201/9781003499718-5

TABLE 5.1
ASTM International Process Categories of AM [8]

Category	Process methodology
Binder jetting (BJ)	Selectively deposit a liquid bonding agent (adhesive) to form powdered materials.
Directed energy deposition (DED)	Utilize concentrated thermal energy to merge depositing materials through the process of melting.
Material extrusion (ME)	Selectively dispense the powder by an orifice or nozzle.
Material jetting (MJ)	Selectively deposit the build material as droplets.
Powder bed fusion (PBF)	Selectively fuse regions of a powder bed utilizing a focused thermal energy source.
Sheet lamination (SL)	Bond pre-existing material sheets to create a three-dimensional object.
Vat photopolymerization (VAT-P)	Employ light-activated polymerization process to cure a liquid photopolymer within a vat.

TABLE 5.2
Key Differences between Direct and Indirect AM Processes

Characteristic	Direct AM	Indirect AM
Consolidation mechanism	Simultaneous shaping and densification through melting and solidification	Shaping with binder, followed by debinding and sintering for densification
Post-processing	Minimal	Debinding and sintering required
Part density	High	Lower than direct AM
Material examples	Metals, some alloys	Limited to materials compatible with binders
Applications	High-performance end-use parts	Prototypes, complex geometries with lower strength requirements

end-use parts. Direct AM for metals could be categorized into two primary technologies: PBF and DED. In contrast, indirect AM techniques depend on a binding agent to initially mould the material powder. These techniques, such as BJ, SLA, and MJ, require subsequent post-processing steps like debinding and sintering in order to remove the binder and achieve full densification.

In all metal additive manufacturing methods, a heat source (laser or electron beam) is used for melting or fusing the initial material (metal powder or wire) in order to produce a dense three-dimensional object. This chapter focuses on the advancement of TMCs using two main direct AM techniques, namely PBF and DED.

5.3 POWDER BED FUSION OF TMCs

PBF is a subset of additive manufacturing, utilizing a heat source, to scan and melt fine metal powder, allowing for precise layer-by-layer deposition to build up a 3D object, as shown in Figure 5.1. After the completion of a layer, a fresh metal powder is spread over the build area, melted, and deposited as a new layer, enabling the production of intricate three-dimensional shapes with high geometric accuracy [9]. It encompasses selective laser sintering (SLS), electron beam melting (EBM), and laser powder bed fusion (L-PBF), also known as SLM or direct metal laser sintering (DMLS) technologies [10]. SLS employs a laser beam to fuse the powder particles at their sintering temperature, whereas L-PBF completely melts the powder particles together. Therefore, SLS technology is beneficial for printing objects made of ceramic, plastic, wax, and sand, while L-PBF technology is more suitable for metal printing. EBM works in the same way as L-PBF; however, it employs an electron beam unit to heat the parts. Table 5.3 provides a comparison between L-PBF and EBM additive manufacturing techniques.

L-PBF processing of Ti-TiB composite materials has gained popularity among AM of TMCs. This is likely because of the favourable characteristics of TiB as a reinforcement, such as the ability to form clean interfaces and strong bonds with the titanium matrix. This TiB stability in titanium is attributed to the extremely limited solubility of boron in Ti, which is less than 0.001 atomic percent compared to 1.8% for nitrogen. Attar et al. fabricated TMCs from Ti and TiB_2 powders using the L-PBF technique [12]. The interaction between titanium and titanium dioxide particles leads to the growth of needle-shaped TiB within the titanium matrix, as presented in Figure 5.2. The microstructural characterization indicated that the L-PBF processed composite had a finer distribution of relatively small-sized TiB precipitates than in the as-cast material. However, the compressive properties

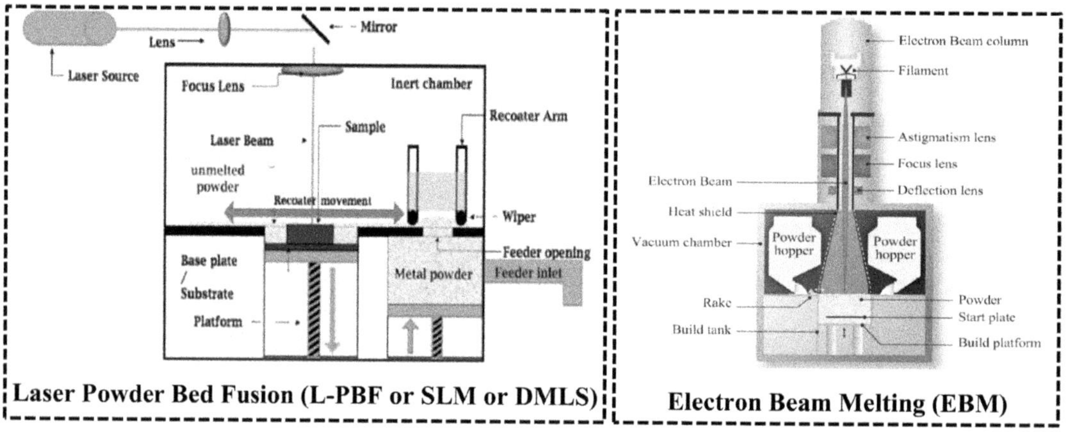

FIGURE 5.1 Working principles of powder bed fusion additive manufacturing techniques. Reproduced from [3, 11].

TABLE 5.3
Comparison of L-PBF and EBM

Parameter	LPBF	EBM
Energy source	High-powered laser	Electron beam
Environment	Inert gas atmosphere (argon or nitrogen)	High vacuum
Energy power (W)	200–1000	3000+
Spot diameter (μm)	50–100	The beam can be concentrated to a highly localized point
Material powder size (μm)	Fine (15–45)	Relatively coarse (45–150)
Layer thickness (μm)	Thin (25–60)	Thicker (50–150)
Advantages	• High resolution and precision • Wide range of compatible materials • Good surface finish • Re-usable powder	• Lower thermal stress • Can process reactive materials • Potential for faster build times due to thicker layers
Disadvantages	• Susceptible to thermal stress cracking • Limited build size due to high laser power • Requires support structures for complex geometries	• Requires vacuum environment • Increased cost and complexity • Limited powder selection due to conductivity requirements • Lower resolution and precision compared to LPBF

and hardness values were observed to be similar in both cases. Similar findings were reported by Kang et al. who utilized CP-titanium and chromium diboride (2wt%) powder in the L-PBF process to produce TMCs containing in-situ TiB particles [13]. The AM-manufactured composite was characterized by improved wear resistance and hardness attributed to the formation of the near α-Ti phase.

The properties of the powder employed in L-PBF play a critical role in defining the quality of the manufactured components. Particle morphology and size distribution are critical factors that impact processability, thus ultimately affecting the relative density and mechanical characteristics of the end product. Additionally, the method of powder production, such as mechanical milling, significantly impacts powder morphology. While mechanical milling is effective for alloying constituent powders, optimizing process parameters like ball-to-powder ratio or milling time is crucial. Attar et al. investigated the milling time effects on mechanical properties and densification of L-PBF fabricated Ti-5wt%TiB composites [14]. It was found that the composite prepared from powder milled for 2 hours exhibited better mechanical properties to those of the 4-hour milled powder. This indicates that the spherical morphology of the 2-hour milled powder allows for more effective powder deposition during L-PBF, leading to improved powder flowability and improved mechanical performance of the fabricated composites. In another research study conducted by Gu et al. the effect of variation in TiC reinforcement content on the microstructural evolution and mechanical performance of L-PBF AM fabricated Ti-TiC was studied [15]. It was observed that, as the content of TiC increases, densification becomes more challenging due to the difficulty of melting and consolidating the Ti-matrix with TiC reinforcement. Ti-TiC composite with less than 15wt% TiC demonstrated better distribution of reinforcement within the fine layered Ti-matrix, whereas, when its content increased to 17.5wt% and 22.5%, clustering and a network with dendritic morphology was observed, respectively, as shown in Figure 5.3. The researchers also examined the

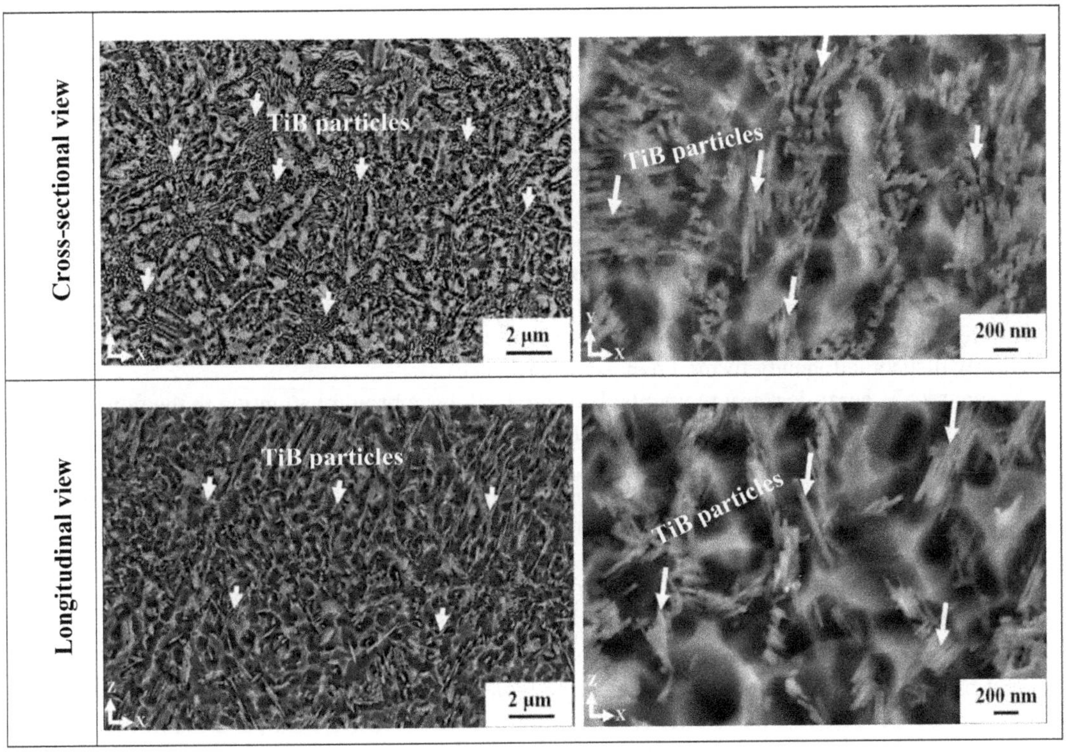

FIGURE 5.2 SEM images depicting the microscopic structures of the Ti-TiB composites fabricated using the L-PBF method. Reproduced from [12].

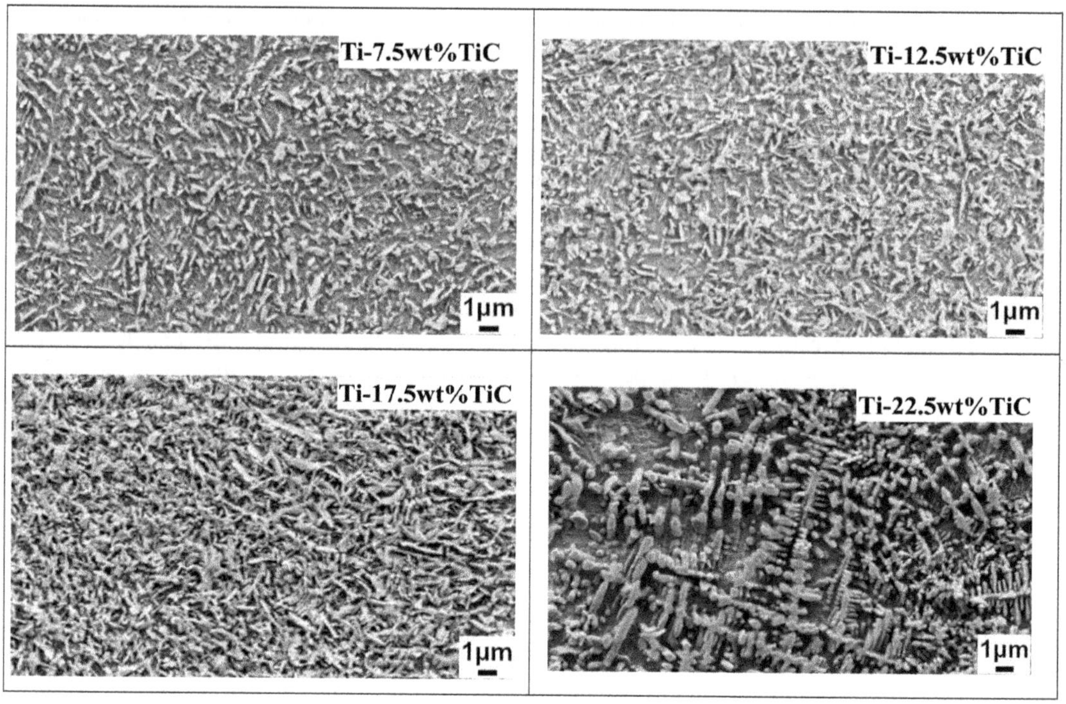

FIGURE 5.3 SEM images showing the microstructures of the L-PBF processed Ti-x%TiC composite. Reproduced from [15].

influence of volumetric energy density using L-PBF process for Ti-15wt%TiC nanocomposites [16]. The Ti-15wt%TiC composite exhibited a significant impact of energy density on the microstructure and mechanical properties. Lower energy densities (≤ 120 J/mm^3) resulted in a uniform dispersion of TiC particles, resulting in a fine microstructure. Conversely, higher energy densities (350 J/mm^3) resulted in the formation of coarse dendrites and reinforcement agglomerates. Furthermore, mechanical properties were improved with the increase of energy density from 90 J/mm^3 to 120 J/mm^3 and declined at an elevated value of 360 J/mm^3. He et al. also observed an increase in the tensile strength of the Ti-5wt%TiC composite in comparison to the CP-Ti, which was attributed to the uniformly distributed, newly formed nanoscale TiC [17].

As regards the ceramic reinforcements, B_4C and CrB_2 have brought attention to the fabrication of TMCs as a result of the formation of in-situ TiC and TiB particles within the titanium matrix. Xia et al. conducted a study where they used the L-PBF process to fabricate TMCs containing in-situ TiB and TiC particles using a mixture of Ti and B_4C powder [18]. In this case, L-PBF processing of the Ti-B_4C composite triggered the formation of dispersed TiB and granular TiC phases within the Ti-matrix. These precipitates were uniformly distributed, contributing to the overall hardness and strength improvement. Furthermore, larger TiB and TiC phases were observed with the increase in energy density. Kang et al. used the L-PBF process to reinforce CP-Ti with 2wt% CrB_2 powder. CrB_2 was selected for its exceptional wear and corrosion resistance properties, in addition to its compatibility with titanium due to chromium being a β phase stabilizer [13]. In-situ reactions during the L-PBF process resulted in the formation of titanium boride (TiB) particles in the titanium matrix. It was found that the composite hardness varied based on the processing conditions. The higher energy density resulted in both improved wear resistance and increased hardness. This was attributed to the increased content of TiB particles.

TABLE 5.4
Relationship between Reinforcement Type and Content and the Mechanical Properties of the TMCs, Fabricated by Powder Bed Fusion (PBF) AM Techniques

	Reinforcement				
Matrix	Material	Content	Technique	Properties	Reference
CP-Ti	–	–	L-PBF	Hardness: 261±13 HV CYS: 560±5 MPa UCS: 1136±15 MPa TYS: 555 MPa UTS: 757 MPa El: 19.5%	[12, 19]
	TiB_2	2.5wt%	L-PBF	Hardness: 333±14 HV	[20]
		5wt%	L-PBF	Hardness: 392±21 HV	[20]
		7.5wt%	L-PBF	Hardness: 435±15 HV	[20]
	TiB	8.35vol%	L-PBF	Hardness: 402±7 HV CYS: 1,103±20 MPa UCS: 1,421±47 MPa	[12]
	TiC	5wt%	L-PBF	UTS: 914 MPa El: 18.3%	[17]
		7.5wt%	L-PBF	Hardness: 553 HV	[15]
		12.5wt%	L-PBF	Hardness: 577 HV	[15]
		17.5wt%	L-PBF	Hardness: 487 HV	[15]
		22.5wt%	L-PBF	Hardness: 412 HV	[15]
	B_4C	5wt%	L-PBF	Hardness: 577.1 HV	[18]
	CrB_2	2wt%	L-PBF	Hardness: 320–380 HV	[13]
Ti-6Al-4V	–	–	L-PBF	Hardness: 375±4 HV UCS: 1,384±30 MPa Max. strain: 33%	[21]
	–	–	EBM	Hardness: 4.1 GPa TYS: 950 MPa UTS: 1,110 MPa El: 13% UCS: 1,300 MPa	[6]
	SiC	0.5wt%	L-PBF	TYS: 1,288 MPa UTS: 1,322 MPa El: 9.9%	[22]
			EBM	TYS: 1,085 MPa UTS: 1,159 MPa El: 13.6%	[22]
	HA	5wt%	EBM	Hardness: 6.8 GPa UTS: 123 MPa El: 5.5% UCS: 875 MPa	[6]
	B_4C	0.5wt%	L-PBF	Hardness: 458±5 HV UCS: 1,535±18 MPa Max. strain: 19%	[21]
		1wt%	L-PBF	Hardness: 546±7 HV UCS: 1,747±42 MPa Max. strain: 14%	[21]

Where *TYS*—tensile yield strength, *UTS*—ultimate tensile strength, *CYS*—compressive yield strength, *UCS*—ultimate compressive strength, *El*—maximum elongation in tension, *Max. strain*—maximum compressive strain.

5.4 DIRECTED ENERGY DEPOSITION OF TMCs

DED is a powder-fed or wire-fed AM process that utilizes a concentrated thermal energy source (laser, electron beam) to melt and deposit feedstock material in a layer-by-layer manner onto a substrate plate, building a three-dimensional object as shown in Figure 5.4 [23]. The key feature of DED is the formation of a molten pool where the feedstock material interacts with the energy source and melts. This molten pool then solidifies rapidly upon cooling, bonding with the previously deposited layer and contributing to the overall structure. DED is the most common AM integration in the hybrid additive-subtractive manufacturing processes (HAPs), which is used with other conventional manufacturing processes to provide adequate accuracy and desired surface roughness in large-scale manufacturing of TMCs [24].

DED offers several advantages for the fabrication of titanium and its alloys [25, 26]:

- High-power density and high deposition rate: The focused energy source allows for deep penetration and complete melting of titanium powder or wire, minimizing porosity and ensuring good mechanical properties
- Offers the ability to not only fabricate components from pure metals and functionally graded components but also to repair and clad existing valuable parts
- DED employs a synchronized feedstock delivery system, introducing either metal powder or wire concurrently with a focused thermal energy source
- Can obtain in-situ alloying in the metal pool as multiple powder feedstock can be fed at the same time
- Functionally graded materials of desired chemical composition can be fabricated through tailored microstructure, consequently attained by controlling the process parameters, including laser power, nozzle feed rate, scan speed, and hatch spacing. The selection of appropriate raw materials further contributes to the desired final chemical composition
- DED manufactured components have minimal distortion or cracks, excellent metallurgical bonding, a small heat-affected zone, and accurate deposition. Therefore, DED can be considered as a more effective tool for high-value parts repair, especially in the medical, aerospace, and defence fields of engineering.

There are few setbacks to which researchers are still finding a solution such as final rough surface and low powder efficiency. Several studies have reported challenges in achieving uniform microstructures, which can subsequently lead to variations in the mechanical characteristics of

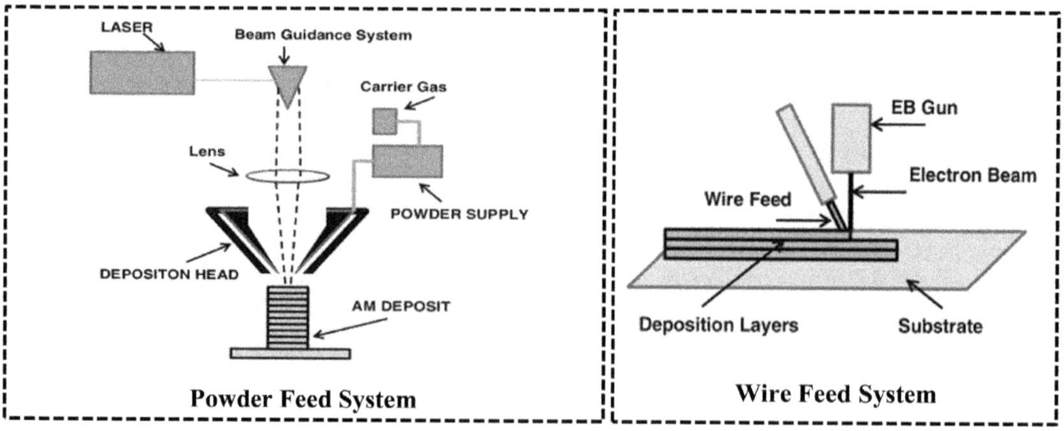

FIGURE 5.4 Working principles of directed energy deposition (DED) additive manufacturing techniques. Reproduced from [23].

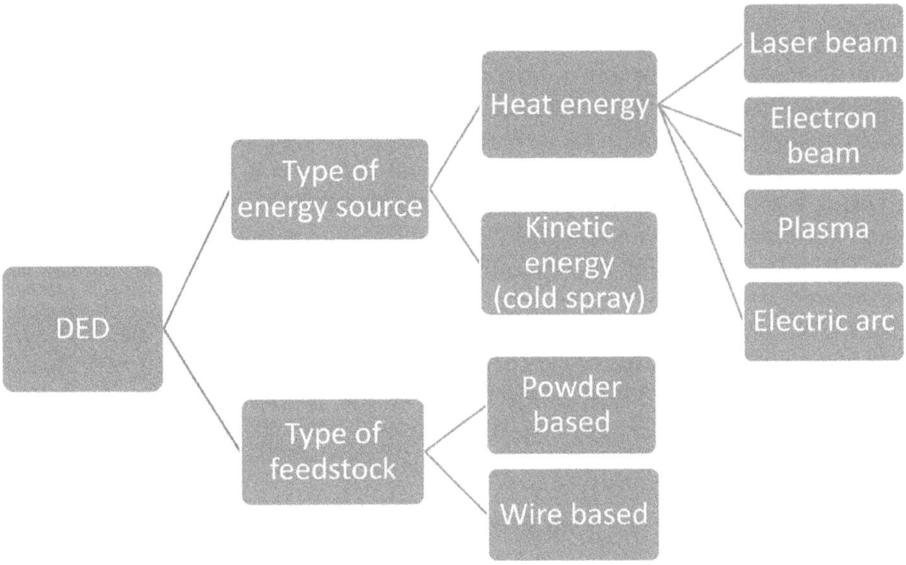

FIGURE 5.5 Classification of directed energy deposition (DED) technique. Reproduced from [28].

DED-fabricated parts. This can be attributed to the coarser nature of DED feedstock material (wire or powder) and large melt pool, which lead to uneven cooling rates and segregation during solidification [27]. Theoretically, DED has been classified as shown in Figure 5.5.

In general, there are several variations of each classification of DED process among which the commercialized ones can be listed as [25]: laser cladding (LC), laser powder deposition (LPD), direct metal deposition (DMD) or electron beam metal deposition (EBMD), direct light fabrication (DLF), shape deposition manufacturing (SDM), laser forming (Lasform), laser engineer net shaping or direct laser fabrication (LENS), laser powder fusion (LPF), laser-aided direct-metal deposition (LADMD) or laser metal deposition (LMD), laser direct casting (LDC), laser-based multi-directional metal deposition (LBMDMD), and laser aided manufacturing process (LAMP).

A broad perspective on the effect of specific DED methods applications on the mechanical performance of TMCs was summarized in Table 5.5. The table focuses on the TMCs including CP-Ti and Ti-6Al-4V matrices reinforced with TiB, TiC, SiC, TiN, and CaP. The properties of these DED-fabricated TMCs were compared to those of unalloyed commercially pure (CP) titanium.

From the research study of Zhang et al. it was observed that the change in the proportion of TiC reinforcement content marginally affects the YS and UTS of the Ti-TiC composites [29], whereas, with the decrease in the concentration of the TiC reinforcement, the ductility increases and the hardness decreases. The reason behind this is the uniform distribution of TiC particles, with a larger amount of it being unmelted, in the alloy matrix of the Ti-TiC composites [29]. For Ti-TiB composite the changes in mechanical properties were studied with respect to different laser powers, keeping the volume of Boron reinforcement constant and equal to 1.6wt%. It was concluded that the laser power increase from 125W to 200W resulted in a decrease in the non-uniform distribution of the reinforcement because, under high power, fine TiB particles were formed, resulting in the formation of 3-dimensional quasi-continuous network (3DQCN) microstructure [30, 31]. The presence of the 3DQCN microstructure significantly hinders the coarsening of titanium grains, which results in improved mechanical performance of the composites. Das et al. proposed their research on defect-free Ti-SiC composite with a variation of two process parameters including laser power and scanning speed [32]. The change in microstructure from Ti/SiC to Ti_5Si_3/TiC due to the chemical reaction is the reason behind a significant increase in wear resistance and hardness (to a max. 1167 HV). Apart from DLD, laser cladding was used in the manufacturing of Ti-SiC composite as reported by Li et al. [33]. Characterization of the formed Ti_5Si_3 and TiC particles revealed average sizes of about 100–400 nm and 1 μm, respectively. The clad layer consisting of 80%Ti and 20%SiC by volume significantly enhanced the hardness of the Ti64 matrix (339 HV) to 932 HV. This improvement can be primarily attributed to TiC and Ti_5Si_3 particles, which act as effective load-bearing reinforcements within the matrix. Consequently, the cladded material exhibited a significant improvement in wear resistance. Furthermore, Ti6Al4V–TiB composites have been also produced using 2 wt% B (elemental) [34]. Due to the chemical reaction and rapid solidification a B27 orthorhombic microstructure was characterized by homogenously dispersed TiB precipitates within the Ti6Al4V matrix. This microstructure results in improved creep resistance, wear resistance, and fracture toughness at elevated temperatures.

Titanium composites are also commercially used in implants to provide a lubricating effect between two surfaces. A decrease in the coefficient of friction and improved wear resistance are achieved in Ti-CaP and Ti6Al4V-CaP, where CaP is calcium phosphate. Bandyopadhyay et al. produced these composites with different wt% of CaP as shown in Table 5.5 [35]. As the concentration of CaP increased in the base matrix, the influence of CaP tribo-film amplified. This resulted in a consequent increase in mechanical properties. The yield strength, compressive strength, and hardness of the Ti-CaP composite increased from 780 MPa, 1,300 MPa, and 540 HV to 1,470 MPa, 2,070 MPa and 930 HV, respectively, whereas the hardness of the Ti6Al4V-CaP composite exhibited a moderate improvement, increasing from 440 HV to 540 HV. For Ti6Al4V-TiC composite fabrication, Wang et al. used wire-fed Ti6Al4V base matrix and powder-fed TiC reinforcement, which resulted in an increase of about 100% and 95%, in YS and UTS, respectively, in comparison to those of the Ti-TiC composites fabricated by the laser powder deposition [29, 36]. The TiC concentration was systematically adjusted, in the range of 8 vol%–74 vol%. Although improved YS and UTS with the addition of 8 vol% TiC was achieved, the ductility was negatively impacted. However, the addition of 15 vol% TiC significantly enhanced the wear resistance of TMCs.

TMCs with a contact surface of high wear resistance useful for loading bearing implants were manufactured by Balla et al. who reinforced Ti6Al4V alloy by incorporating TiN particles using

TABLE 5.5
Relationship between Reinforcement Type and Content and the Mechanical Properties of the TMCs, Fabricated by Directed Energy Deposition AM Techniques

Matrix	Reinforcement Material	Reinforcement Content	Technique	Properties	Reference
CP-Ti	–	–	LPD	Hardness: 260 HV TYS: 520 MPa UTS: 585 MPa El: 19%	[29]
	–	–	LENS	Hardness: 345.5±5.4 HV CYS: 773–787 MPa UCS: 836–851 MPa Max. strain: 54.9–56.5%	[30, 31]
	TiC	10vol%	LPD	Hardness: 301 HV TYS: 520 MPa UTS: 575 MPa El: 2.5%	[29]
		20vol%	LPD	Hardness: 336 HV TYS: 515 MPa UTS: 590 MPa El: 1.5%	[29]

TABLE 5.5 *(Continued)*
Relationship between Reinforcement Type and Content and the Mechanical Properties of the TMCs, Fabricated by Directed Energy Deposition AM Techniques

Matrix	Reinforcement		Technique	Properties	Reference
	Material	Content			
	B	1.6wt%	LENS	Hardness: 392.6±8.9 HV CYS: 940–1010 MPa UCS: 1016–1092 MPa Max. strain: 26.5–36.4%	[30, 31]
	SiC		LENS	Hardness: 976–1167 HV	[32]
	CaP	3.3wt%	LENS	Hardness: 540±20 HV CYS: 780±20 MPa UCS: 1300±30 MPa	[35]
		6.8wt%	LENS	Hardness: 760±30 HV CYS: 1220±20 MPa UCS: 1450±70 MPa	[35]
		10wt%	LENS	Hardness: 930±50 HV CYS: 1470±90 MPa UCS: 2070±70 MPa	[35]
Ti-6Al-4V	–	–	LMD	UTS: 959.4–1091 MPa El: 11.45–5.52%	[38]
	SiC	10wt%	Laser cladding	Hardness: 527.3 HV	[33]
		20wt%	Laser cladding	Hardness: 932.2 HV	[33]
	CaP	2wt%	LENS	Hardness: 440±70 HV	[35]
		5wt%	LENS	Hardness: 540±50 HV	[35]
	TiC	8vol%	Wire and powder fed LENS	TYS: 1038–1041.7 MPa UTS: 1075–1118.1 MPa El: 0.3–1.05%	[36]
		5vol%	LMD	UTS: 1225.5 MPa El: 1.31%	[38]
		10vol%	LMD	UTS: 1100.4 MPa El: 0.49%	[38]
		15vol%	LMD	UTS: 997.2 MPa El: 0.3%	[38]
		20vol%	LMD	UTS: 907.3 MPa El: 0.15%	[38]
		15wt%	Wire and powder fed EBM	Hardness: 7.3 GPa UTS: 950 MPa El: 6%	[39]
		35wt%	Wire and powder fed EBM	Hardness: 7.5 GPa UTS: 930 MPa El: 2%	[39]
	TiN	10wt%	LENS	Hardness: 527±7 HV	[37]
		15wt%	LENS	Hardness: 489±31 HV	[37]
		20wt%	LENS	Hardness: 514±24 HV	[37]
		40wt%	LENS	Hardness: 1138±61 HV	[37]
	AlB_2	20wt%	Laser cladding	Hardness: 804 HV	[40]

Where *TYS*—tensile yield strength, *UTS*—ultimate tensile strength, *CYS*—compressive yield strength, *UCS*—ultimate compressive strength, *El*—maximum elongation in tension, *Max. strain*—maximum compressive strain.

the Laser Engineered Net Shaping (LENS™) technique [37]. The in vitro biocompatibility analysis demonstrated that these composites had non-toxic properties in comparison to the Ti6Al4V substrate. In this study, the weight percentage of TiN was increased from 10% to 40 %, which resulted in an exponential increment in the hardness of the composite from 527 HV to 1138 HV. The addition of TiN reinforcement in Ti6Al4V alloy also remarkably improved the in-vitro wear performance, exhibiting a wear rate of 3.74×10^{-6} mm^3/Nm. This represents a notable reduction compared to the most preferable CoCrMo alloy, which exhibits a wear rate of 1.04×10^{-5} mm^3/Nm.

5.5 SUMMARY, CHALLENGES, AND OUTLOOK

In this chapter, the research on the fabrication of titanium-based matrix composites (TMCs) by using different additive manufacturing approaches was summarized. The literature review suggests that few reinforcements including as TiC, TiB_2, TiB, B_4C, CrB_2, SiC, B, TiN, CaP, and hydroxyapatite (HA) have been used mostly utilized to fabricate TMCs. Although there are other reinforcements available that are suitable for the Ti and Ti-alloys, only a few attempts were made to assess their effectiveness in AM.

It is worth mentioning that AM offers a powerful tool for manufacturing titanium-based matrix composites with superior properties compared to unreinforced Ti-alloys. By selecting appropriate reinforcements and optimizing process parameters, one can produce components with enhanced wear resistance, high-temperature performance, and biocompatibility for diverse applications in the transport industry, biomedical, and other critical sectors. With continuous advancements in AM research, particularly in the field of TMCs, we can anticipate even more innovative and high-performance TMCs emerging from the world of AM.

Despite the numerous advantages, additive manufacturing of TMCs faces certain challenges:

- Porosity Control: Managing porosity is crucial for achieving good mechanical properties. Careful selection of feedstock powder and optimization of process parameters is essential to minimize voids and ensure a dense structure
- Residual Stresses: The rapid solidification process during AM processes can generate residual stresses in the final part. These stresses can affect dimensional accuracy and potentially lead to cracking. Process optimization and post-processing techniques like stress relief annealing are employed to mitigate this issue
- Feedstock Development: Developing high-quality feedstock materials like pre-alloyed powders with good flowability and laser absorptivity is crucial for the successful processing of TMCs

In the near future, several promising developments are expected to occur:

- Novel research efforts focus on topology optimization and the development of innovative designs for additive manufacturing of TMCs. These advancements aim to achieve superior structural and functional performance in the resulting components
- Optimization of process parameters for utilization of diverse reinforcements in TMCs for producing parts without defects and with the desired surface quality, microstructure, and mechanical properties
- Current efforts are underway to develop real-time process control strategies that involve the use of machine vision or other sensor technologies. These control systems enhance the assurance of quality and stability of the manufacturing process for TMCs
- Advancements in additive manufacturing (AM) technology involve the development of AM machines and the exploration of hybrid AM processes, to enable the fabrication of TMCs through both in-situ and ex-situ methods

REFERENCES

[1] Dubey, V.P., Kopec, M., Łazińska, M. and Kowalewski, Z.L., 2023. Yield surface identification of CP-Ti and its evolution reflecting pre-deformation under complex loading. *International Journal of Plasticity*, *167*, p. 103677.

[2] Cooke, K. and Alhubaida, A., 2022. Microstructural response and wear behaviour of Ti-6Al-4V impregnated with Ni/Al_2O_3 + TiO_2 nanostructured coating using an electric arc. *Scientific Reports*, *12*(1), p. 21978.

[3] Singh, N., Hameed, P., Ummethala, R., Manivasagam, G., Prashanth, K.G. and Eckert, J., 2020. Selective laser manufacturing of Ti-based alloys and composites: Impact of process parameters, application trends, and future prospects. *Materials Today Advances*, *8*, p. 100097.

[4] Attar, H., Ehtemam-Haghighi, S., Kent, D. and Dargusch, M.S., 2018. Recent developments and opportunities in additive manufacturing of titanium-based matrix composites: A review. *International Journal of Machine Tools and Manufacture*, *133*, pp. 85–102.

[5] Kujur, M.S., Dubey, V.P., Mallick, A. and Gupta, M., 2021. Tensile characteristics of metal matrix composites. In *Encyclopedia of Materials: Composites*, D. Brabazon, ed., Oxford: Elsevier, pp. 298–312.

[6] Terrazas, C.A., Murr, L.E., Bermudez, D., Arrieta, E., Roberson, D.A. and Wicker, R.B., 2019. Microstructure and mechanical properties of Ti-6Al-4V-5% hydroxyapatite composite fabricated using electron beam powder bed fusion. *Journal of Materials Science & Technology*, *35*(2), pp. 309–321.

[7] Shishkovsky, I., Kakovkina, N. and Sherbakov, V., 2017. Graded layered titanium composite structures with TiB_2 inclusions fabricated by selective laser melting. *Composite Structures*, *169*, pp. 90–96.

[8] ASTM International, 2012. *F2792, Standard Terminology for Additive Manufacturing Technologies* [Online]. Available: www.astm.org/f2792-10e01.html. [Accessed: 28-March-2024].

[9] Dev Singh, D., Mahender, T. and Raji Reddy, A., 2021. Powder bed fusion process: A brief review. *Materials Today: Proceedings*, *46*, pp. 350–355.

[10] Mehrpouya, M., Tuma, D., Vaneker, T., Afrasiabi, M., Bambach, M. and Gibson, I., 2022. Multimaterial powder bed fusion techniques. *Rapid Prototyping Journal*, *28*(11), pp. 1–19.

[11] Ghibaudo, C., Maculotti, G., Gobber, F., Saboori, A., Galetto, M., Biamino, S. and Ugues, D., 2023. Information-rich quality controls prediction model based on non-destructive analysis for porosity determination of AISI H13 produced by electron beam melting. *The International Journal of Advanced Manufacturing Technology*, *126*(3), pp. 1159–1173.

[12] Attar, H., Bönisch, M., Calin, M., Zhang, L.-C., Scudino, S. and Eckert, J., 2014. Selective laser melting of in situ titanium–titanium boride composites: Processing, microstructure and mechanical properties. *Acta Materialia*, *76*, pp. 13–22.

[13] Kang, N., Coddet, P., Liu, Q., Liao, H. and Coddet, C., 2016. In-situ TiB/near α Ti matrix composites manufactured by selective laser melting. *Additive Manufacturing*, *11*, pp. 1–6.

[14] Attar, H., Prashanth, K.G., Zhang, L.-C., Calin, M., Okulov, I.V., Scudino, S., Yang, C. and Eckert, J., 2015. Effect of powder particle shape on the properties of in situ Ti–TiB composite materials produced by selective laser melting. *Journal of Materials Science & Technology*, *31*(10), pp. 1001–1005.

[15] Gu, D., Meng, G., Li, C., Meiners, W. and Poprawe, R., 2012. Selective laser melting of TiC/Ti bulk nanocomposites: Influence of nanoscale reinforcement. *Scripta Materialia*, *67*(2), pp. 185–188.

[16] Gu, D., Hagedorn, Y.-C., Meiners, W., Wissenbach, K. and Poprawe, R., 2011. Nanocrystalline TiC reinforced Ti matrix bulk-form nanocomposites by selective laser melting (SLM): Densification, growth mechanism and wear behavior. *Composites Science and Technology*, *71*(13), pp. 1612–1620.

[17] He, B., Chang, K., Wu, W. and Zhang, C., 2017. The formation mechanism of TiC reinforcement and improved tensile strength in additive manufactured Ti matrix nanocomposite. *Vacuum*, *143*, pp. 23–27.

[18] Xia, M., Liu, A., Hou, Z., Li, N., Chen, Z. and Ding, H., 2017. Microstructure growth behavior and its evolution mechanism during laser additive manufacture of in-situ reinforced (TiB+TiC)/Ti composite. *Journal of Alloys and Compounds*, *728*, pp. 436–444.

[19] Cui, Y.-W., Wang, L. and Zhang, L.-C., 2024. Towards load-bearing biomedical titanium-based alloys: From essential requirements to future developments. *Progress in Materials Science*, *144*, p. 101277.

[20] Jin, J., Zhou, S., Zhao, Y., Zhang, Q., Wang, X., Li, W., Chen, D. and Zhang, L.-C., 2021. Refined microstructure and enhanced wear resistance of titanium matrix composites produced by selective laser melting. *Optics & Laser Technology*, *134*, p. 106644.

[21] Li, H., Yang, Z., Cai, D., Jia, D. and Zhou, Y., 2020. Microstructure evolution and mechanical properties of selective laser melted bulk-form titanium matrix nanocomposites with minor B_4C additions. *Materials & Design*, *185*, p. 108245.
[22] Portolés, L., Blasco, J.-R., Martín, J.-M., Burgos, N., Borghetto, M., Zoz, A., Ludwig, N., Maccio, T., Marcos, M.D., García-Manrique, J.A. and Solano, L., 2024. In situ synthesis of titanium alloy powders reinforced with nanoparticles for powder bed fusion: A step towards safer and more sustainable manufacturing. *Additive Manufacturing*, *82*, p. 104032.
[23] Frazier, W.E., 2014. Metal additive manufacturing: A review. *Journal of Materials Engineering and Performance*, *23*(6), pp. 1917–1928.
[24] Merklein, M., Junker, D., Schaub, A. and Neubauer, F., 2016. Hybrid additive manufacturing technologies—an analysis regarding potentials and applications. *Physics Procedia*, *83*, pp. 549–559.
[25] Saboori, A., Gallo, D., Biamino, S., Fino, P. and Lombardi, M., 2017. An overview of additive manufacturing of titanium components by directed energy deposition: Microstructure and mechanical properties. *Applied Sciences*, *7*(9), p. 883.
[26] Barwinska, I., Kopec, M., Łazińska, M., Brodecki, A., Durejko, T. and Kowalewski, Z.L., 2021. Suitability of laser engineered net shaping technology for Inconel 625 based parts repair process. *Materials*, *14*(23), p. 7302.
[27] Selcuk, C., 2011. Laser metal deposition for powder metallurgy parts. *Powder Metallurgy*, *54*(2), pp. 94–99.
[28] Dass, A. and Moridi, A., 2019. State of the art in directed energy deposition: From additive manufacturing to materials design. *Coatings*, *9*(7), p. 418.
[29] Zhang, Y., Wei, Z., Shi, L. and Xi, M., 2008. Characterization of laser powder deposited Ti–TiC composites and functional gradient materials. *Journal of Materials Processing Technology*, *206*(1), pp. 438–444.
[30] Hu, Y., Ning, F., Wang, H., Cong, W. and Zhao, B., 2018. Laser engineered net shaping of quasi-continuous network microstructural TiB reinforced titanium matrix bulk composites: Microstructure and wear performance. *Optics & Laser Technology*, *99*, pp. 174–183.
[31] Hu, Y., Cong, W., Wang, X., Li, Y., Ning, F. and Wang, H., 2018. Laser deposition-additive manufacturing of TiB-Ti composites with novel three-dimensional quasi-continuous network microstructure: Effects on strengthening and toughening. *Composites Part B: Engineering*, *133*, pp. 91–100.
[32] Das, M., Balla, V.K., Basu, D., Bose, S. and Bandyopadhyay, A., 2010. Laser processing of SiC-particle-reinforced coating on titanium. *Scripta Materialia*, *63*(4), pp. 438–441.
[33] Li, N., Xiong, Y., Xiong, H., Shi, G., Blackburn, J., Liu, W. and Qin, R., 2019. Microstructure, formation mechanism and property characterization of Ti + SiC laser cladded coatings on Ti6Al4V alloy. *Materials Characterization*, *148*, pp. 43–51.
[34] Genç, A., Banerjee, R., Hill, D. and Fraser, H.L., 2006. Structure of TiB precipitates in laser deposited in situ, Ti-6Al-4V–TiB composites. *Materials Letters*, *60*(7), pp. 859–863.
[35] Bandyopadhyay, A., Dittrick, S., Gualtieri, T., Wu, J. and Bose, S., 2016. Calcium phosphate–titanium composites for articulating surfaces of load-bearing implants. *Journal of the Mechanical Behavior of Biomedical Materials*, *57*, pp. 280–288.
[36] Wang, F., Mei, J., Jiang, H. and Wu, X., 2007. Laser fabrication of Ti6Al4V/TiC composites using simultaneous powder and wire feed. *Materials Science and Engineering: A*, *445–446*, pp. 61–466.
[37] Balla, V.K., Bhat, A., Bose, S. and Bandyopadhyay, A., 2012. Laser processed TiN reinforced Ti6Al4V composite coatings. *Journal of the Mechanical Behavior of Biomedical Materials*, *6*, pp. 9–20.
[38] Wang, J., Li, L., Tan, C., Liu, H. and Lin, P., 2018. Microstructure and tensile properties of TiCp/Ti6Al4V titanium matrix composites manufactured by laser melting deposition. *Journal of Materials Processing Technology*, *252*, pp. 524–536.
[39] Builuk, A., Kazachenok, M. and Martynov, S., 2019. Electron beam additive manufacturing of TiCx/Ti-6Al-4V composite. *AIP Conference Proceedings*, *2167*(1), p. 020039.
[40] Feng, Y., Feng, K., Yao, C., Li, Z. and Sun, J., 2018. Microstructure and properties of in-situ synthesized (Ti3Al + TiB)/Ti composites by laser cladding. *Materials & Design*, *157*, pp. 258–272.

6 A Review on Selective Laser Melting of Aluminium Metal Matrix Composites

Kumar Ramar, Gedion Habtay Gebremicheal, Ghebremlak Andemariam, Lewhat Michael Weldegebriel, Hurui Tesfalem Woldu, Vijay Ananth Suyamburajan, and Ajith Arul Daniel

6.1 INTRODUCTION

Nowadays, MMCs are widely used in the structure manufacturing industry. Especially aluminium-based MMCs are highly used in emerging applications of military vehicles, space, and aircraft industries [1–4]. Commercially used matrix materials are aluminium, magnesium, cobalt, copper, nickel, and titanium alloys. But here, the most wanted matrix materials are aluminium and titanium alloys for advanced applications [5]. Along with commonly used reinforcements are aluminium oxides, titanium carbide, silicon carbide, tungsten carbide, etc., in which silicon carbide, aluminium oxide, and titanium carbide are repeatedly used in the manufacturing of composites [6]. The concept of reinforcing ceramic particles into the matrix alloy leads to enhancing the strength, stiffness, and stability of the materials, which are highly demanded by the aerospace industry. At the same time, other properties like thermal stability, corrosion and wear resistance, damping effect, and fatigue strength are also significantly enhanced [7].

6.1.1 AMMCs

Aluminium alloy has delightful properties like corrosion resistance, being lightweight, heat treatment ability, good thermal conductivity, etc., which are used in different sections of construction such as automotive, aerospace, electrical and electronics, sports, and recreational components [7]. However, low strength and wear resistance have been challenges in its application. The concept of reinforcing particles into the aluminium-based alloy matrix is the greatest indication for producing superior-strength materials [8, 9]. Mainly, good moduli and resistance to higher temperatures are obtained from aluminium metal matrix composites (AMMCs) [10–14]. Manufacturing of AMMCs is done through two main techniques: casting methods and powder metallurgy, but it offers a reduction in mechanical properties due to the problem of heterogeneous distribution of reinforcement particles. Also, the fabrication of components through those methods has more porosity and segregation of particles and is limited to simple-shaped parts [15, 16].

6.1.2 Additive Manufacturing (AM) Processes

An account of highly complex geometry in parts is needed in the aerospace industry, so manufacturing those components through a customary method is thought-provoking or sometimes impossible [17, 18]. To rectify this issue, advanced processing routes are essential for manufacturing these complex ingredients. AM is the paramount alternative technique used to produce geometry in complex shapes.

DOI: 10.1201/9781003499718-6

It indicates the group of manufacturing processes in which a constituent thought to be fictitious is completed through layer-by-layer addition of cured materials like metallic or non-metallic powder, liquid in resin form, and solid form (wire, laminated sheets). The machines used for this technique require a 3D part model in standard tessellation format (STL.file) and convert it into a slicing format for the manufacture of the desired component through the layer-by-layer bonding method. Nowadays, different AM systems, stereolithography apparatus, are available on the market for treating various forms of materials [19]. Classifications of the proposed methods are based on the initial state of the material, energy source, layer thickness, curing time, built size, etc. The properties of AMMCs are decided based on the distribution level of reinforcement materials into the aluminium alloy matrix throughout the composite. Due to the advancement of AM methods, several hurdles are faced in the traditional process, including dispersion of reinforcement into the matrix, difficulties in wettability, and chemical reactions between matrix and reinforcement [20–23]. Selective laser melting (SLM) stands out as the optimal additive manufacturing (AM) system for producing advanced metal matrix composites (AMMCs), effectively avoiding the issues associated with traditional methods. SLM is the metal powder-based AM technique in which components are fabricated through a layer-by-layer bonding mechanism. It has the advantages of low segregation and net-shape processing with more complex geometries. In this chapter, brief information about the fabrication of AMMCs using the SLM technique is discussed.

6.2 PROCESSING OF SLM

6.2.1 SLM Process

SLM is a powder-based fused deposition AM process in which the product is produced from loosely packed powder through a selective laser beam [24]. Table 6.1 shows the previously used process parameters for the processing of AMMCs by SLM. A 3D CAD model is first designed with the modelling software on a separate computer and then converted into a 2D slicing format. The layer thickness is around 30–100 μm thick and defines the geometric cross-section of the parts to be produced during the slicing period [25]. The data from the pre-processing steps is transferred into the controlling system computer of the AM machine.

Figure 6.1 shows the schematic representation of the SLM processes. Initially, the powders with the homogenously mixed combination of aluminium alloy and reinforcement material are filled in the powder feed bed container. Then the powder feed platform is raised to the required level through the control of the servomotor, and after that, a roller mechanism is operated to spread the powder uniformly over the part-build platform for a thickness of one layer. The excess powder during the rolling process is collected in a separate container. A laser beam with 70–100 μm intensity is used to provide heat energy over the powder layer for selectively fusing based on data defined in the slicing process [25]. Once the fabrication of the first layer is finished, the build platform is moved downward at a depth of pre-defined layer thickness. Again, repeat the same procedure until the finishing of the desired products is defined by the 3D model.

TABLE 6.1
Process Parameters Used for SLM Process

		SLM process parameters			
Authors	Composite	Laser power	Scanning speed	Layer thickness	Hatch Spacing
Jue et al. [35]	$Al\text{-}Al_2O_3$	130 W	450–750 mm/s	–	–
Han et al. [41]	$Al\text{-}Al_2O_3$	200 W	100–600 mm/s	30μm	70, 100μm
Du et al. [34]	$AlSi_{10}Mg\text{-}Al_2O_3$	350 W	100–1500 mm/s	–	0.1–0.2μ m
Liao et al. [39]	$AlSi_{10}Mg\text{-}Al_2O_3$	200 W	250–350 mm/s	–	0.05, 0.1μm
Liao et al. [38]	$AlSi_{10}Mg\text{-}Al_2O_3$	200 W	100–300 mm/s	20μm	60–160μm

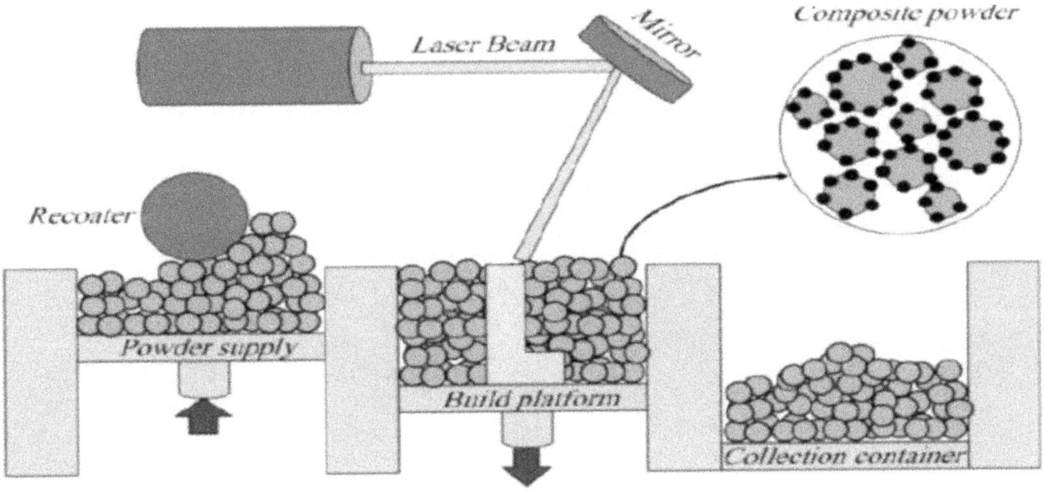

FIGURE 6.1 SLM processes [17].

6.2.2 Composite Build Powders for SLM Process

Before manufacturing AMMCs by the SLM operation, a feed stock of composite powder as an initial material with a combination of matrix and reinforcement materials is prepared [26, 27]. The preparation methods and characteristics of the mixed composite powder are very important because they affect the performance and accuracy of the parts to be fabricated. During the mixing of reinforcement ceramic particles with the metallic powder, the following factors should be considered: particle size, material type, volumetric percentage of powder, and morphology structure. Those factors determine the mechanical properties of AMMC parts, which include microstructure, homogeneity, and process ability [28, 29]. Recently, several processes have been available for making composite powders that permit the fabrication of AMMCs.

Regular mixing of ceramic particles with metal powders manually is one of the methods for preparing composite powder. Figure 6.2 shows the arrangement of mixing process of ceramic powders. During this process, ceramic reinforcement particles are slightly stuck to the surface of metal powders, leading to a lack of decoration. In the ball milling method, the distribution of reinforced ceramic particles into metallic powders is achieved through mechanical alloying. In this process, due to the high impact induced by the balls on the mixture of particles, it creates cold-welding of particles, and particles fracture and repeatedly deform [30]. The process parameters required for ball milling are speed of rotation, time for mixing, and weight ratio [31]. The composite powder provided by this method offers good powder flow ability and uniform spreading over the powder bed during the SLM process of AMMC fabrication. For instance, Han et al. [32, 33] used the ball-milled method for the preparation of Al-Al_2O_3 powder for SLM applications, and in this research, 200g of Al with 4 vol% of Al_2O_3 were mixed with a 5:1 ball-to-powder physical property ratio. They observed the undesirable effect of cold-welding and oxidation is prevented by the add-on of stearic acid.

Du et al. [34] prepared AMMCs with 2 and 5 wt% of nano-sized Al_2O_3 using the SLM process, in which the ball milling process is used for the preparation of composite powders. The gas-atomized AlSi10Mg was successfully mixed with 270 nm Al_2O_3 powder with a mixing time of 5 hours at 600 rpm rotational speed and a 1:1 ball-to-powder ratio. Therefore, due to its time-cost effectiveness and suitability for a wide range of materials, the ball milling method has highly attracted researchers for composite powder preparation. Table 6.2 shows the initial metal powder prepared in the observed previous research work for the SLA process.

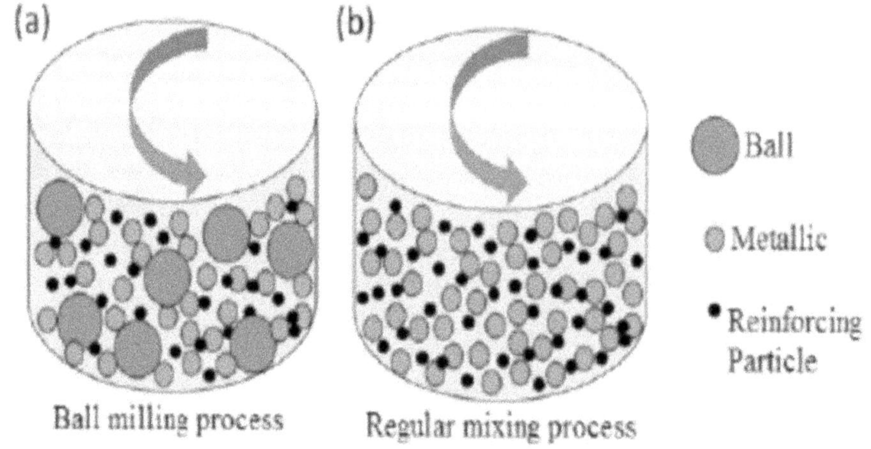

FIGURE 6.2 **a)** Ball milling, **b)** regular mixing processes [17].

TABLE 6.2
Metal Powder Prepared by Ball Milling

Authors	Composite metal powder	Weight ratio	Particle size	Milling processing time and speed
Jue et al. [35]	$Al\text{-}Al_2O_3$	$Al{:}Al_2O_3 = 80{:}20$	1.02 μm (Al), 9.04 μm (Al_2O_3)	16 hours and 200 rpm
Wang et al [36], Gu et al [37]	$AlSi_{10}Mg\text{-}Al_2O_3$	$AlSi_{10}Mg{:}\ Al_2O_3 = 80{:}\ 20$	30 μm ($AlSi_{10}Mg$), 9.04 μm (Al_2O_3)	8 hours and 200 rpm
Liao et al [38]	$AlSi_{10}Mg\text{-}Al_2O_3$	$AlSi_{10}Mg{:}\ Al_2O_3 = 85{:}\ 15$	33.1 μm ($AlSi_{10}Mg$), 26.6 μm (Al_2O_3)	4 hours and 70 rpm
Liao et al [39]	$AlSi_{10}Mg\text{-}Al_2O_3$	$AlSi_{10}Mg{:}Al_2O_3 = 1{:}1$	33 μm ($AlSi_{10}Mg$), 26.6 μm (Al_2O_3)	–
Du et al [40]	$AlSi_{10}Mg\text{-nano}Al_2O_3$	-	42 μm ($AlSi_{10}Mg$), 270 nm (nAl_2O_3)	5 hours and 600 rpm

6.3 PROPERTIES OF AMMCs DEVELOPED BY SLM

Nowadays, a lot of SLM machines are available on the market, and each machine has unique process variations in the ratio of metal powders, layer thickness, etc. The mechanical properties of AMMCs are shown in Table 6.3. So the comparison of various properties of the produced AMMC material based on the SLM process parameters is unreasonable [44]. The ceramic particle Al_2O_3 is a good reinforcement phase in aluminium alloys to achieve that aim since it has characteristics of high strength, withstanding elevated temperatures, higher hardness, and better wear resistance [45]. Therefore, it can be used as reinforcement material to improve the mechanical strength by acting as a barrier for dislocation movement [46]. The SLM system is the advanced AMMCs production machine; it has an effective arrangement for part design and no need for machining after the process to produce a produce a good net-shape product and, especially, an ultrafine microstructure can be obtained [47, 48].

Microstructure and relative density variation were observed in the SLM processed powder as shown in Figure 6.3. Jue et al. [35] investigated the hardness of $Al\text{-}Al_2O_3$ composite with a weight ratio of 80:20 developed by SLM. It addressed the greater hardness value in the range of 148–175

HV and also obtained a superior wear rate of 4.75 × 10–5 $mm^3N^{-1}m^{-1}$ and a coefficient of friction of 0.11. Figure 6.4 shows the variation in WR and COF for SLM processed parts. Besides, Gu et al. [37] investigated the coefficient of friction, surface roughness, and wear rate of the Al-Al_2O_3 composites from the SLM process. Han et al [40, 42] did a series of research studies for the fabrication

TABLE 6.3
Mechanical Properties of AMMCs Fabricated by the SLM Process

		Mechanical properties of AMMCs			
Authors	**Composite material**	**Coefficient of friction**	**Wear rate, $mm^3N^{-1}m^{-1}$**	**Hardness**	**Strength, MPa**
Jue et al. [35]	Al-Al_2O_3	0.11	4.75×10^{-5}	148–175 HV	×
Gu et al [37], Jue et al [41]	$AlSi_{10}Mg$-Al_2O_3	0.3	2.94×10^{-5}	×	×
Han et al [40, 42]	Al-Al_2O_3	×	×	48.5 HV	Yield strength: 109, ultimate tensile strength: 160
Nalivaiko et al [43]	Al-Al_2O_3	×	×	58.3 HB	–

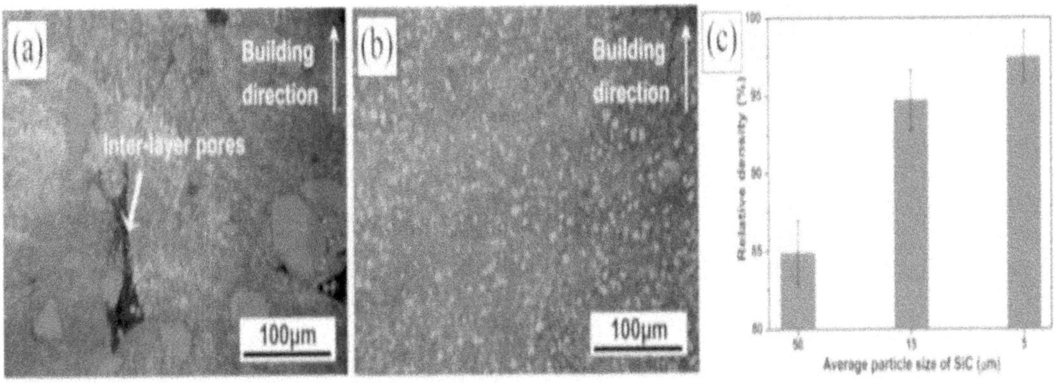

FIGURE 6.3 Microstructure of SLM-processed powder: **a)** SiC coarse grain, **b)** SiC fine powder particles; **c)** the change in relative density of parts [17].

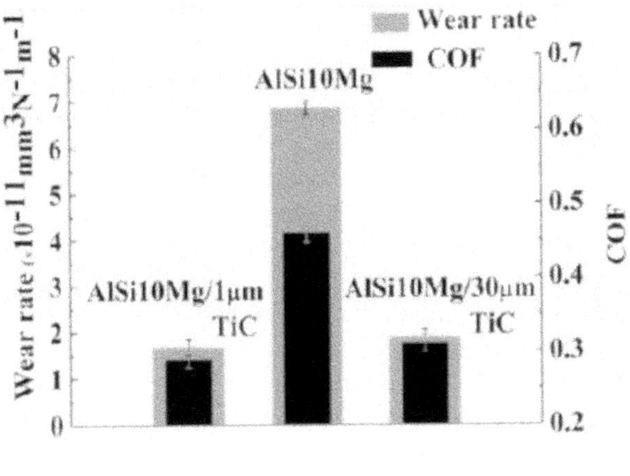

FIGURE 6.4 Wear rate and coefficient of friction [17].

of Al-Al_2O_3 composites using SLM, and then wear performance and hardness (48.5 HV) were studied. Due to the dispersion of Al_2O_3 reinforcement, the hardness of the composites has increased up to 17.5 %. Also, the tensile performance was reported in [40], and compared with pure aluminium the yield strength of 80 MPa was increased to 109 MPa in Al-Al_2O_3 composites through SLM, an improvement of 36.3%. Also, an ultimate tensile strength improvement of 45.5% was obtained when comparing pure aluminium of 110 MPa with the Al-Al_2O_3 composites of 160 MPa. However, the UTS of 160 MPa was comparable to less than the AMMCs produced by some other methods (stir casting, die casting) [48, 49], although this value is greater than the outcomes from other research works [50, 51] on AMMCs through the microwave sintering process.

Gu et al. [52] fabricated the AlSi10Mg/TiC nanocomposites through SLM, in which the mechanical properties were obtained for the fabricated specimens. The results of this research work denoted an betterment in UTS of 486 MPa with 10.9% elongation and a microhardness of 188.3 HV. The combination of two mechanisms of grain reinforcement and boundary strengthening in developed AMMCs was due to the enhancement in material behaviour. It's provided a novel ring-like microstructure in AlSi10Mg/TiC nanocomposites by altering the laser energy density. Also, with enhancing laser energy density, the microstructural change in TiC reinforcement from the average size of 77–93 nm to a coarsened submicron structure particle size of 154 nm with enhancing denseness was observed. Homogeneous distribution of the TiC particle in the AlSi10Mg matrix was obtained due to increases in laser energy density of SLM and low COF of (0.36) with reduced wear.

Ghosh et al. [53] analysed the wear resistance properties of Al4.5Cu3Mg/SiC developed through the SLM process. The wear resistance of AMMCs was enhanced with the addition of reinforcement up to 20 wt% of SiC particles in matrix alloy by the way of increased hardness. However, these features were impacted because to cracking and abrasive wear characteristics following a 20 wt%. of SiC particles to Al4.5Cu3Mg. Thermal expansion and contraction residual stresses were induced during the SLM process of AMMCs because cracks created at the interfacial bonding between the Al4.5Cu3Mg/SiC materials. Hence, due to limited plastic deformation at the surface of AMMCs for the composites having greater than 20 wt% addition of SiC particles, the wear resistance was reduced.

Building processes with reinforcement of hard ceramic SiC particles into aluminium in laser processing of AMMCs leads to poor wettability at the interface of reinforcement and matrix due to low absorption rate capacity. This is the main disadvantage of producing AMMC parts through SLM when SiC disperses into aluminium. For that particular situation, an in-situ reaction between the interface of SiC and aluminium allows the formation of the reaction product of Al4SiC4, which happens above the elevated temperature of 1,670 K. Zhou et al. [54] addressed the wettability of Al and SiC.

Chang et al. [55] focused on the developed AMMCs of in-situ Al4SiC4 with different proportions of reinforcement of SiC contents. They investigated the consequence of SiC on the AMMCs developed by SLM, in which mechanical properties, phase transformation, and microstructure features were studied. The in-situ response between the interface of aluminium matrix and SiC particles was improved by a reduction in the size of the SiC particle. The dispersion of reduced-size 3 μm SiC particles into the matrix alloy betters the homogeneity of the microstructure of the AMMCs. Also, a density level of 97.2% of fine SiC dispersion was achieved by the SLM process, which led to a 50% improvement in microhardness of about 218 HV compared to an unreinforced alloy. Further, it was noted that the wear resistance was improved by 66% and the coefficient of friction was reduced by 19% for the developed AMMCs by SLM.

Li et al. [56] fabricated AMMC through SLM with two varied reinforcements of SiC and TiN contents into the Al12Si alloy. They systematically studied the phenomenon of SLM process parameters on the produced AMMC parts. It resulted in the reinforcement of SiC contents breaking down into matrix alloys during the SLM process at higher energy densities. However, this breakdown effect does not take place during the dispersion of TiN into matrix alloys with high density. It offered higher microhardness of the produced SLM parts.

Dadbakhsh et al. [57] developed AMMCs with various proportions of Fe_2O_3 in-situ reinforcement into a pure aluminium matrix by the SLM process. This research was concentrated on controlling the interfacial bonding in Al/Fe_2O_3 composites. By the way, they reported the influence of reinforcement particles in a pure aluminium matrix on the hardness and microstructure formation of the AMMCs. It was found that the well-bonded hard phase's formation includes Fe_3Al, $Al1_3Fe_4$, and Al_2Fe with homogeneous dispersion in the microstructure of Al/Fe_2O_3 composites for different percentages of Fe_2O_3 content and laser parameters. The porosity due to entrapped gases during the SLM process was decreased while increasing the content of Fe_2O_3 particles in the matrix.

Kang et al. [58] studied the wear characteristics and microstructure of quasicrystal Al91Fe4Cr5 reinforced into an aluminium matrix by the SLM process. The outcomes of this work demonstrated that the increasing laser power intensity altered the particle size of quasicrystals from nanometres to micrometers. Also, it was identified as having a higher wear resistance of $2.64 \times 10^{-4} mm^3/Nm$ for the developed composite parts. Then, the cracks on the surface of the composites were easily found due to their low ductility.

Even though the nano-sized reinforced particles disturbed on the matrix alloy provide the beneficial effects of increased hardness and mechanical strength of AMMCs, it should be noted that the high percentage of coarse reinforcing particle addition in the matrix can decline the mechanical properties. Xue et al. [59] analyzed the interruption subsurface of the AlSi10Mg with 15 wt% SiC particle composites manufactured by the SLM process. The average size of 46 μm SiC was distributed into the matrix alloy, leading to enhanced hardness by 47%, but the elongation to fracture and ultimate tensile strength were noted to be significantly lower. This reduction in mechanical properties is due to premature failure caused by interior porosities and coarse-grained SiC particles in the structure of AMMCs.

Zhou et al. [60] processed the AlSi10Mg with TiC particles by SLM, which offered a lower COF and loss than the unreinforced matrix alloy. Here, two different TiC particle sizes of 1 μm and 30 μm were treated on the AiSi10Mg matrix alloy, and a lesser variation in wear resistance resulted in both composites. Mainly, higher densification levels, enhanced hardness, and homogeneity in the microstructure are associated with AMMCs with a fine size of TiC, which improves wear resistance.

Kimura et al. [61] reported the performance of SLM during the production of two different AMMCs with a combination of carbon fiber-reinforced aluminium matrix and copper-plated carbon fibre-reinforced aluminium composites. The densified SLM composites offered low thermal conductivity and higher hardness due to the formation of an Al_4C_3 phase at the interface. The laser scanning rate of greater than 100 J/mm^3 showed a higher relative density of 99.75% in carbon fibre AMMCs. During the part-build period by SLM on copper-plated carbon fibre, AMMCs could control the reflection of the laser beam and avoid the chemical reaction of carbon fibre with aluminium. Consequently, the hardness of the carbon fibre-reinforced aluminium composite was less than that of copper-plated carbon fibre AMMCs, but both composites had the same thermal conductivity.

6.4 APPLICATIONS OF AMMCs

Due to the innovation of modern technology, it is often essential to get the desired properties through the process of producing composite materials. The composite materials are adjoining two or more different phases of material leads to superior properties than the pure or single-phase material [62–67]. Table 6.4 shows the applications of AMMCs used in different industrial sectors. Aluminium alloys are specially utilized for good wear and corrosion resistance characteristics. Here, AMMCs play an important role by combining different phases of material to provide a good functional material for advanced applications, particularly for automobile, aerospace, and marine industries.

TABLE 6.4
List of Applications of AMMCs

Industries	Applications
Aerospace	Structure in wings, fuselage, airline body, military vehicles, and cargo
Train	Structure in train for getting good fuel efficiency
Automobile	Components like pistons, connecting rods, engine structure, rotors, brake disc, etc.
Ship building	Structure in maritime applications to achieve higher fuel efficiency
Sports	Development of a variety sports goods like Al/SiC, Al/B_4C
Electrical transmission	Due to high corrosion resistance, it is used for making electrical conductors.
Packaging	Especially used for food containers and beverage bottles and cans

6.5 CONCLUSIONS

In this section, the feasibleness of using the SLM operation to manufacture AMMCs for different uses such as aerospace, automobiles, marine industries, and others was successfully reviewed. Among the different AM methods, SLM offers a better relationship to the distribution of reinforcement material into the aluminium metal matrices. AMMCs are still of greater importance in various industries as higher functional materials since they have outstanding mechanical properties with greater wear resistance and a higher strength-to-weight ratio. So AMMCs present good prospects and possibilities for researchers and design engineers to develop new technologies for industrial development. Several prior studies discussed here thoroughly examine the challenges and obstacles in producing AMMCs. By incorporating mico-sized reinforcement particles into the aluminium matrix, the consistency and strength are enhanced, although this comes with a decrease in ductility. However, nano-sized reinforced particles benefit from an improvement in ductility along with higher hardness and strength. This chapter, informed by previous investigators of AMMCs, can help address the problems in the manufacturing of composites by the SLM process.

REFERENCES

[1] Sahin, Y. and Murphy, S., 1996. The effect of fibre orientation of the dry sliding wear of borsic-reinforced 2014 aluminium alloy. *Journal of Material Science*, *31*, pp. 5399–5407.

[2] Seo, Y.H. and Kang, C.G., 1995. The effect of applied pressure on particle-dispersion characteristics and mechanical properties in melt-stirring squeeze-cast SiCp/Al composites. *Journal of Materials Processing Technology*, *55*, pp. 370–379.

[3] Purazrang, K., Kainer, K.U. and Mordike, B.L., 1991. Fracture toughness behaviour of a magnesium alloy metal-matrix composite produced by the infiltration technique. *Composites*, *22*, pp. 456–462.

[4] Rahimian, M., Ehsani, N., Parvin, N. and Baharvandi, H., 2009. The effect of particle size, sintering temperature and sintering time on the properties of Al–Al_2O_3 composites, made by powder metallurgy. *Journal of Materials Processing Technology*, *209*, pp. 5387–5393.

[5] Li, N., Liu, W., Wang, Y., Zhao, Z., Yan, T., Zhang, G. and Xiong, H., 2021. Laser additive manufacturing on metal matrix composites: A review. *Chinese Journal of Mechanical Engineering*, pp. 34–38.

[6] Hu, Y.B., Cong, W.L. and Wang, X.L., 2018. Laser deposition additive manufacturing of TiB-Ti composites with novel three-dimensional quasi continuous network microstructure: Effects on strengthening and toughening. *Composites Part B: Engineering*, pp. 91–100.

[7] Kok, M., 2005. Production and mechanical properties of Al_2O_3 particle-reinforced 2024 aluminium alloy composites. *Journal of Materials Processing Technology*, *161*, pp. 381–387.

[8] Valiev, R.Z. and Langdon, T.G., 2006. Principles of equal-channel angular pressing as a processing tool for grain refinement. *Progress in Materials Science*, *51*, pp. 881–981.

[9] Gu, D.D., Meiners, W., Wissenbach, K. and Poprawe, R., 2012. Laser additive manufacturing of metallic components: Materials, processes and mechanisms. *International Materials Review*, *57*, pp. 133–164.

[10] Singh, J. and Chauhan, A., 2016. Overview of wear performance of aluminium matrix composites reinforced with ceramic materials under the influence of controllable variables. *Ceramics International, 42*, pp. 56–81.
[11] Rahimian, M., Parvin, N. and Ehsani, N., 2011. The effect of production parameters on microstructure and wear resistance of powder metallurgy Al-Al_2O_3 composite. *Materials & Design, 32*, pp. 1031–1038.
[12] Prasad, B.K., Das, S., Jha, A.K., Modi, O.P., Dasgupta, R. and Yegneswaran, A.H., 1997. Factors controlling the abrasive wear response of a zinc-based alloy silicon carbide particle composite. *Composites Part A: Applied Science and Manufacturing, 28*, pp. 301–308.
[13] Wang, H.M., Wang, C.M. and Cai, L.X., 2003. Wear and corrosion resistance of laser clad Ni2Si/NiSi composite coatings. *Surface and Coating Technology, 168*, pp. 202–208.
[14] Miracle, D.B., 2005. Metal matrix composites—from science to technological significance. *Composites Science and Technology, 65*, pp. 2526–2540.
[15] Groover, M.P., 2007. *Fundamentals of Modern Manufacturing: Materials Processes, and Systems.* John Wiley & Sons, pp. 1–5.
[16] Kainer, K.U., 2006. *Metal Matrix Composites: Custom-Made Materials for Automotive and Aerospace Engineering*. Weinheim, Germany: John Wiley & Sons, pp. 1–4.
[17] Eskandar, F., Ali, G. and Mohamed, E., 2020. Selective laser melting of aluminum and titanium matrix composites: Recent progress and potential applications in the aerospace industry. *Aerospace, 7*, pp. 70–77.
[18] Sajjadi, S.A., Ezatpour, H.R. and Beygi, H., 2011. Microstructure and mechanical properties of Al–Al_2O_3 micro and nano composites fabricated by stir casting. *Material Science and Engineering: A, 528*, pp. 8765–8771.
[19] Gu, D., Wang, H. and Dai, D., 2016. Laser additive manufacturing of novel aluminum based nanocomposite parts: Tailored forming of multiple materials. *Journal of Manufacturing Science and Engineering, 138*, pp. 1–11.
[20] Khaing, M., Fuh, J. and Lu, L., 2001. Direct metal laser sintering for rapid tooling: Processing and characterisation of EOS parts. *Journal of Materials Processing Technology, 113*, pp. 269–272.
[21] Mangano, C., Chambrone, L., Van, N.R., Miller, C., Hatton, P.V. and Mangano, C., 2014. Direct metal laser sintering titanium dental implants: A review of the current literature. *International Journals of Biomaterials*, pp. 1–11.
[22] Grünberger, T. and Domröse, R., 2015. Direct metal laser sintering. *Laser Technik Journal, 12*, pp. 45–48.
[23] Pohl, H., Simchi, A., Issa, M. and Dias, H.C., 2001. Thermal stresses in direct metal laser sintering. In *Proceedings of the 12th Solid Freeform Fabrication Symposium*, Austin, TX, USA, pp. 6–8.
[24] Li, M., Fang, A., Martinez-Franco, E., Alvarado-Orozco, J.M., Pei, Z. and Ma, C., 2019. Selective laser melting of metal matrix composites: Feedstock powder preparation by electroless plating. *Materials Letters, 247*, pp. 115–118.
[25] Ethan, M.P. and Saba, Z.S., 2022. Additive manufacturing of aluminum metal matrix composites: Mechanical alloying of composite powders and single track consolidation with laser powder bed fusion. *Additive Manufacturing, 50*, pp. 1–5.
[26] Fereiduni, E., Ghasemi, A. and Elbestawi, M., 2019. Selective laser melting of hybrid ex-situ/in-situ reinforced titanium matrix composites: Laser/powder interaction, reinforcement formation mechanism, and non-equilibrium microstructural evolutions. *Materials Design, 184*, pp. 1–5.
[27] Simchi, A. and Godlinski, D., 2011. Densification and microstructural evolution during laser sintering of A356/SiC composite powders. *Journal of Material Science, 46*, pp. 1446–1454.
[28] Ma, C., Gu, D., Dai, D., Chen, W., Chang, F., Yuan, P. and Shen, Y., 2015. Aluminum-based nanocomposites with hybrid reinforcements prepared by mechanical alloying and selective laser melting consolidation. *Journals of Material Research, 30*, pp. 2816–2828.
[29] Yu, W., Sing, S., Chua, C., Kuo, C.N. and Tian, X., 2019. Particle-reinforced metal matrix nanocomposites fabricated by selective laser melting: A state of the art review. *Progress in Materials Science, 104*, pp. 330–379.
[30] Suryanarayana, C., 2001. Mechanical alloying and milling. *Progress in Materials Science, 46*, pp. 1–184.
[31] Fereiduni, E., Ghasemi, A. and Elbestawi, M., 2019. Characterization of composite powder feedstock from powder bed fusion additive manufacturing perspective. *Materials, 12*, pp. 36–73.
[32] Han, Q., Setchi, R. and Evans, S.L., 2016. Synthesis and characterisation of advanced ball-milled Al-Al_2O_3 nanocomposites for selective laser melting. *Powder Technology, 297*, pp. 183–192.

[33] Han, Q., Setchi, R. and Evans, S.L., 2017. Characterisation and milling time optimisation of nanocrystalline aluminium powder for selective laser melting. *International Journal of Advanced Manufacturing Technology*, *88*, pp. 1429–1438.

[34] Du, Z., Chen, H.C., Tan, M.J., Bi, G. and Chua, C.K., 2020. Effect of nAl_2O_3 on the part density and microstructure during the laser-based powder bed fusion of AlSi10Mg composite. *Rapid Prototypying Journal*, *26*, pp. 727–735.

[35] Jue, J., Gu, D., Chang, K. and Dai, D., 2017. Microstructure evolution and mechanical properties of Al-Al_2O_3 composites fabricated by selective laser melting. *Powder Technology*, *310*, pp. 80–91.

[36] Wang, L., Jue, J., Xia, M., Guo, L., Yan, B. and Gu, D., 2016. Effect of the thermodynamic behavior of selective laser melting on the formation of in situ oxide dispersion-strengthened aluminum-based composites. *Metals*, *6*, pp. 280–286.

[37] Gu, D., Jue, J., Dai, D., Lin, K. and Chen, W., 2018. Effects of dry sliding conditions on wear properties of al-matrix composites produced by selective laser melting additive manufacturing. *Journal of Tribology*, *140*, pp. 1–5.

[38] Liao, H., Zhu, H., Xue, G. and Zeng, X., 2019. Alumina loss mechanism of Al_2O_3-AlSi10 Mg composites during selective laser melting. *Journal of Alloys and Compounds*, *785*, pp. 286–295.

[39] Liao, H., Zhu, H., Zhu, J., Chang, S. and Zeng, X., 2019. Effect of process parameters on selective laser melting Al_2O_3-Al cermet material. In *Proceedings of the Solid Freeform Fabrication Symposium—an Additive Manufacturing Conference*, Austin, TX, USA, pp. 12–14.

[40] Han, Q., Setchi, R., Lacan, F., Gu, D. and Evans, S.L., 2017. Selective laser melting of advanced Al-Al_2O_3 nanocomposites: Simulation, microstructure and mechanical properties. *Material Science and Engineering: A*, *698*, pp. 162–173.

[41] Jue, J. and Gu, D., 2017. Selective laser melting additive manufacturing of in situ $Al_2Si_4O_{10}$/Al composites: Microstructural characteristics and mechanical properties. *Journal of Composite Materials*, *201*, pp. 10–12.

[42] Han, Q., Geng, Y., Setchi, R., Lacan, F., Gu, D. and Evans, S.L., 2017. Macro and nanoscale wear behaviour of Al-Al_2O_3 nanocomposites fabricated by selective laser melting. *Composiets Part B: Engineering*, *127*, pp. 26–35.

[43] Nalivaiko, A.Y., Ozherelkov, D.Y., Arnautov, A.N., Zmanovsky, S.V., Osipenkova, A.A. and Gromov, A.A., 2020. Selective laser melting of aluminum-alumina powder composites obtained by hydrothermal oxidation method. *Applied Physics: A Materials Science and Processing*, *126*, pp. 1–6.

[44] Yu, P., Deng, C.J., Ma, N.G. and Ng, D.H.L., 2004. A new method of producing uniformly distributed alumina particles in Al-based metal matrix composite. *Materials Letters*, *58*, pp. 679–682.

[45] Dieter, G.E., 1976. *Mechanical Metallurgy*, 3rd edition. New York, NY, USA: McGraw-Hill.

[46] Gu, D. and Yuan, P., 2015. Thermal evolution behavior and fluid dynamics during laser additive manufacturing of Al-based nanocomposites: Underlying role of reinforcement weight fraction. *Journal of Applied Physics*, *118*, pp. 1–9.

[47] Li, R., Liu, J., Shi, Y., Wang, L. and Jiang, W., 2012. Balling behavior of stainless steel and nickel powder during selective laser melting process. *International Journal of Advanced Manufacturing and Technology*, *59*, pp. 1025–1035.

[48] Yadav, S., Chichili, D.R. and Ramesh, K.T., 1995. The mechanical response of a 6061-T6 A1/$A1_2O_3$ metal matrix composite at high rates of deformation. *Acta Metallurgica et Materialia*, *43*, pp. 4453–4464.

[49] Ezatpour, H.R., Sajjadi, S.A., Sabzevar, M.H. and Huang, Y., 2014. Investigation of microstructure and mechanical properties of Al6061-nanocomposite fabricated by stir casting. *Material Design*, *55*, pp. 921–928.

[50] Reddy, M.P., Ubaid, F., Shakoor, R.A., Parande, G., Manakari, V., Mohamed, A.M.A. and Gupta, M., 2017. Effect of reinforcement concentration on the properties of hot extruded Al-Al_2O_3 composites synthesized through microwave sintering process. *Material Science Engineering: A*, *696*, pp. 60–69.

[51] Kok, M., 2005. Production and mechanical properties of Al_2O_3 particle-reinforced 2024 aluminium alloy composites. *Journal of Material Processing Technology*, *161*, pp. 381–387.

[52] Gu, D., Wang, H., Chang, F., Dai, D., Yuan, P., Hagedorn, Y.C. and Meiners, W., 2014. Selective laser melting additive manufacturing of TiC/AlSi10Mg bulk-form nanocomposites with tailored microstructures and properties. *Physics Procedia*, *56*, pp. 108–116.

[53] Ghosh, S.K. and Saha, P., 2011. Crack and wear behavior of SiC particulate reinforced aluminium based metal matrix composite fabricated by direct metal laser sintering process. *Material Design, 32*, pp. 139–145.
[54] Zhou, X.B. and De Hosson, J.Th.M., 1996. Reactive wetting of liquid metals on ceramic substrates. *Acta Metallurgica et Materialia, 44*, pp. 421–426.
[55] Chang, F., Gu, D., Dai, D. and Yuan, P., 2015. Selective laser melting of in-situ Al4SiC4 + SiC hybrid reinforced Al matrix composites: Influence of starting SiC particle size. *Surface & Coatings Technology, 272*, pp. 15–24.
[56] Li, X., Astfalack, L., Drew, P., Schliwa, Y. and Sercombe, T.B., 2016. Selective laser melting of aluminium metal matrix composites. In *Proceedings of the 2nd International Conference on Progress in Additive Manufacturing*, pp. 433–438.
[57] Dadbakhsh, S. and Hao, L., 2014. Effect of Fe_2O_3 content on microstructure of Al powder consolidated parts via selective laser melting using various laser powers and speeds. *International Journal of Advanced Manufacturing and Technology, 73*, pp. 1453–1463.
[58] Kang, N., Fu, Y. and Coddet, P., 2017. On the microstructure, hardness and wear behavior of Al-Fe-Cr quasicrystal reinforced Al matrix composites prepared by selective laser melting. *Materials & Design, 132*, pp. 105–111.
[59] Xue, G., Ke, L., Zhu, H., Liao, H., Zhu, J. and Zeng, X., 2019. Influence of processing parameters on selective laser melted SiCp/AlSi10Mg composites: Densification, microstructure and mechanical properties. *Materials Science and Engineering: A, 764*, pp. 138–155.
[60] Zhou, S.Y., Wang, Z.Y., Su, Y., Wang, H., Liu, G., Song, T.T. and Yan, M., 2020. Effects of micron/submicron TiC on additively manufactured AlSi10Mg: A comprehensive study from computer simulation to mechanical and microstructural analysis. *Journal of the Minerals Metals & Materials Society*, pp. 1–12.
[61] Takahiro, K., Takayuki, N., Takeshi, S. and Takao, M., 2020. In-process fabrication of carbon-dispersed aluminum matrix composite using selective laser melting. *Metals, 10*, pp. 1–9.
[62] Nturanabo, F., Masu, L.M. and Govender, G., 2015. Automotive light-weighting using aluminium metal matrix composites. *Materials Science Forum, 828*, pp. 1–5.
[63] Skillingberg, M. and Green, J., 2007. Aluminum applications in the rail industry. *Light Metal Age-Chicago, 65*, pp. 1–8.
[64] Goede, M. and Stehlin, M., 2009. Super light-car project—an integrated research approach for lightweight car body innovations. In *Innovative Developments for Lightweight Vehicle Structures: Conference Proceedings*, Wolfsburg, pp. 26–27.
[65] Rohatgi, P.K., Ray, S. and Yun, L., 1992. Tribological properties of metal matrix-graphite particle composites. *International Materials Review, 37*, pp. 129–152.
[66] Stephens, J., 1987. High temperature metal matrix composites for future aerospace systems, In *24th Joint Propulsion Conference*, p. 3059.
[67] Sharma, A.K., Bhandari, R., Aherwar, A., Rimašauskienė, R. and Bretotean, C.P., 2020. A study of advancement in application opportunities of aluminum metal matrix composites. *Materials Today Proceedings, 26*, pp. 2419–2424.

7 Investigation on Properties of Metal Matrix Composites Fabricated Through Additive Manufacturing

Aravindh Raj Babu Rudrakotti and Velmurugan Chandrasekaran

7.1 INTRODUCTION

In recent decades, MMC has ascended to the forefront of material science, representing a paradigm shift in the realm of structural integrity, thermal regulation, and acoustic insulation. The evolution of MMC has been propelled by their remarkable attributes, which surpass those of conventional lightweight materials on numerous fronts. Notably, MMC exhibits heightened levels of resistance to deformation, bolstered compressive strength, and enhanced crashworthiness, setting them apart as the premier choice for countless applications.

This comprehensive review article serves as a compendium of the manufacturing methodologies employed in the production of MMC, ranging from conventional to new techniques encompassing both direct and indirect foaming methodologies. Delving into the depths of innovation, the review explains unconventional approaches to metal foam fabrication, including but not limited to additive manufacturing, pioneering methodologies, and the burgeoning field of metal syntactic foaming. Each method is dissected accurately, offering a comprehensive understanding of its principles, applications, and potential for advancement.

7.2 METAL MATRIX COMPOSITES MATERIALS

Composites represent synergistic blends of two or more distinct materials, harmonizing to manifest a superior amalgamation of various properties like hardness, strength, stiffness, toughness, and ductility, characteristics often unattainable within a solitary material. Within these complicated blends, the matrix functions as a channel, distributing applied loads to the reinforcement, while the reinforcement, in turn, endows the composites with the desired mechanical strength. Depending on their morphology, reinforcement materials can be categorized as either particles or fibres.

MMC is inclusive of two or more materials in a metallic matrix [1]. In current industrial requirements MMCs have attained more attention in manufacturing components due to their promising properties like high strength, hardness, ductility, toughness, temperature resistance, corrosion resistance, etc. [2]. MMC is a promising material for manufacturing lightweight components [3]. MMC can be manufactured by conventional production process based on three states of materials: solid-state material, semi-solid-state material, and liquid-state material processes [4].

 DOI: 10.1201/9781003499718-7

7.2.1 Solid State of MMC

7.2.2 Physical Vapor Deposition

The manufacturing of metal matrix composites (MMC) via the physical vapor deposition (PVD) process involves the deposition of reinforcing phases onto a metallic substrate in a vacuum environment. In this method, the reinforcing material is vaporized and subsequently deposited onto the metallic matrix, forming a thin film composite layer. The PVD process for MMC production typically begins with the preparation of the metallic substrate, followed by the deposition of the reinforcing material using techniques such as physical vapor deposition, magnetron sputtering, or electron beam evaporation. The method is followed in a controlled vacuum environment to ensure uniform deposition and adhesion between the reinforcing phase and the metallic matrix.

One common application of MMC manufactured by the PVD method is the aerospace industry, where these materials are used to increase the mechanical and thermal properties of components such as turbine blades and structural elements. Additionally, MMC produced via PVD finds applications in electronics, automotive, and biomedical industries due to their enhanced strength, hardness, and wear resistance [5–7].

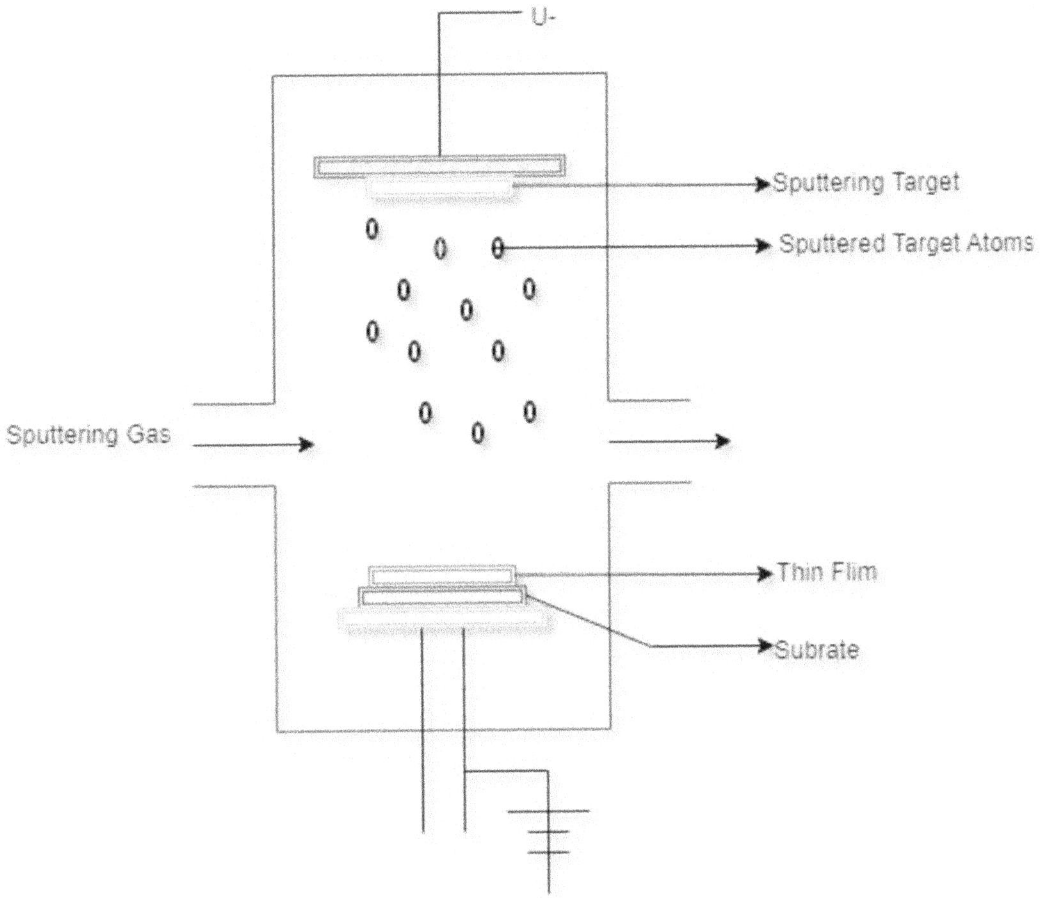

FIGURE 7.1 Shows diagrammatic view of physical vapor deposition.

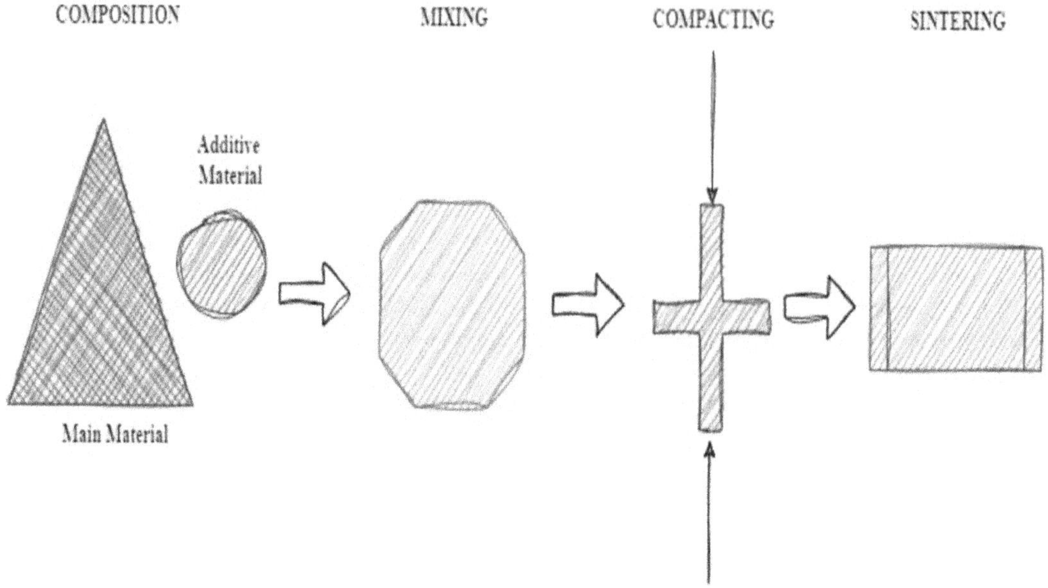

FIGURE 7.2 Powder metallurgy process [8].

7.2.3 Powder Metallurgy

The blended powder undergoes a cold pressing method in addition to the sintering and forging process. Pressed powders are called a "green material." Further, the green material undergoes a post-processing process depending on its preparation methods. The next step involves working with cold plastic if the green material is mostly sintered. On the other hand, if the green product is made using cold pressure, hot plastic work is applied to it in the later stages of processing [9, 10].

7.2.4 Spark Plasma Sintering

Using SPS, high-density composites with exceptionally good grain growth or size may be created [11]. The SPS process schematic design is displayed in Figure 7.3. Many contemporary compounds may be developed with the use of this SPS technique [12]. The SPS process occurs at atmospheric pressure in a shielded gas or empty space environment [13, 14].

7.2.5 Diffusion Bonding

Diffusion bonding is the joining method of manufacturing material in a solid state to prepare MMC by bonding together similar or dissimilar materials at elevated temperatures. This involves cleaning the surfaces to remove any contaminants and ensuring proper alignment of the materials. The matrix (base metal matrix powder) with reinforcement powder is then mixed in the required proportions. Once the blending is complete, the mixed powders are subjected to cold pressing to form a green compact. The green compact is then heated to an elevated temperature in a controlled atmosphere, typically in a vacuum or under partial pressure of an inert gas such as Argon [16]. Depending on the specific application and desired properties, the bonded MMC may undergo further processing steps such as machining, heat treatment, or surface finishing [17].

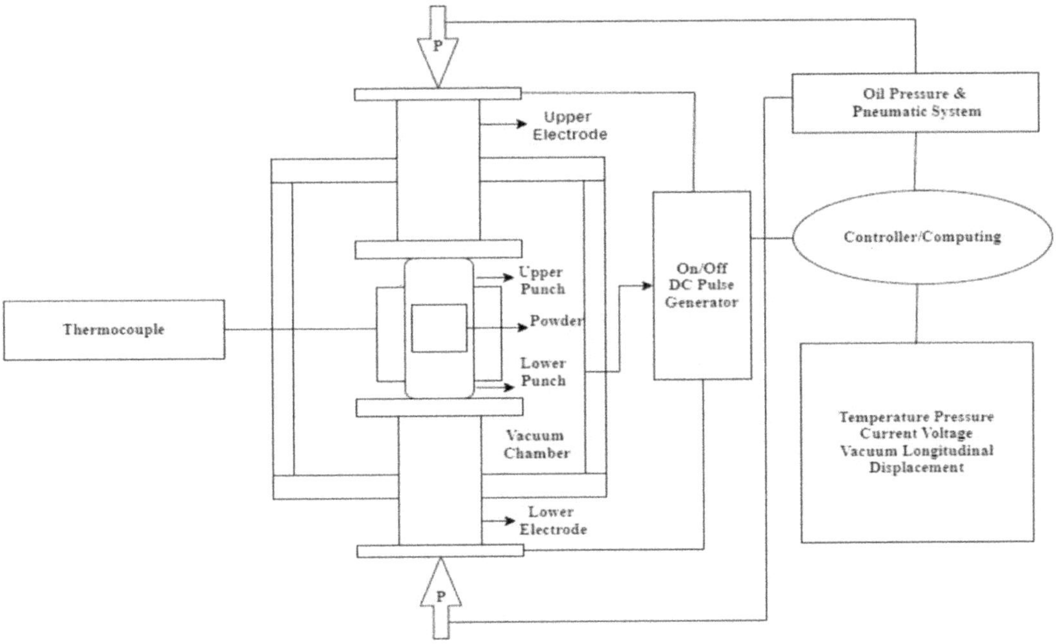

FIGURE 7.3 Spark plasma sintering [15].

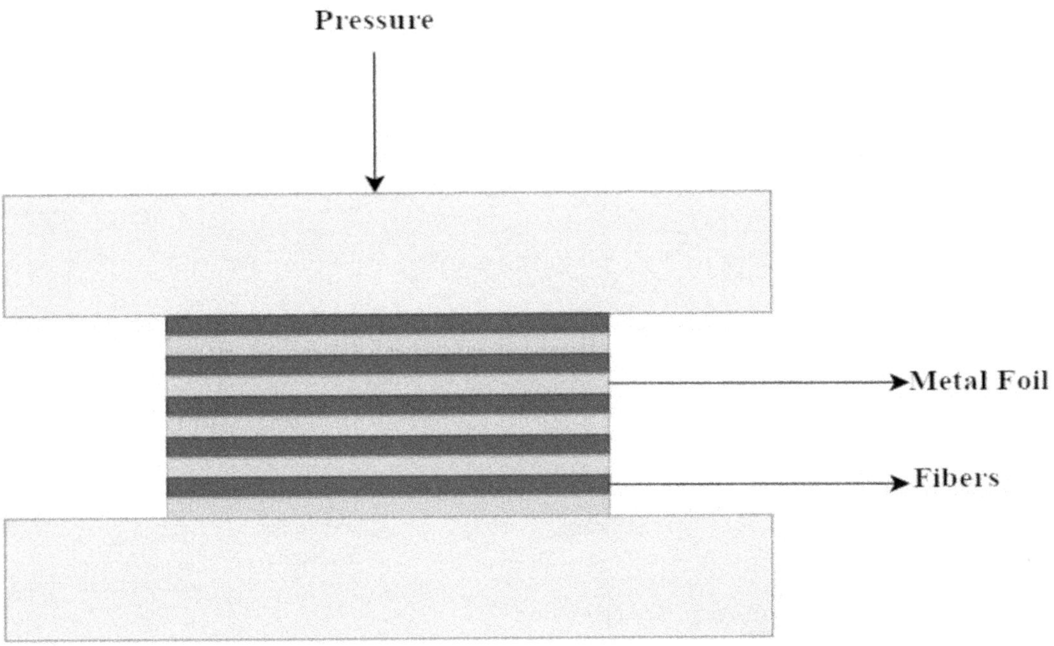

FIGURE 7.4 Schematic diagram of diffusion bonding process.

7.3 SEMI SOLID MMC

7.3.1 Compocasting Method

PVD compocasting is a semi-solid processing technique in which mechanical agitation is used to insert ceramic reinforcing particles into a semi-solid matrix alloy. Two separate versions of this technology are referred to as semisolid-liquid (SL) and semisolid-semisolid (SS) [18]. Figure 7.5 displays the compocasting process diagram.

7.3.2 Thixoforging Method

By using these techniques, the alloy is heated to a semi-solid or thixotropic state, where it combines the properties of a liquid and a solid. Ceramic reinforcing particles are added to the semi-solid metal matrix during the thixoforging process. Better dispersion and distribution of the reinforcing particles are made possible by the semi-solid condition of the matrix, which enhances the mechanical qualities of the finished MMC [19]. Figure 7.6 shows the schematic process of the thixoforging process.

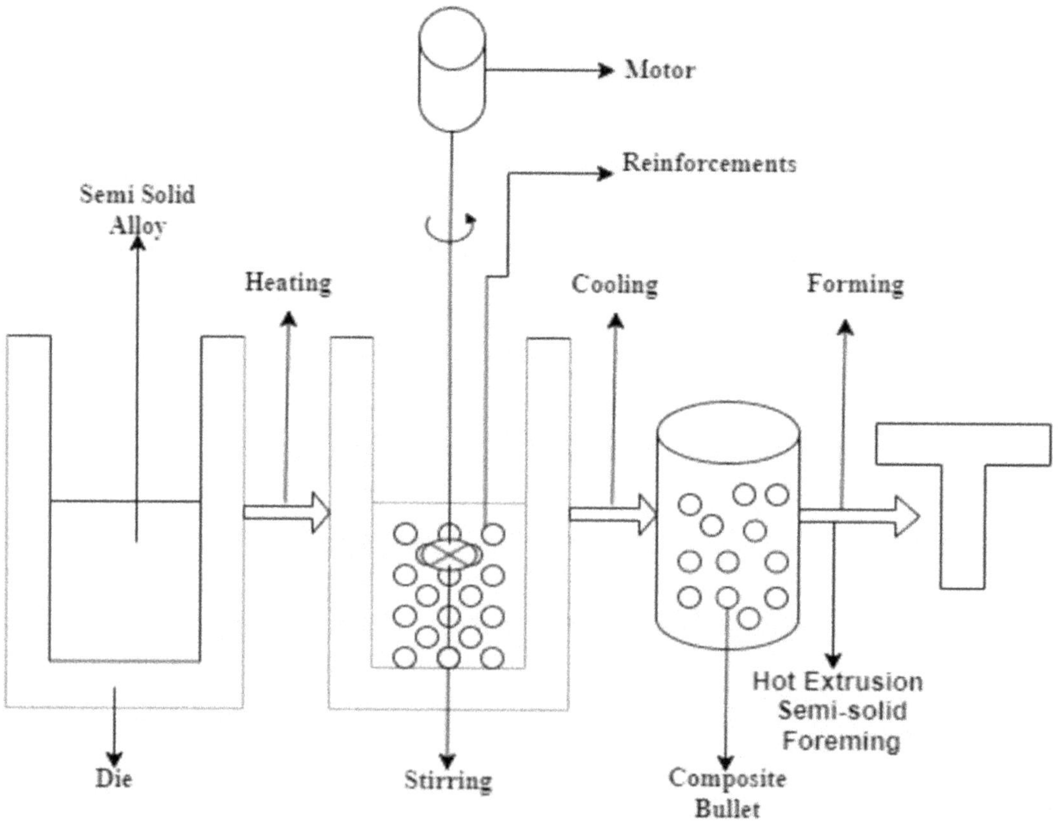

FIGURE 7.5 Shows process diagram of compocasting.

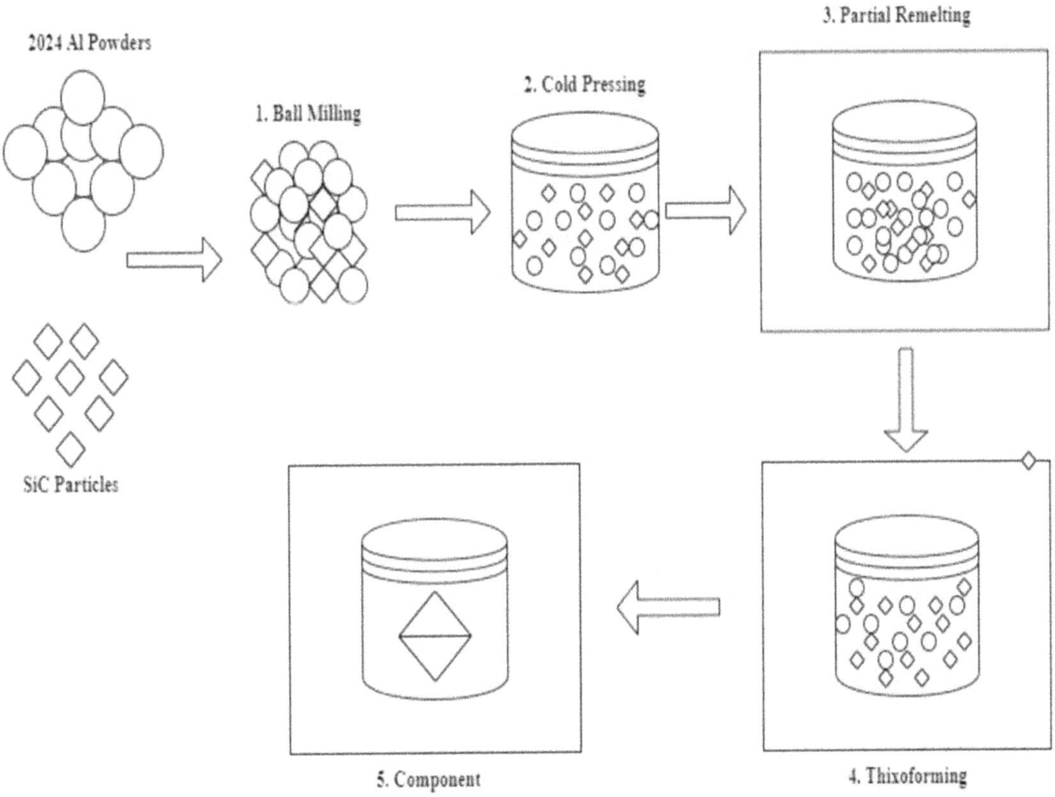

FIGURE 7.6 Shows schematic diagram for preparation of 2024 Al powders by thixoforging process.

7.4 LIQUID-BASED MMC

New liquid-based manufacturing techniques, such as ultrasonic-assisted casting, stirring casting, and pressure infiltration methods spray deposition have emerged in addition to conventional methods.

7.4.1 Stirring Casting

Stirring casting involves the mechanical stirring of a liquid metal alloy with ceramic reinforcing particles to achieve uniform dispersion. This method facilitates the incorporation of particles into the molten alloy, enhancing the mechanical properties of the resulting composite material [20]. Figure 7.7 shows the diagrammatic view of the stir-casting process

7.4.2 Ultrasonic-Assisted Casting

Using ultrasonic vibrations, ultrasonic-assisted casting (UAC) improves the dispersion of reinforcing particles in a molten metal matrix. Improved mechanical characteristics and microstructural homogeneity are the outcomes of the ultrasonic energy's assistance in dispersing agglomerates and encouraging uniform dispersion [22]. Figure 7.8 displays the ultrasonic-assisted casting schematic design.

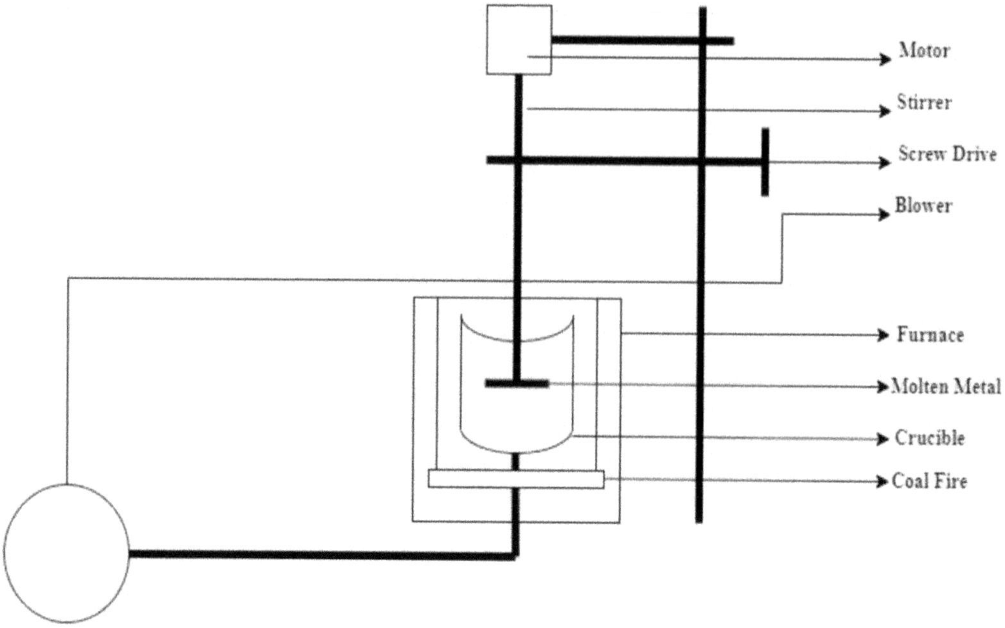

FIGURE 7.7 Diagrammatic view of stir casting process [21].

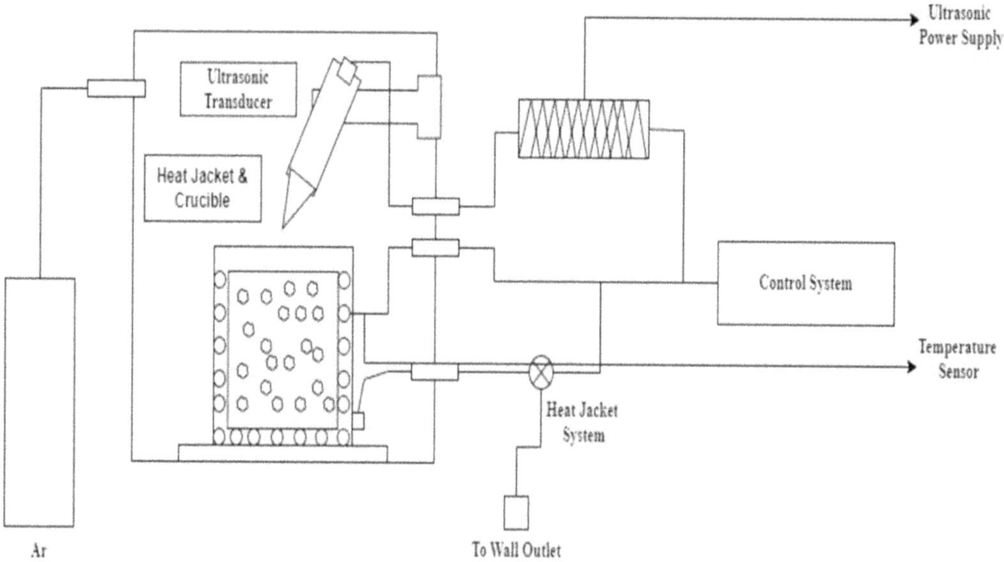

FIGURE 7.8 Shows a schematic diagram of ultrasonic-assisted casting.

7.4.3 Spray Deposition

Spray deposition involves atomizing a molten metal alloy into fine droplets, which are then deposited onto a substrate to form a thin film or coating. This method allows for the fabrication of uniform and dense coatings with controlled microstructures and tailored properties [23].

7.4.4 Pressure Infiltration

Pressure infiltration utilizes external pressure to force a molten metal alloy into a preform or porous structure containing reinforcing particles. The molten metal infiltrates the void spaces within the preform, resulting in a composite material with enhanced mechanical properties and improved interfacial bonding between the matrix and reinforcement phases [24].

7.5 LIMITATIONS AND CHALLENGES FACED IN CONVENTIONAL FABRICATION TECHNIQUES

Certain domestic MMC technologies have progressed to an advanced state, such as exhaust pressure infiltration technology in atmospheric conditions. The technique used in stirring casting and powder metallurgy is equivalent to those of other nations. All things considered, conventional material powerhouses like the United States, Japan, Canada, and the United Kingdom still have more advanced MMC technology than China. There are numerous places where the industrialization gap is very noticeable:

- The material production equipment, precision machining, precision forming, specific alloy design for metal matrix, connection technology, and reinforcement preparation are all still missing links in the MMC industrial chain. Specifically, China lacks specific matrix alloys; it also has high failure rates in domestic production equipment; it battles with uneven quality in domestic reinforcement and matrix alloy powders; and it has not conducted enough research on welding technology and precise machining
- Poor material selection and possible financial losses for customers are caused by China's unfinished fundamental MMC database, flawed national standard system, and ambiguous material classification

7.6 ADDITIVE MANUFACTURING TECHNIQUES FOR THE DEVELOPMENT OF MMC

Composite powder combinations have been assembled using two main traditional production techniques: hot pressing (HP) and spark plasma sintering (SPS) [25, 26]. However these approaches have several difficulties, especially when it comes to how the matrix and reinforcement interact. Problems including uneven reinforcement dispersion in the matrix and poor interfacial bonding provide major challenges, particularly in the production of intricately shaped components [27]. To improve MMC applicability across sectors, it is also necessary to solve the significant economic hurdle posed by their high manufacturing costs [28, 29].

7.6.1 Industrial Need of Metal AM

Metal additive manufacturing is witnessing substantial growth concerning its industrial relevance and market performance. It achieved an income of 1.51 billion EUR in 2018, and forecasts an expected 5.4 billion United States Dollars by 2027 [30, 31]. The metal-based AM market is primarily attributed to the various fields of automotive sectors, dental, surgical instruments, and aerospace sectors, where AM are highly favoured, and additively manufactured components are already utilized for final applications. Additionally, AM has proven advantageous in various repairment procedures, on-demand spare part production, and custom or low-volume manufacturing [32].

Currently, additive manufacturing technologies are deeply embedded across diverse sectors, including the production of medical-related implants and aerospace components, which require high performance [33].

However, there's still an urgent demand for AM, particularly metal AM, to be more widely incorporated into various industrial sectors [34]. The disruptive nature of metal-based AM processes has caused many industries to prefer to perceive their incorporation as potentially risky [35, 36]. The two primary process categories used in metal AM at the moment are directed energy deposition (DED) AM process and powder bed fusion (PBF) AM process [37].

Furthermore, differences in metallurgical properties have been observed between components fabricated using additive techniques and those manufactured conventionally. The considerable residual stresses and integrity shortcomings present notable obstacles for high-performance applications, particularly in situations requiring superior resistance property to high-temperature fatigue [38], while essential components like turbine blades are still undergoing early development stages [39].

7.6.2 Powder Bed Fusion

Figure 7.9 illustrates a diagrammatic depiction of the laser-based powder bed fusion procedure. In this method, a laser beam traverses predetermined sections of the powder bed at a regulated pace, causing the powder to fuse onto the fundamental solid material via either complete fusion, termed the selective laser melting AM process (SLM) or partial fusion, known as the selective laser sintering AM process (SLS). Upon completion of laser exposure for one layer, the powder bed descends by a specified layer thickness, and a fresh layer of powder is evenly distributed. This process repeats until the component is fabricated. Maintaining consistent gas flow crossways to the build area is essential for guaranteeing the quality and attributes of components manufactured through selective laser melting (SLM) [40].

7.6.3 Direct Energy Deposition

Figures 7.10 and 7.11 illustrate the core concept of the DED process, which amalgamates principles from cladding and welding methodologies [41]. Thermal energy, be it from a laser source,

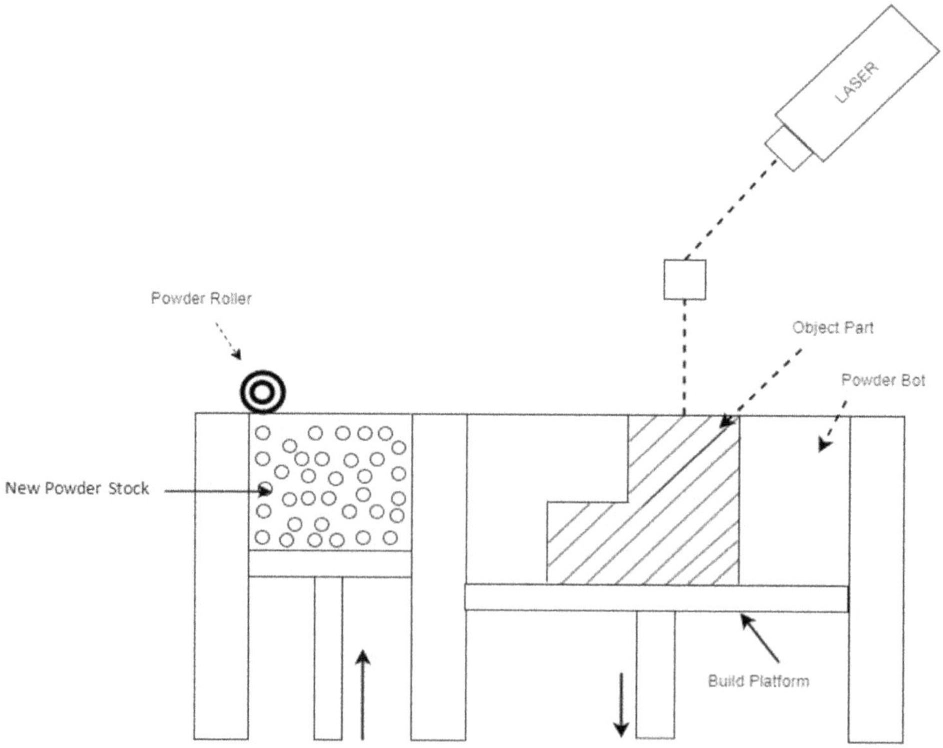

FIGURE 7.9 Illustrates the diagrammatic view of powder bed fusion process.

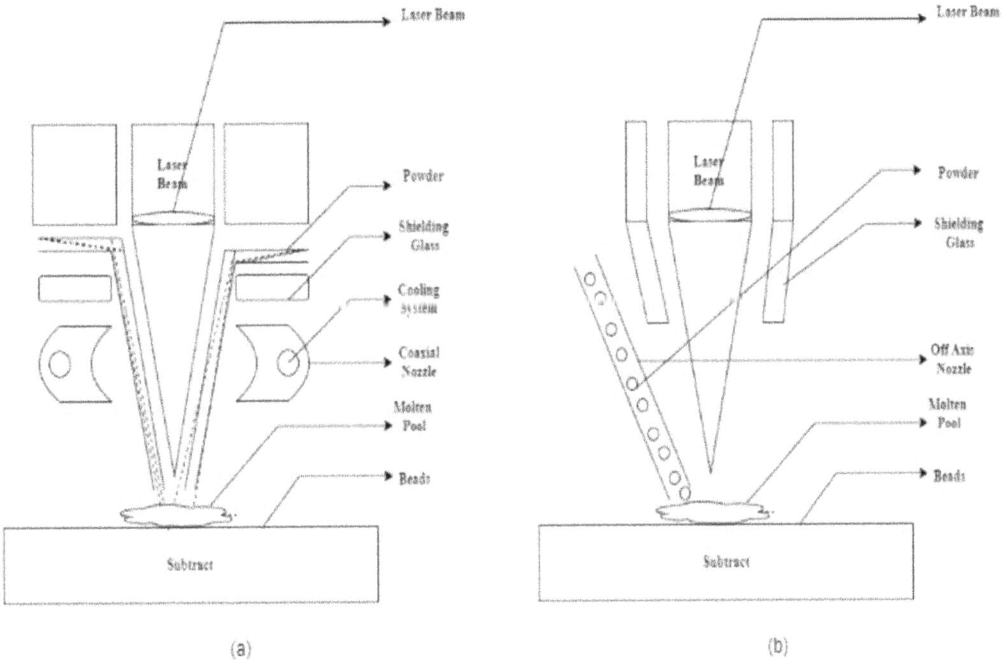

FIGURE 7.10 Powder based AM method of DED process **a)** co-axial feeding type, **b)** off-axis feeding type.

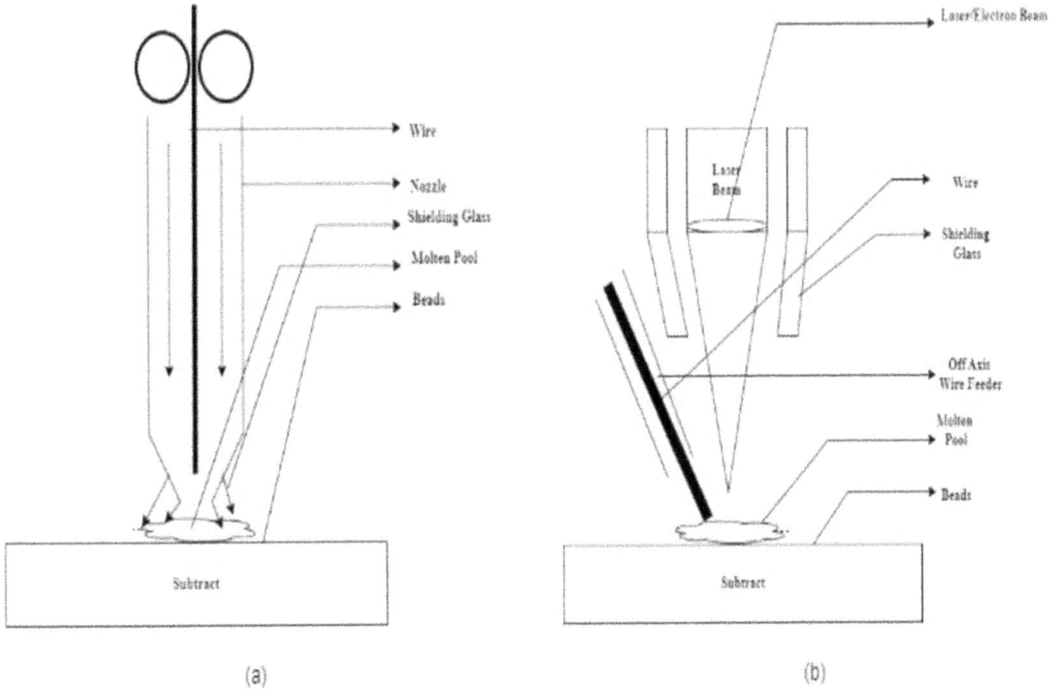

FIGURE 7.11 Wire based method of DED AM process **a)** co-axial feeding type, **b)** off-axis feeding type.

electron ray, or fusing heat flux, is engrossed onto the preceding layer, as outlined in Table 7.1. Simultaneously, a variety of wire and powder types constituting the feedstock are directed to the centre of the thermal energy field [42]. This leads to the creation of a melted pool formation by the fusion of the prior layer and the feedstock within and the concentrated thermal energy area [43]. Subsequent cooling of the molten pool results in the formation of a deposition drop. By repeating this process iteratively, a three-dimensional metallic component is manufactured. The DED process typically demonstrates cooling rates ranging from 10^3 to 10^5 °C/s [44].

The DED process can be classified into two types based on the feedstock: powder and wire feeding, as shown in Figure 7.12 [45]. In the powder feeding stage of DED processes, a laser serves as

TABLE 7.1
Characteristics Features and Applications of Common MMC

Material	Characteristics	Application
Titanium alloys	• High fracture toughness • High corrosion resistance • At higher temperatures it gives good mechanical properties • Lower thermal expansions	Automotive, aerospace, and bio-medical application
Stainless steel	• High in ductile • Good corrosion resistance • High strength	Corrosion resistance and construction applications
Copper alloys	• Good mechanical properties • High thermal conductivity • High electrical conductivity	Electrical, rocket engine manufacturing applications

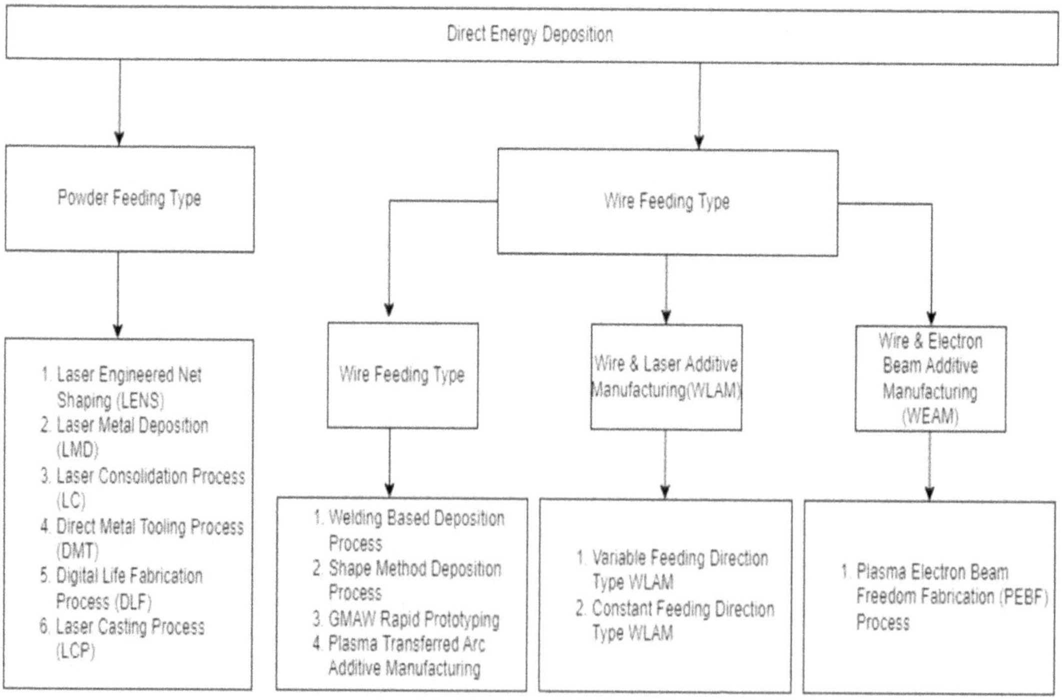

FIGURE 7.12 Different types of DED process.

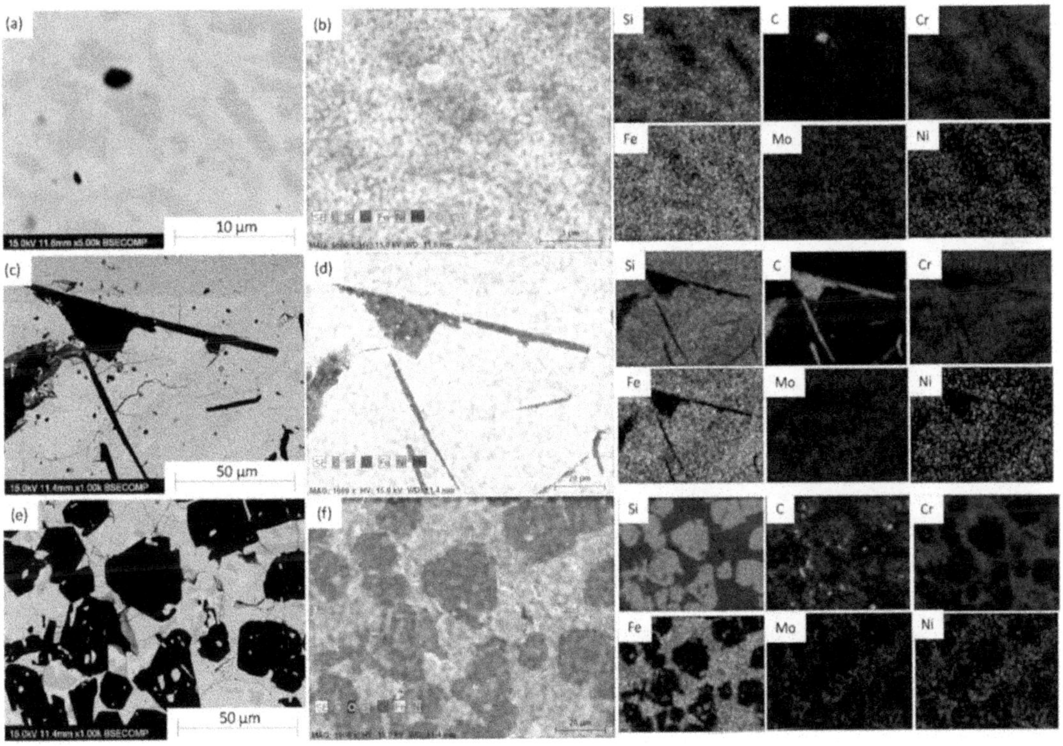

FIGURE 7.13 Scanning electron microscope images depict multilayer-wall samples produced with varying proportions of SiC particles, along with their respective Energy-Dispersive X-ray (EDX) elemental maps: **a)** 316 L/40SiC, **b)** EDX map for (a), **c)** 316 L/60SiC, **d)** corresponding EDX map for (c), **e)** 316 L/80SiC, **f)** corresponding EDX map for (e).

the thermal-based energy source, while in wire suckling types, the thermal energy can be provided by an electric arc, plasma arc, laser, or electron beam, as outlined in Figure 7.13 [46]. Co-axial and lateral feeding technologies are utilized in the DED process to introduce the feedstock into the molten pool, as shown in Figure 7.13. Combining a powder-feeding device with a laser results in the DED process [47]. Wire-feeding DED processes are further categorized into WAAM, WLAM, and WEAM processes, depending on the thermal energies employed [48].

7.7 INVESTIGATION OF THE PROCESS PARAMETERS ON PROPERTIES OF MMC FABRICATED THROUGH ADDITIVE MANUFACTURING

This area of study focuses on improving the physical, chemical, and mechanical properties of ceramic and non-ceramic materials by the use of fibres or particles. Interestingly, alumina, boron nitride, graphite, and silicon carbide particles are the most often employed reinforcing elements in metal matrix composites (MMC) to boost hardness and strength [49]. The production of highly reinforced SiC/316 L composites has shown considerable promise when employing DLD additive manufacturing. With this method, composites with particle contents ranging from 20 to 80 weight percent may be created without significantly changing the production process. This presents several options for creating low-density, high-hardness components, especially when combined with additive manufacturing's capabilities.

However, augmentation of SiCp content results in decreased hardness, reaching approximately 890 HV0.1, as the matrix becomes soaked with carbides and larger, additional brittle graphite

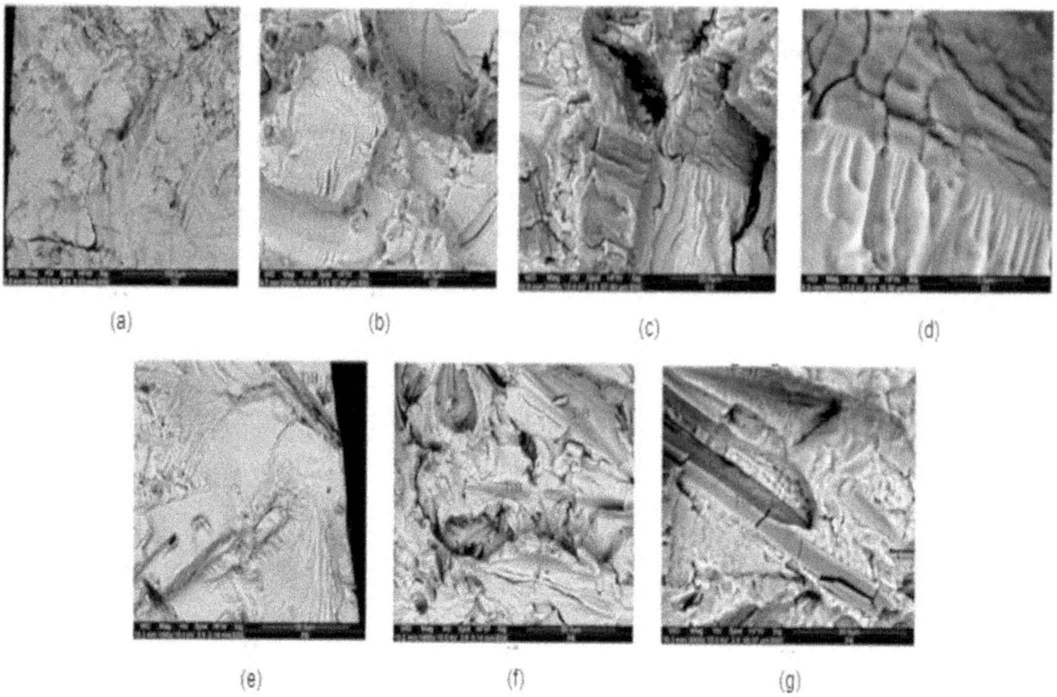

FIGURE 7.14 BSE images of 45° additive build orientation samples that show fracture characteristics such as **a)** cleavage in alpha Ti matrix, **b)** region of striations in alpha Ti matrix, **c)** fractured TiB particles surrounded by features of blocked pores during fracture, **d)** an enlarged portion of (c) showing striations in broken TiB particle. The following are BSE images of 90° AM orientation samples that show fracture features such as **e)** alpha Ti matrix cleavage, **f)** striations in alpha Ti matrix, **g)** fractured TiB particles surrounded by features of blocked pores during fracture [53].

precipitates form. For developed SiCp concentrations within the 316 L composition, the ideal manufacturing conditions shift toward reduced scanning speeds. This demonstrates the feasibility of producing cermet materials derived from steel with decreased density and increased hardness compared to the original steel [50]. The development of second, sixth, and seventh series Al alloys and their MMC poses challenges owing to its elevated thermal-related conductivity and issues associated with hardening, such as cracks, porosity, and thermal distortion. The MMC exhibited a notable enhancement in microhardness and ultimate tensile strength, with improvements of 11.3% and 22.3% and 30.5% and 31.5% respectively, compared to Al 6061 alloys [51]. The maximum strength is seen in the 90° direction and the lesser in the AM build's perpendicular (0°) direction [52].

A comparative investigation revealed that composites which are heat treated produced at 1,090°C acquired a stronger ultimate tensile strength (UTS) of around 992 MPa, with only a 0.8% loss in uniform elongation (UE), in contrast to SDSSs which are heat treated (UTS: ~839 MPa; UE: ~20.6%). Foam stir additive manufacturing (FSAM) was used to manufacture a laminated sheet made of Al5083–AZ31B alloys. It looked at how the speed in rotational motion of the tool affected the lamination's bonding strength and microstructural properties. Strong interfacial materials like Mg_2Al_3 and Al_4Fe help to increase the hardness and strength of the bonding [54]. Using laser-based PBF, Inconel 625 alloys reinforced with nano-additives were created. TiC nanoparticles and/or nickel-coated multi-walled carbon nanotubes were added. The reinforced samples showed lower elongation to failure but better yield strength, ultimate tensile strength, and hardness when compared to PBF-manufactured In625 [55]. The maximum ultimate tensile strength of 415 MPa, or 74.3% of the base metal's strength, was attained by sample L9 (1,000 rpm, 90 mm/min, and 2 NPs) thanks to the uniform dispersion of precipitates with a low dissolving density. The production of seven-layered Al-7075 laminates utilizing friction stir additive manufacturing led to the discovery that the tool-workpiece speed was the most important input parameter. For the hardness, UTS, and percentage elongation, TWS was responsible for 56.94%, 52.33%, and 47.33% of the total [56].

7.8 CONCLUSION

The production of particle-reinforced MMC via PBF demands careful consideration in both material selection and process design to mitigate defects and improve properties. Hence, this paper discusses a strategy for preparing, designing, and manufacturing MMC as follows:

Material preparation and selection

- Promoting the usage of alloys in the powder form and managing the moisture level of such materials to guarantee ideal powder dispersion and avoid oxidation
- Looking through a material database to find appropriate matrices and reinforcement combinations suited to certain application needs
- Examining if reinforcements and matrix alloys are compatible to prevent structural flaws and early failure brought on by inadequate bonding

Process optimization

- Gain a basic grasp of the physical metallurgical phenomena related to the material and process in order to eliminate defects such as pores, fractures, deformation, and delamination
- One important factor in achieving the desired phase in MMC during the PBF process is the type of process atmosphere, such as N_2 and Ar atmospheres
- Methodically creating different PBF process parameters depending on powder shape, sizes, and melt pool characterization, such as recoating speed, layer thickness, and laser-related factors

Microstructure characterization

- Recognizing the factors driving the anisotropic and inhomogeneous microstructural development of MMC manufactured using PBF
- Acknowledging that decomposed elements from reinforced particles can alter the matrix's chemical composition during PBF, significantly impacting MMC properties; acknowledging that the inclusion of nanoparticles and submicron- or micron-sized particles can aid in matrix grain recrystallization, grain refinement, and defect reduction

Postprocessing

- Complying with a postprocessing workflow including wire EDM, heat treatments, machining, and surface finishing for MMC manufactured using PBF
- Employing heat treatments as necessary to obtain desired phases and alleviate anisotropy, inhomogeneity, and residual stress induced by high cooling rates
- Carefully selecting heat treatments tailored to the chosen materials to enhance properties and manage phenomena like grain refinement, elemental segregation, and texture alterations resulting from particle reinforcements

One of DED's most attractive qualities is its capacity to generate multi-material components, which has garnered acclaim from academic and industry groups alike. Multi-material DED can be used to overcome the current challenges in combining different materials, making it possible to create components made of different materials in a single process. Alloy design, metal matrix composites, and the development of functionally graded materials are common uses for multi-material DED.

Within the field of alloy design, DED offers a far faster and more economical method for conducting investigational trials when investigating novel alloys. This is especially crucial for the creation of novel high-entropy alloys, as there is a wide variety of possible compositions, and making experimental testing simpler is crucial. Surface engineers, on the other hand, may benefit greatly from metal matrix composite coatings, particularly with regard to their remarkable wear resistance.

The fact that DED has the innate capacity to regulate feedstock arrangement and modify it as a separate process parameter is compelling evidence that this technique is appropriate for multi-material manufacturing. It is already thought to be the most advanced technique available for creating multi-material structures. Before this technology may be easily incorporated into industrial applications, a few issues need to be resolved. These challenges span both process engineering and materials science domains, as outlined later.

DED's inherent ability to regulate feedstock composition as a distinct process parameter and control it is strong proof that this approach is suitable for multi-material production. It is now regarded as the most sophisticated method for building multi-material buildings. A few problems must be fixed before this technology may be simply used in industrial applications. The following list of issues pertains to the fields of materials science and process engineering.

Furthermore, it's critical to look at how process factors affect the thermal history since thermal cycles during processing have a substantial impact on the characteristics and integrity of multi-material L-DED. These process variables will eventually be adjusted to produce a particular heat cycle. Furthermore, to balance the interactions between the metallic and ceramic phases in MMC, the temperature history will be meticulously regulated.

A greater comprehension of how temperature cycles impact component interactions in multi-material powder mixes is crucial for materials engineering. Given the frequency of non-equilibrium synthesis in L-DED, large datasets must be created in order to optimize the setup procedure and target or steer clear of certain microstructures. Furthermore, extensive testing is required to determine the in-service behaviour of multi-material constructions, including structural behaviour and tribological performance. This information will act as a basis for developing broad guidelines for material selection and processing condition.

REFERENCES

[1] Singh, L., Singh, B. and Saxena, K.K., 2020. Manufacturing techniques for metal matrix composites (MMC): An overview. *Advances in Materials and Processing Technologies*, *6*(2), pp. 441–457.

[2] Senthil, S., Raguraman, M. and Thamarai Manalan, D., 2021. Manufacturing processes & recent applications of aluminium metal matrix composite materials: A review. *Materials Today: Proceedings*, *45*, pp. 5934–5938.

[3] Samal, P., Vundavilli, P.R., Meher, A. and Mahapatra, M.M., 2020. Recent progress in aluminum metal matrix composites: A review on processing, mechanical and wear properties. *Journal of Manufacturing Processes*, *59*, pp. 131–152.

[4] Bhoi, N.K., Singh, H. and Pratap, S., 2020. Developments in the aluminum metal matrix composites reinforced by micro/nano particles—a review. *Journal of Composite Materials*, *54*(6), pp. 813–833.

[5] Vasile, B.S., Birca, A.C., Surdu, V.A., Neacsu, I.A. and Nicoară, A.I., 2020. Ceramic composite materials obtained by electron-beam physical vapor deposition used as thermal barriers in the aerospace industry. *Nanomaterials*, *10*(2), p. 370.

[6] Jayappa, K., Kumar, V. and Purushotham, G.G., 2021. Effect of reinforcements on mechanical properties of nickel alloy hybrid metal matrix composites processed by sand mold technique. *Applied Science and Engineering Progress*, *14*(1), pp. 44–51.

[7] Zhao, C., Zhou, J., Zhong, K., Bai, Y. and Qi, L., 2024. Enhancing understanding metal matrix composites through molecular dynamics simulation: A comprehensive review. *Computational Materials Science*, *239*, p. 112993.

[8] Banerjee, S., Poria, S., Sutradhar, G. and Sahoo, P., 2020. Abrasive wear behavior of WC nanoparticle reinforced magnesium metal matrix composites. *Surface Topography: Metrology and Properties*, *8*(2), p. 025001.
[9] Shi, J. and Wang, Y., 2020. Development of metal matrix composites by laser-assisted additive manufacturing technologies: A review. *Journal of Materials Science*, *55*(23), pp. 9883–9917.
[10] Chen, W., Yang, T., Dong, L., Elmasry, A., Song, J., Deng, N., Elmarakbi, A., Liu, T., Lv, H.B. and Fu, Y.Q., 2020. Advances in graphene reinforced metal matrix nanocomposites: Mechanisms, processing, modelling, properties and applications. *Nanotechnology and Precision Engineering*, *3*(4), pp. 189–210.
[11] Oguntuyi, S.D., Johnson, O.T. and Shongwe, M.B., 2021. Spark plasma sintering of ceramic matrix composite of TiC: Microstructure, densification, and mechanical properties: A review. *The International Journal of Advanced Manufacturing Technology*, *116*(1), pp. 69–82.
[12] Tokita, M., 2021. Progress of spark plasma sintering (SPS) method, systems, ceramics applications and industrialization. *Ceramics*, *4*(2), pp. 160–198.
[13] Ogunmefun, O.A., Bayode, B.L., Jamiru, T. and Olubambi, P.A., 2023. A critical review of dispersion strengthened titanium alloy fabricated through spark plasma sintering techniques. *Journal of Alloys and Compounds*, p. 170407.
[14] Dudina, D.V., Georgarakis, K. and Olevsky, E.A., 2023. Progress in aluminium and magnesium matrix composites obtained by spark plasma, microwave and induction sintering. *International Materials Reviews*, *68*(2), pp. 225–246.
[15] Jana, P., Oza, M.J., Schell, K.G., Bucharsky, E.C., Laha, T. and Roy, S., 2022. Study of the elastic properties and thermal shock behavior of Al–SiC-graphite hybrid composites fabricated by spark plasma sintering. *Ceramics International*, *48*(4), pp. 5386–5396.
[16] Shirzadi, A.A. and Wallach, E.R., 1997. New approaches for transient liquid phase diffusion bonding of aluminium based metal matrix composites. *Materials Science and Technology*, *13*(2), pp. 135–142.
[17] Muratoğlu, M., Yilmaz, O. and Aksoy, M., 2006. Investigation on diffusion bonding characteristics of aluminum metal matrix composites (Al/SiCp) with pure aluminum for different heat treatments. *Journal of Materials Processing Technology*, *178*(1–3), pp. 211–217.
[18] Singh, R.P., Gupta, G.K. and Paliwal, M., 2018, September. Synthesis, consolidation and modelling study of AA2014-TiB_2 composite prepared by powder metallurgy (P/M) method. In *Materials Science Forum*, Trans Tech Publications Ltd, vol. 928, pp. 45–50.
[19] Inegbenebor, A.O., Bolu, C.A., Babalola, P.O., Inegbenebor, A.I. and Fayomi, O.S.I., 2018. Aluminum silicon carbide particulate metal matrix composite development via stir casting processing. *Silicon*, *10*, pp. 343–347.
[20] Poovazhagan, L., Kani, K., Balaji, V.R., Ganesh, P. and Avudaiappan, A.K., 2014. Development of AA6061/SiCp metal matrix composites by conventional stir casting and ultrasonic assisted casting routes—a comparative study. *Advanced Materials Research*, *984*, pp. 384–389.
[21] Srivatsan, T.S. and Lavernia, E.J., 1992. Use of spray techniques to synthesize particulate-reinforced metal-matrix composites. *Journal of Materials Science*, *27*, pp. 5965–5981.
[22] Etemadi, R., Wang, B., Pillai, K.M., Niroumand, B., Omrani, E. and Rohatgi, P., 2018. Pressure infiltration processes to synthesize metal matrix composites—a review of metal matrix composites, the technology and process simulation. *Materials and Manufacturing Processes*, *33*(12), pp. 1261–1290.
[23] Li, N., Liu, W., Wang, Y., Zhao, Z., Yan, T., Zhang, G. and Xiong, H., 2021. Laser additive manufacturing on metal matrix composites: A review. *Chinese Journal of Mechanical Engineering*, *34*, pp. 1–16.
[24] Dadkhah, M., Mosallanejad, M.H., Iuliano, L. and Saboori, A., 2021. A comprehensive overview on the latest progress in the additive manufacturing of metal matrix composites: Potential, challenges, and feasible solutions. *Acta Metallurgica Sinica (English Letters)*, *34*, pp. 1173–1200.
[25] Sharma, S.K., Saxena, K.K., Dixit, A.K., Singh, R. and Mohammed, K.A., 2022. Role of additive manufacturing and various reinforcements in MMCs related to biomedical applications. *Advances in Materials and Processing Technologies*, pp. 1–18.
[26] Griffiths, R.J., Perry, M.E., Sietins, J.M., Zhu, Y., Hardwick, N., Cox, C.D., Rauch, H.A. and Yu, H.Z., 2019. A perspective on solid-state additive manufacturing of aluminum matrix composites using MELD. *Journal of Materials Engineering and Performance*, *28*, pp. 648–656.
[27] Mertens, A.I., 2021. Metal matrix composites processed by laser additive manufacturing: Microstructure and properties. In *Additive Manufacturing*, Elsevier, pp. 409–425.
[28] Armstrong, M., Mehrabi, H. and Naveed, N., 2022. An overview of modern metal additive manufacturing technology. *Journal of Manufacturing Processes*, *84*, pp. 1001–1029.

[29] Yoozbashizadeh, M., Yavari, P. and Khoshnevis, B., 2018. Novel method for additive manufacturing of metal-matrix composite by thermal decomposition of salts. *Additive Manufacturing*, *24*, pp. 173–182.

[30] Martin, J.H., Yahata, B.D., Clough, E.C., Mayer, J.A., Hundley, J.M. and Schaedler, T.A., 2018. Additive manufacturing of metal matrix composites via nanofunctionalization. *MRS Communications*, *8*(2), pp. 297–302.

[31] Fereiduni, E., Yakout, M. and Elbestawi, M., 2019. Laser-based additive manufacturing of lightweight metal matrix composites. *Additive Manufacturing of Emerging Materials*, pp. 55–109.

[32] Karabulut, Y. and Ünal, R., 2022. Additive manufacturing of ceramic particle-reinforced aluminum-based metal matrix composites: A review. *Journal of Materials Science*, *57*(41), pp. 19212–19242.

[33] Attar, H., Ehtemam-Haghighi, S., Kent, D. and Dargusch, M.S., 2018. Recent developments and opportunities in additive manufacturing of titanium-based matrix composites: A review. *International Journal of Machine Tools and Manufacture*, *133*, pp. 85–102.

[34] Chaudhary, B., Patel, M., Jain, N.K., Murugesan, J. and Patel, V., 2023. Friction stir powder additive manufacturing of Al 6061/FeCoNi and Al 6061/Ni metal matrix composites: Reinforcement distribution, microstructure, residual stresses, and mechanical properties. *Journal of Materials Processing Technology*, *319*, p. 118061.

[35] Memarzadeh, A., Safaei, B., Tabak, A., Sahmani, S. and Kizilors, C., 2023. Advancements in additive manufacturing of polymer matrix composites: A systematic review of techniques and properties. *Materials Today Communications*, p. 106449.

[36] Das, M. and Balla, V.K., 2015. Additive manufacturing and innovation in materials world. In *Additive Manufacturing*, Taylor & Francis, p. 297.

[37] Kumar, M.B. and Sathiya, P., 2021. Methods and materials for additive manufacturing: A critical review on advancements and challenges. *Thin-Walled Structures*, *159*, p. 107228.

[38] Gardner, L., 2023, January. Metal additive manufacturing in structural engineering—review, advances, opportunities and outlook. *Structures*, Elsevier, *7*, pp. 2178–2193.

[39] Mahmood, M.A., Popescu, A.C. and Mihailescu, I.N., 2020. Metal matrix composites synthesized by laser-melting deposition: A review. *Materials*, *13*(11), p. 2593.

[40] Liu, S. and Shin, Y.C., 2017. The influences of melting degree of TiC reinforcements on microstructure and mechanical properties of laser direct deposited Ti6Al4V-TiC composites. *Materials & Design*, *136*, pp. 185–195.

[41] Novichenko, D., Marants, A., Thivillon, L., Bertrand, P.H. and Smurov, I., 2011. Metal matrix composite material by direct metal deposition. *Physics Procedia*, *12*, pp. 296–302.

[42] Ostolaza, M., Arrizubieta, J.I., Lamikiz, A., Plaza, S. and Ortega, N., 2023. Latest developments to manufacture metal matrix composites and functionally graded materials through AM: A state-of-the-art review. *Materials*, *16*(4), p. 1746.

[43] Dadbakhsh, S., Mertens, R., Hao, L., Van Humbeeck, J. and Kruth, J.P., 2019. Selective laser melting to manufacture "in situ" metal matrix composites: A review. *Advanced Engineering Materials*, *21*(3), p. 1801244.

[44] Özel, T., Shokri, H. and Loizeau, R., 2023. A review on wire-fed directed energy deposition based metal additive manufacturing. *Journal of Manufacturing and Materials Processing*, *7*(1), p. 45.

[45] Koczak, M.J. and Premkumar, M.K., 1993. Emerging technologies for the in-situ production of MMCs. *JOM*, *45*, pp. 44–48.

[46] Zhang, C., Li, Z., Zhang, J., Tang, H. and Wang, H., 2023. Additive manufacturing of magnesium matrix composites: Comprehensive review of recent progress and research perspectives. *Journal of Magnesium and Alloys*, *11*(2), pp. 425–461.

[47] Fang, Y., Kim, M.K., Zhang, Y., Duan, Z., Yuan, Q. and Suhr, J., 2022. Particulate-reinforced iron-based metal matrix composites fabricated by selective laser melting: A systematic review. *Journal of Manufacturing Processes*, *74*, pp. 592–639.

[48] Yang, H., Chen, X., Huang, G., Song, J., She, J., Tan, J., Zheng, K., Jin, Y., Jiang, B. and Pan, F., 2022. Microstructures and mechanical properties of titanium-reinforced magnesium matrix composites: Review and perspective. *Journal of Magnesium and Alloys*, *10*(9), pp. 2311–2333.

[49] Riquelme, A., de Rojas Candela, C.S., Rodrigo, P. and Rams, J., 2022. Influence of process parameters in additive manufacturing of highly reinforced 316L/SiCp composites. *Journal of Materials Processing Technology*, *299*, p. 117325.

[50] Li, J., Zong, B.Y., Wang, Y.M. and Zhuang, W.B., 2010. Experiment and modeling of mechanical properties on iron matrix composites reinforced by different types of ceramic particles. *Materials Science and Engineering: A*, *527*(29–30), pp. 7545–7551.

[51] Chen, D., Pan, Q., Liu, Z., Zeng, S., Shi, Q., Peng, J. and Li, Y., 2024. The strengthening mechanism of mechanical and tribological properties of TiC-reinforced 316L matrix composites fabricated by selective laser melting. *Materials Today Communications*, *39*, p. 108916.

[52] Balakumar, T., Riahi, R.A. and Edrisy, A., 2024. Experimental and analytical study of directional isothermal fatigue in additively manufactured Ti-TiB metal matrix composites. *Metals*, *14*(4), p. 408.

[53] Fang, Y., Zhang, Y., Kim, M.K., Kim, T., Duan, Z., Yuan, Q. and Suhr, J., 2024. Multi-scale hybrid reinforced super duplex stainless steel matrix composites with high strength and ductility via laser powder bed fusion and an in-situ synthesis strategy. *Additive Manufacturing*, *85*, p. 104152.

[54] Krishna, K.V., Patil, S.M., Sharma, S., Yuqi, J., Joshi, S.S., Radhakrishnan, M. and Dahotre, N.B., 2024. Additive friction stir deposition of Al 6061-B_4C composites: Process parameter, microstructure and property correlation. *Materials Science and Engineering, 910.*

[55] Wei, S., Tang, J.J., Jha, A., Lau, K.B., Wuu, D., Soh, V., Zhu, Q., Tan, C.C., Wang, P. and Chi, D., 2024. Individual and collective effects of titanium carbides and multi-walled carbon nanotubes on reinforcing the Inconel 625 alloy fabricated by laser powder bed fusion. *Materials Characterization*, *207*, p. 113555.

[56] Liu, Y., Chen, H., Han, R., Peng, Y., Wang, K. and Li, X., 2024. Investigating the microstructure and mechanical properties of 316L/TiB2 composites fabricated by laser cladding additive manufacturing. *Journal of Materials Research and Technology*, *29*, pp. 28–39.

8 Advancement in Fabrication of Metal Matrix Composites

Sudhagar Sukkasamy, Vijayakumar Palanivel, Suresh Subramanian, Yessian Robinson, and Duraisamy Velmurugan

8.1 INTRODUCTION

In the present industrial landscape, metal matrix composites (MMC) have emerged as highly sought-after materials due to their exceptional properties and versatility. Over the years, extensive research has been focused on the development and optimization of MMC, with numerous studies contributing to a deeper understanding of their properties and applications [1]. Despite the wealth of research already conducted, many researchers continue to explore novel approaches and advancements in MMC technology, underscoring the ongoing interest and significance of these materials in various industrial sectors. One of the key factors driving the widespread adoption of MMC is their unique combination of properties, which surpass those offered by their individual constituents. These properties include high strength, stiffness, dimensional stability, moisture resistance, thermal and electrical conductivity, and toughness of fracture [2]. The impressive array of characteristics exhibited by MMC makes them well-suited for a diverse industrial application, from aerospace and automotive to electronics and infrastructure. As industries strive for materials with enhanced performance and reliability, MMCs stand out as a compelling solution, offering a balance of strength, durability, and adaptability that meets the demands of modern manufacturing and engineering challenges [3].

Extensive research has been conducted on MMC, shedding light on their potential superiority over traditional monolithic alloys. One of the defining characteristics of MMC is its composition, which consists of two distinct materials. In MMC, the matrix material, typically a metal, encloses the reinforcement and forms the composite. Serving as the backbone of the composite, the matrix material provides structural integrity and support. On the other hand, the reinforcement material constitutes the dispersed phase within the composite. It is this phase that receives the load transmission and distribution from the metal matrix, enhancing the mechanical properties and performance of the composite. This collaborative interaction between the matrix and reinforcement materials is fundamental to the unique advantages offered by MMC [4].

When dealing with reinforcement in composite materials, it is crucial to carefully describe the forms and quantities of reinforcement utilized. Reinforcement directly influences the mechanical properties of the composites, playing a noteworthy role in defining their overall performance. The categories of reinforcement employed are pivotal in analyzing the attributes of composite, as they ultimately dictate the final properties. The ultimate properties of the composites are profoundly dependent on both the reinforcement and the matrix material properties. In the manufacturing of MMC, a diverse range of materials is employed as reinforcements, including particles, short fibres, whiskers, and continuous fibres [5]. Among these varieties, particle reinforcements stand out for their ability to uniformly disperse within the matrix phase, owing to their enhanced isentropic features. This uniform dispersion contributes to the improved mechanical properties and performance of MMC, making particle reinforcements a preferred choice in many applications.

DOI: 10.1201/9781003499718-8

Despite being recognized as a promising alternative to traditional materials, composites face several challenges that hinder their widespread adoption. Achieving desired material qualities is paramount in composite manufacturing, as these properties determine the performance and suitability of the final product [6]. The attainment of such qualities hinges on a multitude of factors, including the selection of constituent materials, the chosen fabrication approach, process parameters, and the overall composition of the composite [7]. These components are essential in determining the material's attributes and functionality. To overcome these challenges and achieve the necessary material qualities, careful consideration must be given to the selection of suitable materials and fabrication techniques. By strategically choosing materials and employing appropriate fabrication methods, manufacturers can optimize the properties of composites to meet specific application requirements [8].

MMCs are produced using a diverse array of fabrication methods, which can be broadly arranged into two main categories: liquid-state and solid-state procedures [9]. In liquid-state fabrication, the matrix material is heated above its melting point and the reinforcement particles are introduced into the molten metal. This allows for the thorough mixing and spreading of the reinforcement within the liquid molten metal. Stir casting is one of the most widely utilized liquid-state fabrication methods, which is renowned for its versatility and ease of use. In stir casting, stirring of the molten metal is performed while incorporating the reinforcement material, ensuring homogeneity and uniform dispersion throughout the composite [10].

Solid-state manufacturing techniques for MMC involve the consolidation of solid materials without melting them [11]. These methods are often preferred for their ability to maintain the integrity of the reinforcement and matrix materials while forming a composite structure. One of the primary solid-state production methods for MMC is powder metallurgy, which entails blending metal powders with reinforcement particles followed by compaction and sintering. During compaction, pressure is applied to the powder mixture to shape it into a desired form, while sintering involves heating the compacted material to temperatures lower than the melting temperature of the matrix material, allowing for the particles to bond together [12]. Another solid-state manufacturing method is diffusion bonding, where the matrix and reinforcement materials are stacked in layers and subjected to high temperatures and pressures to facilitate atomic diffusion and bonding between the materials.

In recent times, the development of innovative manufacturing methods for MMC fabrication has surged, with a particular emphasis on additive manufacturing techniques. The emergence of additive manufacturing has significantly impacted the fabrication process of MMC, propelling it to new heights. Unlike conventional manufacturing methods, additive manufacturing facilitates the fabrication of near-net-shaped components, which can be utilized directly for end-use applications. This advancement significantly streamlines the manufacturing process, eliminating the need for additional processing steps typically required with traditional methods [13]. MMC fabricated through additive manufacturing techniques can be readily deployed for real-time applications, showcasing their versatility and efficiency. The subsequent section discusses several advanced manufacturing techniques for MMC.

8.2 ACCUMULATIVE ROLL BONDING (ARB)

ARB was pioneered by Tsuji et al. in 1999 as a groundbreaking solid-state multi-stage process for fabricating advanced materials, particularly MMC. This innovative technique employs severe plastic deformation to systematically refine the microstructure of the material, enhancing its mechanical properties. The process involves several stages, including surface preparation, cutting, stacking, and rolling of thin metal foils or sheets [14]. Figure 8.1 schematically illustrates the sequential steps involved in ARB, showcasing the intricate process of stacking and rolling the metal layers to achieve bonding and grain refinement. This method revolutionized the field of material science and engineering, offering a precise and scalable approach for producing MMC with tailored properties for diverse industrial applications [15].

The initial step in ARB is crucial, involving meticulous surface preparation to ensure optimal bonding and material integrity. This process begins with thorough degreasing and the removal of

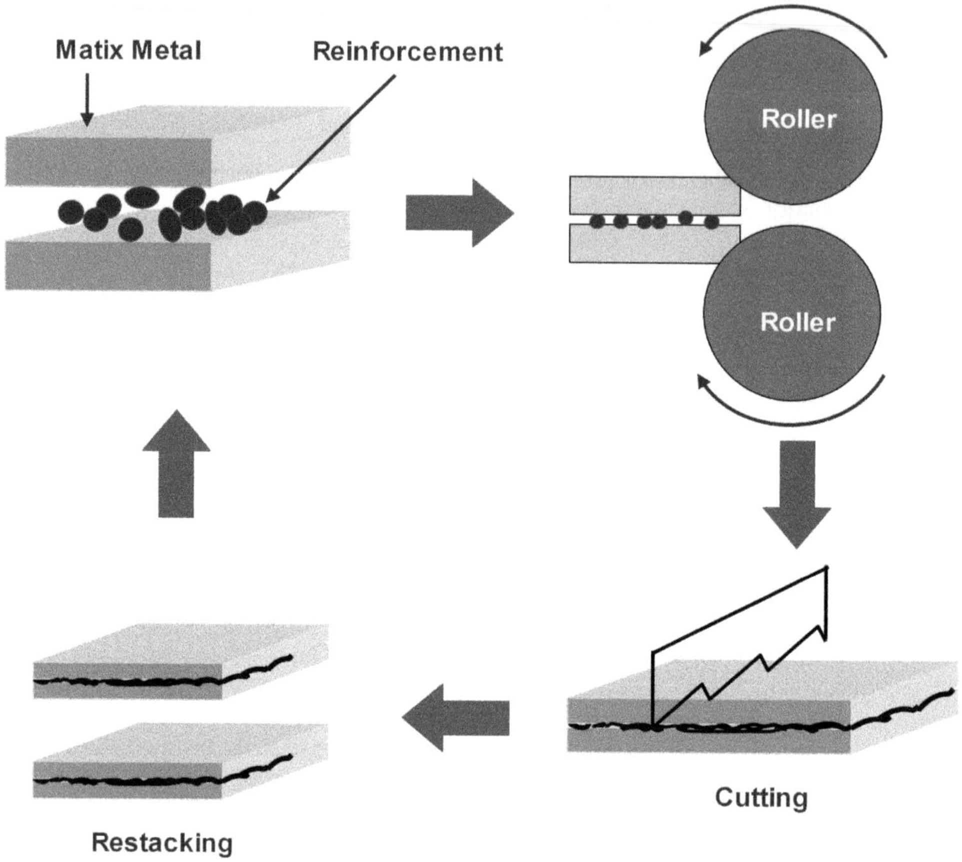

FIGURE 8.1 Schematic of accumulative roll bonding process.

any dirt or oxide layers present on the metal surface, ensuring a clean and receptive surface for subsequent bonding. Following surface preparation, the next step entails the dispersal of reinforcement particles onto the metal surface, strategically positioning them for enhanced mechanical properties [16]. Subsequently, another metal plate is carefully stacked over the reinforcement, creating a layered assembly ready for consolidation. Once packed and stacked, the metal plates undergo the main stage of the ARB process: passing through a set of rollers rotating in opposite directions. The pressure exerted by the rollers induces severe plastic deformation, effectively reducing the thickness of the metal plates while promoting intimate contact between the layers. This intense deformation process facilitates the bonding of the two metal plates, along with the embedded reinforcement particles, creating a robust composite structure with tailored mechanical properties. Through meticulous surface preparation and controlled deformation, ARB offers a reliable and scalable approach for producing advanced metal matrix composites suitable for diverse industrial applications [17].

As the metal plates are compressed and thinned, the material is forced to elongate to accommodate the reduction in thickness while maintaining its overall volume. After the initial rolling process, the elongated plates are cut into halves to facilitate further processing and achieve the desired dimensions. Once the plates are halved, they are carefully restacked, typically in alternating layers, to promote homogeneity and optimize the distribution of reinforcing particles. Subsequently, the stacked plates undergo another round of rolling, wherein they are passed through the rollers once again for further consolidation. This additional rolling stage serves to further reduce the thickness of the material, enhance bonding between the layers, and promote grain refinement.

8.2.1 Bonding Mechanism in ARB

Various studies have delved into comprehending the intricate mechanisms underlying the bonding process in ARB, resulting in the proposal of four major theories: Film theory, energy barrier theory, recrystallization theory, and diffusion bonding theory. Among these, considerable research supports the prominence of the film theory as the primary bonding mechanism during ARB. According to the film theory, when two clean metal surfaces come into contact, they naturally form bonds between them. This process is facilitated by the meticulous surface preparation involved in ARB, where all dirt and oxide layers are meticulously removed from the metal surfaces. As the metal undergoes severe plastic deformation during rolling, the surface layers of the metal plates break, and virgin metal protrudes outwards, establishing contact with the virgin layer of another plate. This phenomenon promotes the formation of bonds between the adjacent metal layers [18]. The validity of this bonding mechanism has been substantiated through detailed studies, including scanning electron microscopic analyses, which provide visual confirmation of the bonding process at the microstructural level. Figure 8.2 illustrates the bonding mechanism in the ARB process.

8.2.2 ARB Process Parameter

The properties of MMC fabricated through the ARB process are determined by numerous parameters. Figure 8.3 shows the major parameters that play a key role in ARB.

8.3 SELECTIVE LASER MELTING (SLM)

SLM represents a groundbreaking advancement within the realm of additive manufacturing. Originally developed to print metals, alloys, and ceramics, SLM has revolutionized the manufacturing landscape with its precision and versatility. In recent years, this innovative technique has extended its capabilities to include the printing of MMC, marking a significant milestone in materials engineering. Unlike traditional manufacturing methods, SLM offers a near-net shape manufacturing process, enabling the creation of intricate geometries and complex structures with

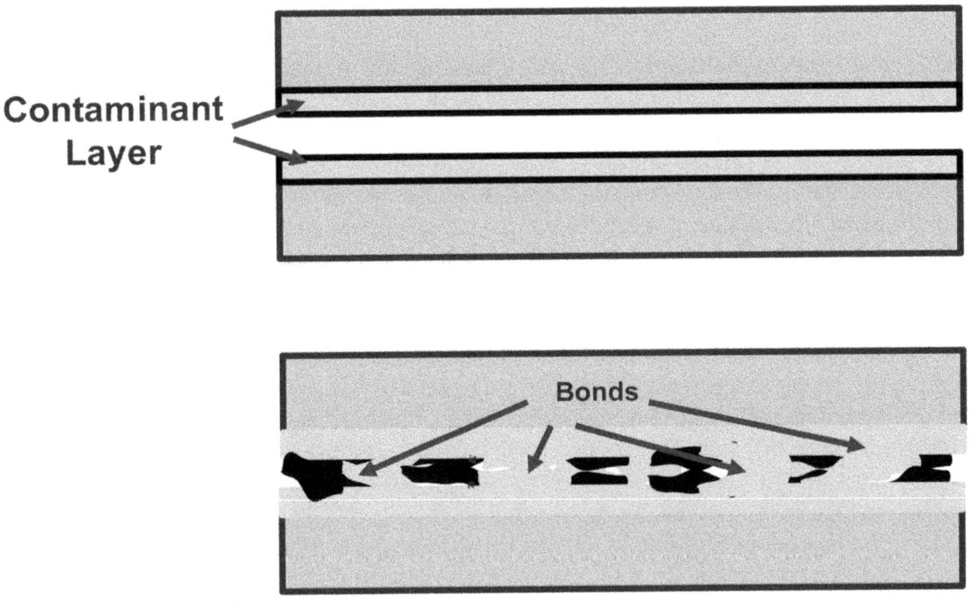

FIGURE 8.2 Mechanism of bonding in ARB.

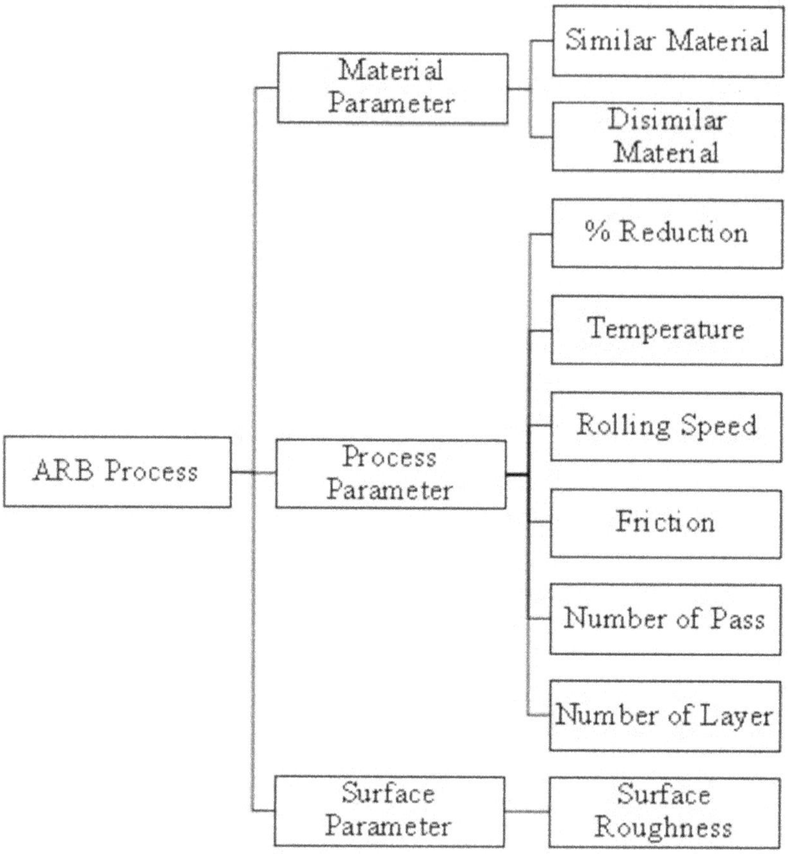

FIGURE 8.3 Process parameters in ARB.

unparalleled accuracy and efficiency. The metal powders are selectively heated and melted using a high-powered laser; SLM facilitates the precise fusion of materials, resulting in components with exceptional mechanical properties and performance characteristics [19].

The core of the SLM process consists few key components, including the laser source, scanning head, powder coater, build platform, and inert gas supply system [20]. The diagrammatic representation of the SLM process shown in Figure 8.4 reveals a meticulous sequence of steps that culminate in the precise fabrication of complex metal components. The process initiates with the coater dispersion of a thin layer of metal powder evenly above the build surface, with the height of the powder layer matching the desired layer height during slicing. A concentrated laser beam with high energy is directed onto the powder, rapidly melting and fusing the metal particles together. Guided by the scanning head, the laser traverses the powder layer, selectively solidifying the material to form the desired cross-sectional shape of the component. After finishing one layer, the coater adds a fresh layer of powder atop the existing one, repeating this process layer by layer until the final part is completely fabricated [21]. The whole process is accomplished in an inert gas environment in order to eliminate oxidation of material during printing.

In SLM, the process of building a 3D part involves a series of intricate phase transformations within the material, transitioning from solid state to liquid state and then returning back to solid state. This complex metamorphosis is governed by a multitude of phenomena that occur during the melting and solidification of metal powders. These phenomena include the assimilation and

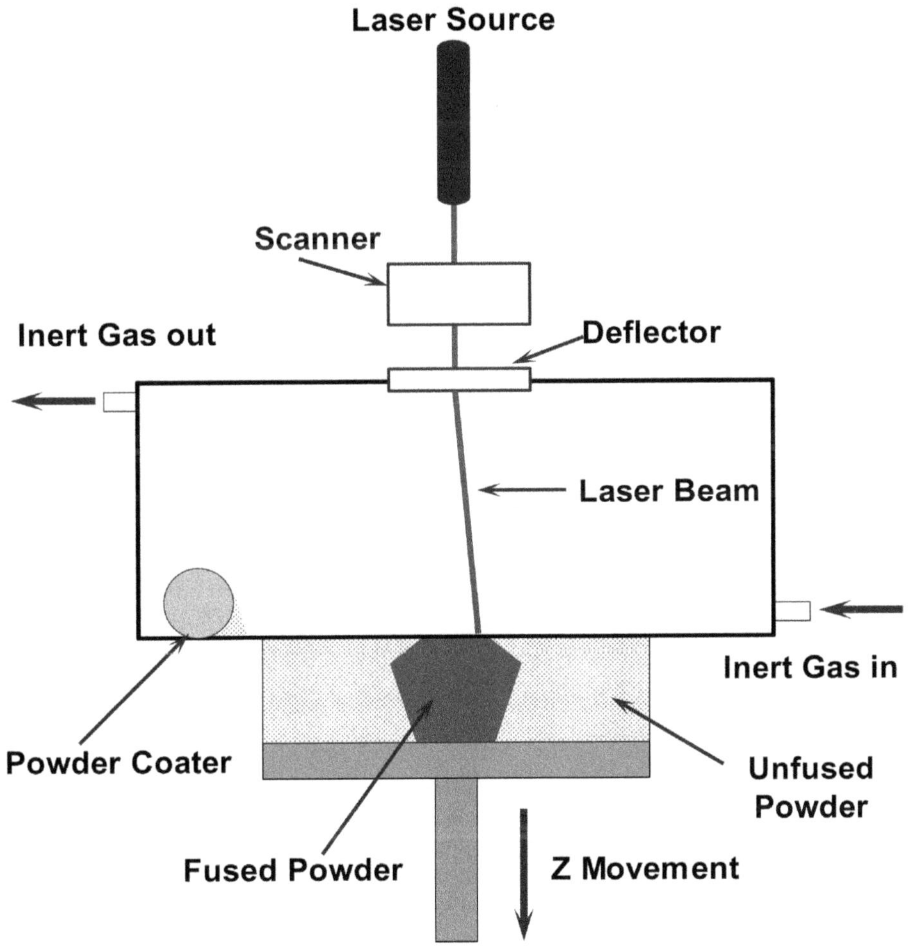

FIGURE 8.4 Schematic of SLM process.

diffusion of laser radiation, thermal diffusion mechanisms, fluid flow dynamics, evaporation and emission of materials, phase transformations, and potential chemical reactions. The precise control and management of these phenomena are paramount in SLM to ensure the production of flawless products with optimal mechanical properties and structural integrity. Central to this control is the collaboration between the powder particles and the laser beam, as the energy absorbed by the particles dictates their behaviour and ultimately influences the perfection of the end product [22]. As the laser beam converges on the powder particles, it undergoes multiple reflections and absorptions, leading to localized heating and subsequent melting of the particles.

8.3.1 SLM Process Parameters

The strength and quality of the final product produced via SLM are subject to the influence of a multitude of parameters, as depicted in Figure 8.5. These parameters are broadly categorized into two main groups: Material parameters and machine parameters. Material parameters encompass critical factors such as the composition of the feedstock powder, particle size distribution, and particle morphology. These characteristics directly impact the material's behaviour during the printing process and ultimately contribute to the properties of the final printed part. However, machine

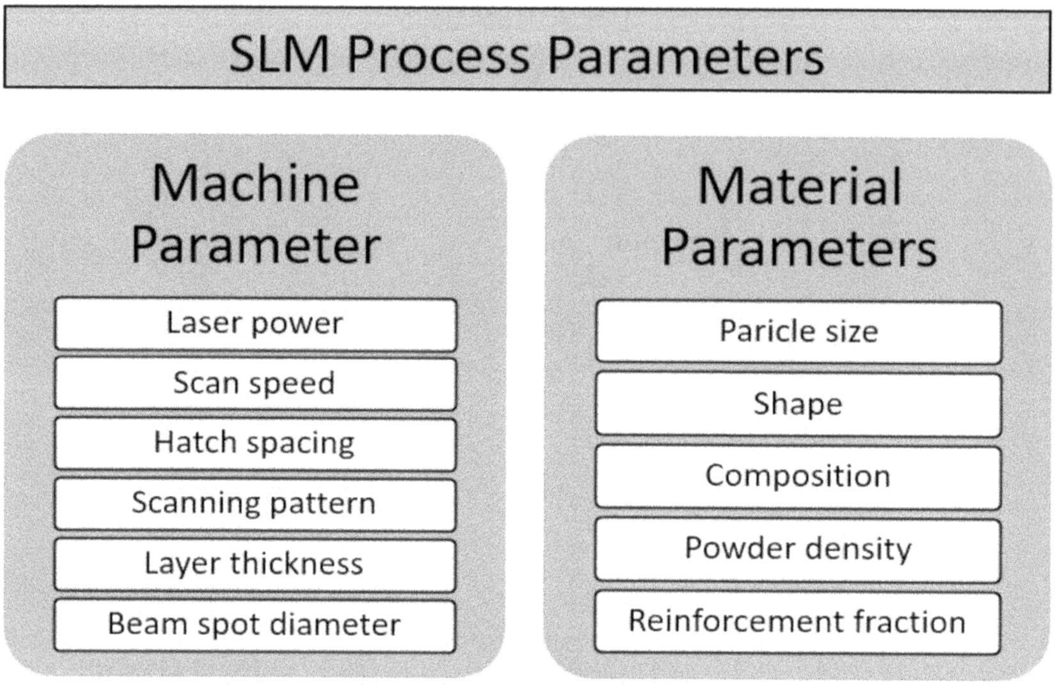

FIGURE 8.5 Process parameters of SLM.

parameters encompass a range of settings and variables inherent to the SLM system itself. These include laser power, scan speed, hatch spacing, scanning pattern, powder layer thickness, and beam spot diameter, among others. Each of these machine parameters plays an important role in dictating the precise deposition of material layers and the consolidation of the printed part. By carefully controlling and optimizing both material and machine parameters, manufacturers can achieve superior outcomes in terms of part quality, mechanical performance, and dimensional accuracy in SLM [23].

8.3.2 SLM Materials

An essential phase in the manufacturing of MMC via SLM is the meticulous preparation of the feedstock material. The feedstock material for MMC comprises both the matrix material and the reinforcement materials, with their particle shape, size, and volume fraction playing pivotal functions in defining the final properties of the composite produced through SLM [24]. Notably, while commercial feedstock of nanocomposite powder specifically tailored for SLM remains unavailable, researchers have explored various preparation methods to effectively blend the matrix and reinforcement powders [25]. Among these methods, the direct mixing approach has garnered significant attention. However, reinforcement agglomeration and poor wettability are unavoidable in the direct mixing method.

To prevent the agglomeration of nanoparticles in MMC, alternative processing methods are necessary. Among these methods, ball milling emerges as the most popular choice, prized for its affordability and suitability to a wide range of powders. In this technique, powder particles are repeatedly fractured, deformed, and cold welded. As the brittle reinforcement particles undergo fracture before adhering to the matrix powder, the predominant phenomenon within the matrix is cold welding, attributed to plastic deformation. This plastic deformation facilitates a dynamic interaction between cold welding and fracture. This equilibrium confirms the stability of powder size and the absence of particle clustering, crucial for the uniform dispersion of nanoparticles.

TABLE 8.1
MMC Fabricated through SLM

Matrix material	Reinforcement material
316L stainless steel	TiB_2
Ti	TiC
Fe	SiC
AlSi10Mg	TiC
Ti	TiC
AlSi10Mg	AlN
316L stainless steel	TiB_2
316L stainless steel	TiC
H13 Steel	TiC
Al	Al_2O_3
Al	Fe_2O_3

SLM offers the capability to reliably process only a limited variety of metal powders, resulting in fully dense components. Among the most popular metal powders suitable for SLM are 316L stainless steel, commercially pure titanium, Ti-6Al-4V, AlSi10Mg, A356 aluminium alloy, Inconel 718, NiTi shape memory alloy, and pure aluminium. These materials are favoured for their compatibility with the SLM process and their ability to produce high-quality, fully dense components. Additionally, various reinforcement materials are commonly employed in conjunction with these matrix materials to fabricate MMC via SLM. These reinforcements include TiC, Al_2O_3, carbon nanotubes, SiC, and Mg_2AlO_4, among others [26]. A comprehensive overview of the matrix materials and reinforcement materials commonly utilized for fabricating MMC via the SLM method is provided in the accompanying Table 8.1.

8.4 BINDER JETTING 3D PRINTING

Binder jetting (BJ), pioneered and patented by MIT in the early 1990s, has gained more attention in the past few years, because of its remarkable productivity and relative simplicity compared to other additive manufacturing techniques. As a powder bed type of additive manufacturing process, BJ involves printing the part by spreading a binder onto the powder bed to fabricate a part layer by layer. This novel method has numerous advantages, including rapid production speeds and the ability to create complex geometries with ease [27].

Binder jetting is a cutting-edge additive manufacturing process that encompasses several key components, including the raw powder chamber, powder spreader, build platform, and binder jetting mechanism. Figure 8.6 illustrates the binder jetting process. The process begins with the powder spreader transferring raw powder from the powder chamber to the build platform, where it is evenly spread to a thickness equivalent to the desired layer thickness during slicing. Next, a binder liquid is precisely injected through a nozzle in the form of fine droplets onto the spread powder. These binder droplets serve to agglomerate the powder particles, causing them to combine and adhere together. Notably, the binder liquids are strategically applied only to the required points as dictated by the CAD model, ensuring precise and efficient material deposition [27].

As each layer is built, the powder spreader once again coats the raw powder above the prior layer, and the binder liquids are injected into the required points. This iterative process continues until the final 3D part is fully fabricated. Remarkably, parts produced via binder jetting are self-supporting within the powder bed, thus the requirement of support structures is eliminated, typically required in other additive manufacturing processes. This inherent self-supporting capability contributes to

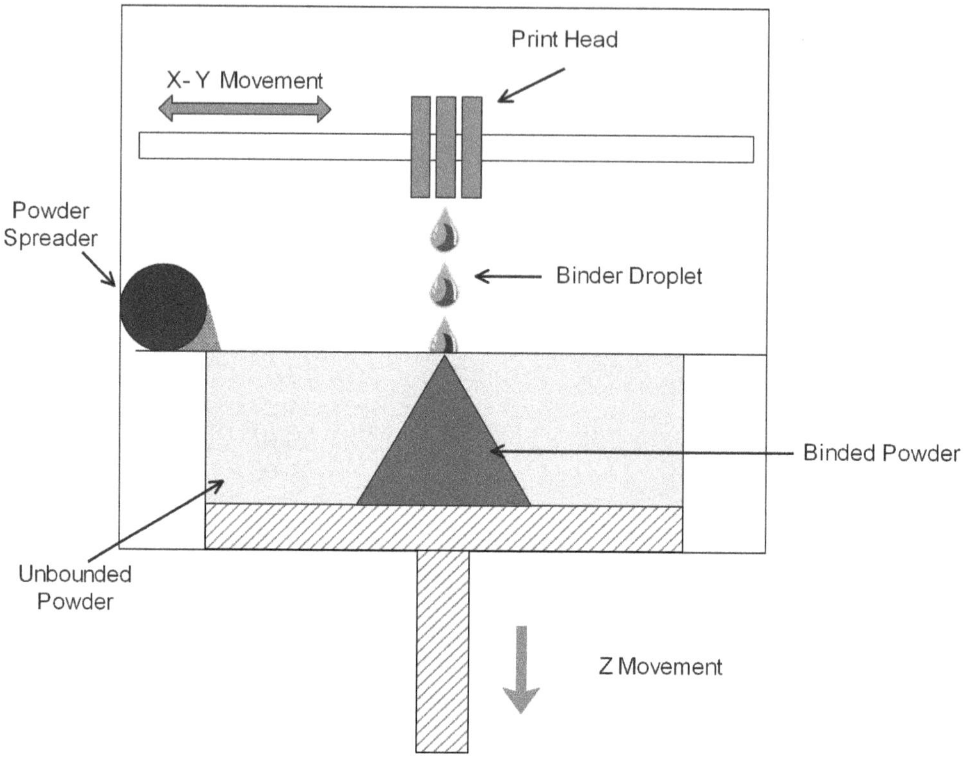

FIGURE 8.6 Schematic of binder jetting process.

the efficiency and simplicity of the BJ process, enabling the fabrication of complex geometries without the constraints imposed by support structures. The printed part is left is inside the print chamber for the binder to fully set, which is known as the green part.

Post-processing plays an important role in the binder jetting, ensuring the final parts meet desired specifications and performance requirements. The post-processing steps typically involve several key tasks. First, the removal of unbound powder is essential to uncover the printed part and prepare it for further processing. This is achieved through the use of pressurized air, which effectively blows away the excess powder without damaging the printed component. Subsequently, the green part, which refers to the printed part in its as-printed state, is carefully removed from the build platform or powder bed. Once removed, the green part is then placed in a furnace for sintering, a crucial step in the post-processing journey. During the sintering process, the binder material used in the printing stage evaporates under controlled heat, leaving behind a porous structure. As the temperature increases, the powder particles within the green part begin to fuse together, resulting in a densification process known as sintering. This fusion of particles ultimately transforms the green part into a solid, bulk component with enhanced mechanical properties and dimensional accuracy [28]. Figure 8.7 depicts the workflow of the BJ process.

8.4.1 BJ Process Parameter

Binder jetting (BJ) is a complex additive manufacturing process that requires careful consideration of numerous variables to produce defect-free, high-strength parts. These variables can be broadly

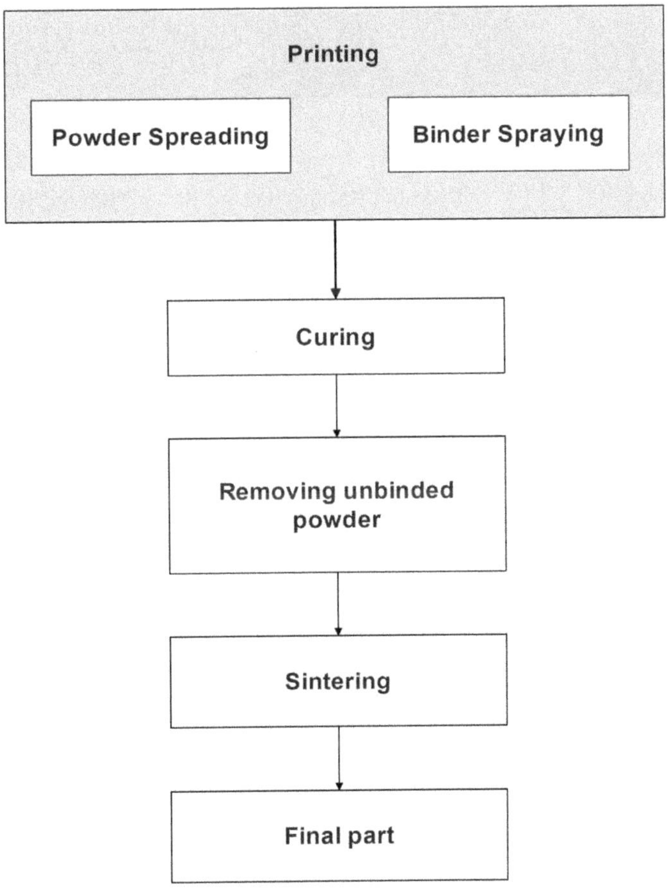

FIGURE 8.7 Workflow of binder jetting process.

categorized into three main groups: material parameters, process parameters, and sintering parameters. Material parameters encompass factors related to the properties of powder materials used in the process, such as powder composition, particle size distribution, and binder formulation. Process parameters, on the other hand, include variables associated with the printing process itself, such as layer thickness, printing speed, and binder droplet size. Sintering parameters include variables associated with sintering such as sintering temperature, time, atmosphere, heating rate, etc. The interaction between these materials and process parameters plays a critical factor in defining the final quality, and performance of the printed parts illustrated in Figure 8.8 [29] are the various factors involved in the binder jetting process.

8.5 CONCLUSION

The fabrication of MMC through advanced methods such as accumulative roll bonding, selective laser melting, and binder jetting processes represents a significant advancement in materials science and manufacturing technology. These innovative techniques offer unique advantages, each contributing to the manufacturing of MMC with custom-made properties and enhanced performance

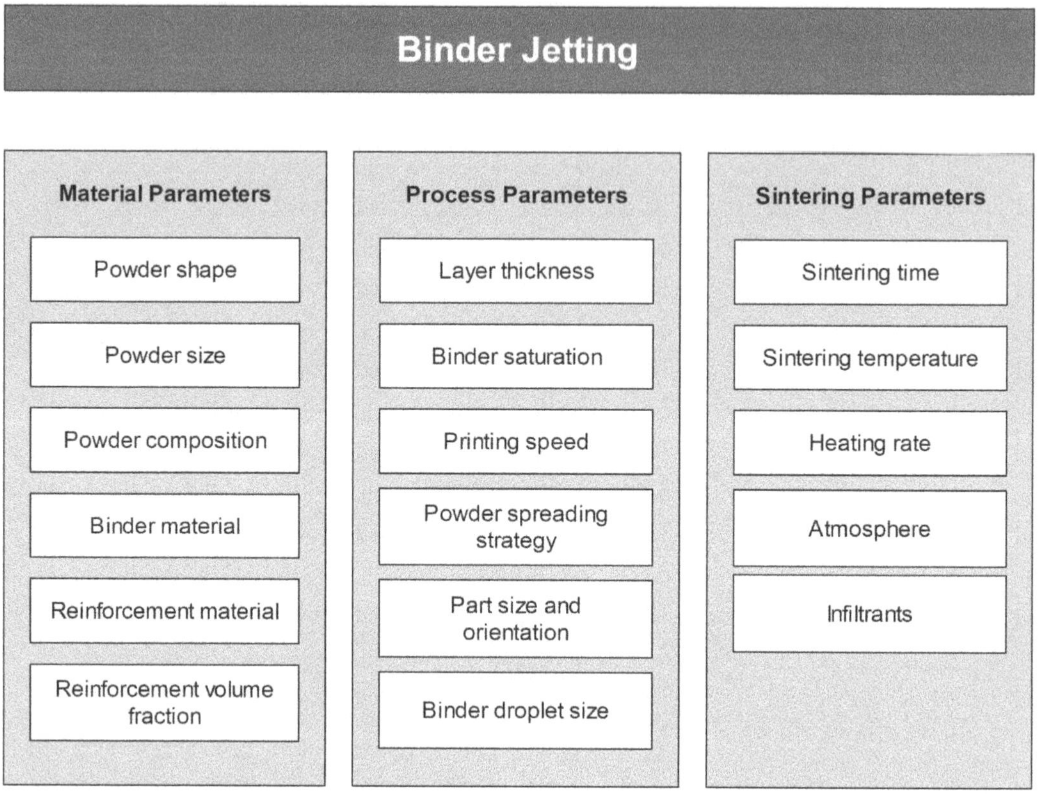

FIGURE 8.8 Process parameters in BJ process.

characteristics. Accumulative roll bonding facilitates the creation of MMC with improved mechanical properties through severe plastic deformation, while selective laser melting enables the precise fabrication of complex geometries with superior control over material composition. Similarly, binder jetting provides a versatile approach for producing MMC with near-net-shaped components, offering cost-effective and efficient manufacturing solutions. By taking advantage of these advanced methods, researchers and engineers can explore new frontiers in MMC development, unlocking the potential for applications across a wide range of industries, from aerospace and automotive to biomedical and beyond.

REFERENCES

[1] Sudhagar, S., Sakthivel, M. and Gopal, P.M., 2015. Parameteric optimization of surface roughness in end milling of aluminium rock dust composite. *Applied Mechanics and Materials*, *813*, pp. 382–387.

[2] Daniel, A.A., Murugesan, S. and Sukkasamy, S., 2017. Dry sliding wear behaviour of aluminium 5059/SiC/MoS_2 hybrid metal matrix composites. *Materials Research*, *20*, pp. 1697–1706.

[3] Kumar, S., Sakthivel, M., Sudhagar, S. and Nivethan, K., 2019. Two body abrasive wear characteristics of Al7068/Si3N4/BN hybrid composite. *Materials Research Express*, *6*(6), p. 066502.

[4] Sudhagar, S., Gopal, P.M., Maniyarasan, M., Suresh, S. and Kavimani, V., 2024. Multi-objective optimization of machining parameters for Si3N4–BN reinforced magnesium composite in wire electrical discharge machining. *International Journal on Interactive Design and Manufacturing*, pp. 1–16.

[5] Preethi, V., Kavimani, V. and Gopal, P.M., 2023. Electrochemical micro-machining of hybrid graphene/silicon nitride-reinforced magnesium composite through integrated entropy-COPRAS approach. *Multiscale and Multidisciplinary Modeling, Experiments and Design*, pp. 1–13.

[6] Balasubramani, V., Pandiarajan, R., Stalin, B. and Kavimani, V., 2023. Mechanical and metallurgical characterization of B4C and SiC reinforced stir casted AA6061 hybrid MMC. *Physica Scripta*, *98*(10), p. 105922.
[7] Sharma, S.K., Saxena, K.K., Salem, K.H., Mohammed, K.A., Singh, R. and Prakash, C., 2022. Effects of various fabrication techniques on the mechanical characteristics of metal matrix composites: A review. *Advances in Materials and Processing Technologies*, pp. 1–18.
[8] Sudhagar, S. and Gopal, P.M., 2022. Investigation on mechanical and tribological characteristics Cu/Si3N4 surface composite developed through friction stir processing. *Silicon*, *14*(8), pp. 4207–4216.
[9] Gopal, P.M., Kavimani, V., Sudhagar, S., Sonar, T. and Venkatesh, S., 2024. Optimization of wire-cut EDM parameters using Taguchi and entropy coupled COPRAS approach for machining of CRT glass powder reinforced magnesium surface composite developed using friction stir processing. *International Journal on Interactive Design and Manufacturing*, pp. 1–12.
[10] Ramanathan, A., Krishnan, P.K. and Muraliraja, R., 2019. A review on the production of metal matrix composites through stir casting—furnace design, properties, challenges, and research opportunities. *Journal of Manufacturing Processes*, *42*, pp. 213–245.
[11] Gopal, P.M. and Kavimani, V., 2021. Influence of silica rich CRT and BN on mechanical, wear and corrosion characteristics of copper-surface composite processed through friction stir processing. *Silicon*, *13*, pp. 3431–3440.
[12] Wu, L., Wu, R., Hou, L., Zhang, J., Sun, J. and Zhang, M., 2017. Microstructure and mechanical properties of CNT-reinforced AZ31 matrix composites prepared using hot-press sintering. *Journal of Materials Engineering and Performance*, *26*, pp. 5495–5500.
[13] Mussatto, A., Ahad, I.U., Mousavian, R.T., Delaure, Y. and Brabazon, D., 2021. Advanced production routes for metal matrix composites. *Engineering Reports*, *3*(5), p.e12330.
[14] Alizadeh, M. and Paydar, M.H., 2009. Fabrication of Al/SiCP composite strips by repeated roll-bonding (RRB) process. *Journal of Alloys and Compounds*, *477*(1–2), pp. 811–816.
[15] Rezayat, M. and Akbarzadeh, A., 2012. Bonding behavior of Al–Al2O3 laminations during roll bonding process. *Materials & Design*, *36*, pp. 874–879.
[16] Karimi, M. and Toroghinejad, M.R., 2014. An alternative method for manufacturing high-strength CP Ti–SiC composites by accumulative roll bonding process. *Materials & Design*, *59*, pp. 494–501.
[17] Liu, C.Y., Wang, Q., Jia, Y.Z., Zhang, B., Jing, R., Ma, M.Z., Jing, Q. and Liu, R.P., 2012. Effect of W particles on the properties of accumulatively roll-bonded Al/W composites. *Materials Science and Engineering: A*, *547*, pp. 120–124.
[18] Pan, D., Gao, K. and Yu, J., 1989. Cold roll bonding of bimetallic sheets and strips. *Materials Science and Technology*, *5*(9), pp. 934–939.
[19] Prashanth, K.G., Shahabi, H.S., Attar, H., Srivastava, V.C., Ellendt, N., Uhlenwinkel, V., Eckert, J. and Scudino, S., 2015. Production of high strength Al85Nd8Ni5Co2 alloy by selective laser melting. *Additive Manufacturing*, *6*, pp. 1–5.
[20] Kruth, J.P., Levy, G., Klocke, F. and Childs, T.H.C., 2007. Consolidation phenomena in laser and powder-bed based layered manufacturing. *CIRP Annals*, *56*(2), pp. 730–759.
[21] Marini, D., Cunningham, D. and Corney, J.R., 2018. Near net shape manufacturing of metal: A review of approaches and their evolutions. *Proceedings of the institution of mechanical engineers, Part B: Journal of Engineering Manufacture*, *232*(4), pp. 650–669.
[22] Ahn, I.H., Moon, S.K., Hwang, J. and Bi, G., 2017. Characteristic length of the solidified melt pool in selective laser melting process. *Rapid Prototyping Journal*, *23*(2), pp. 370–381.
[23] Yu, W.H., Sing, S.L., Chua, C.K., Kuo, C.N. and Tian, X.L., 2019. Particle-reinforced metal matrix nanocomposites fabricated by selective laser melting: A state of the art review. *Progress in Materials Science*, *104*, pp. 330–379.
[24] AlMangour, B., Grzesiak, D. and Yang, J.M., 2017. Selective laser melting of TiB2/316L stainless steel composites: The roles of powder preparation and hot isostatic pressing post-treatment. *Powder Technology*, *309*, pp. 37–48.
[25] AlMangour, B., Grzesiak, D. and Yang, J.M., 2017. In-situ formation of novel TiC-particle-reinforced 316L stainless steel bulk-form composites by selective laser melting. *Journal of Alloys and Compounds*, *706*, pp. 409–418.
[26] Levy, A., Miriyev, A., Elliott, A., Babu, S.S. and Frage, N., 2017. Additive manufacturing of complex-shaped graded TiC/steel composites. *Materials & Design*, *118*, pp. 198–203.

[27] Enneti, R.K., Prough, K.C., Wolfe, T.A., Klein, A., Studley, N. and Trasorras, J.L., 2018. Sintering of WC-12% Co processed by binder jet 3D printing (BJ3DP) technology. *International Journal of Refractory Metals and Hard Materials*, *71*, pp. 28–35.

[28] Snelling, D.A., Williams, C.B., Suchicital, C.T. and Druschitz, A.P., 2017. Binder jetting advanced ceramics for metal-ceramic composite structures. *The International Journal of Advanced Manufacturing Technology*, *92*, pp. 531–545.

[29] Lores, A., Azurmendi, N., Agote, I. and Zuza, E., 2019. A review on recent developments in binder jetting metal additive manufacturing: Materials and process characteristics. *Powder Metallurgy*, *62*(5), pp. 267–296.

9 Fabrication of Intelligent Metal Matrix Composites and Their Applications

Saravanan Rathinasamy, Omkumar Meenakshisundaram, Sathish Thanikodi, and Jothi Arunachalam Solairaju

9.1 INTRODUCTION

9.1.1 Overview of Metal Matrix Composites (MMCs)

Metal matrix composites (MMCs) are engineered materials comprising a metallic base or matrix combined with non-metallic reinforcements. The reinforcements can be in the form of particles, fibres, or other shapes, designed to enhance specific properties of the metal matrix, such as strength, stiffness, or resistance to wear and corrosion. This unique combination allows MMCs to outperform pure metals in various applications, offering tailored mechanical, thermal, and electrical properties. MMCs are commonly used in industries where high strength-to-weight ratios and durability are crucial [1].

9.1.2 Evolution toward Intelligent MMCs

The evolution toward intelligent metal matrix composites (IMMCs) represents a significant leap in materials science, where the emphasis shifts from mere mechanical reinforcement to incorporating adaptive and responsive functionalities. Traditional metal matrix composites (MMCs) focus on enhancing specific properties like strength, toughness, or thermal conductivity by integrating reinforcing materials into a metal matrix. However, the concept of IMMCs builds on these foundations by embedding smart materials and sensor technologies, enabling the composites to respond dynamically to external stimuli [2].

9.1.3 Motivation for Synthesizing Intelligent MMCs

The motivation for synthesizing intelligent metal matrix composites (IMMCs) stems from the growing demand for materials that can meet complex engineering challenges while offering enhanced functionality and adaptability. Traditional metal matrix composites (MMCs) are valued for their high strength-to-weight ratios and durability, but they cannot often adapt to changing conditions or provide real-time feedback [3].

9.2 DESIGN CONSIDERATIONS FOR INTELLIGENT MMCs

9.2.1 Material Selection Criteria

Selecting materials for intelligent metal matrix composites (IMMCs) involves a multi-faceted approach, as these composites must meet both traditional mechanical requirements and the additional demands of smart functionalities. The following are key criteria to consider in material selection. Figure 9.1 shows the production of metal matrix composite.

DOI: 10.1201/9781003499718-9

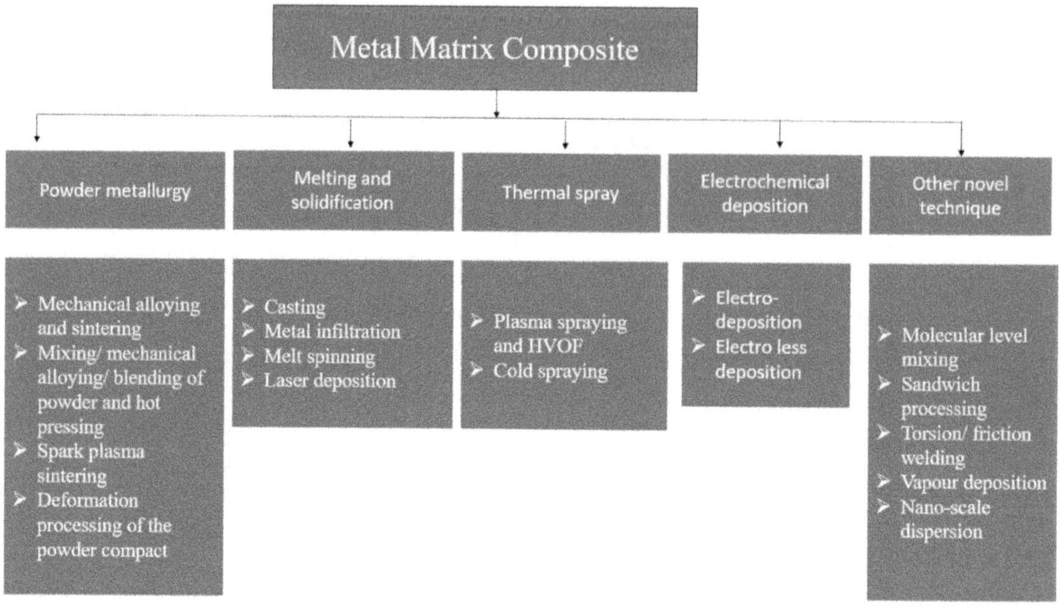

FIGURE 9.1 Basic metal matrix manufacturing process.

9.2.1.1 Matrix Material Properties

The metal matrix forms the primary structure of the IMMC, so it should offer high strength, toughness, and resistance to environmental factors like corrosion and oxidation. Common matrix materials include aluminium, titanium, and magnesium, each chosen for specific properties like light weight, high strength, or heat resistance. Figure 9.2 shows the powder metallurgy process.

9.2.1.2 Smart Material Compatibility

The intelligent components, typically smart materials like piezoelectrics, shape memory alloys, or self-healing polymers, must be compatible with the metal matrix. This compatibility involves not only chemical stability but also mechanical coherence, ensuring that the smart materials can integrate seamlessly without compromising the composite's overall performance.

9.2.1.3 Reinforcement Properties

IMMCs often include reinforcing materials like ceramic particles, fibres, or whiskers to enhance mechanical properties. The reinforcement must be compatible with both the matrix and the smart materials, providing improved strength, stiffness, or wear resistance while allowing the intelligent components to function effectively.

9.2.1.4 Thermal and Electrical Conductivity

Depending on the application, the thermal and electrical conductivity of the composite can be crucial. The choice of materials should ensure that the IMMC has the desired conductivity properties, whether it's for heat dissipation in aerospace applications or electrical responsiveness in sensor-based systems.

9.2.1.5 Processing and Manufacturing Compatibility

The selected materials must be compatible with manufacturing processes such as casting, powder metallurgy, or additive manufacturing. This compatibility ensures that the composite can be produced reliably and with consistent quality.

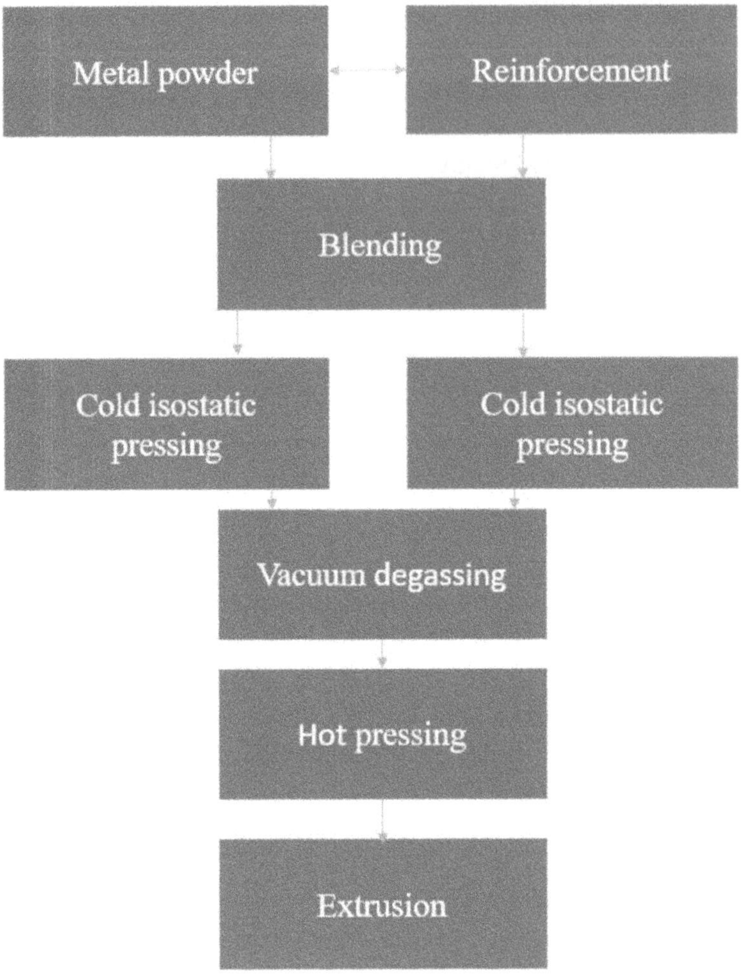

FIGURE 9.2 Powder metallurgy process.

9.2.1.6 Cost and Availability

The economic aspect is also important, with a focus on choosing materials that are cost-effective and readily available in sufficient quantities. The overall production cost of IMMCs can significantly impact their adoption in various industries.

9.2.1.7 Environmental and Safety Considerations

Material selection should consider environmental impact and safety. The chosen materials should comply with regulatory requirements and minimize harmful emissions or waste during manufacturing and end-of-life disposal [4].

9.2.2 Tailoring Microstructural Features

Tailoring the microstructural features of metal matrix composites (MMCs) is a crucial aspect of their design, influencing mechanical properties, thermal behaviour, and the overall performance of the composite.

9.2.2.1 Reinforcement Distribution

A uniform distribution of reinforcement materials, such as particles or fibres, is essential for consistent mechanical properties.

9.2.2.2 Reinforcement Size and Shape

The size and shape of reinforcement materials significantly impact the composite's strength and toughness. Smaller particles generally lead to higher strength and improved wear resistance, while larger particles can increase the ductility of the composite.

9.2.2.3 Matrix Microstructure

The microstructure of the metal matrix plays a pivotal role in determining the composite's overall characteristics. Factors such as grain size, grain boundary distribution, and phase composition need careful control.

9.2.2.4 Interfacial Bonding

The strength of the bond between the reinforcement and the matrix is critical for the integrity of the composite. Proper bonding ensures efficient load transfer and prevents delamination or debonding.

9.2.2.5 Orientation of Reinforcements

For fibre-reinforced MMCs, the orientation of the fibres relative to the applied loads is crucial. Aligning the fibres in the direction of anticipated stresses can significantly increase strength and stiffness.

9.2.2.6 Incorporation of Smart Materials

In Intelligent MMCs, the integration of smart materials requires precise control over microstructural features [5].

9.2.3 Integration of Functional Components

The integration of functional components in metal matrix composites (MMCs) introduces an additional layer of complexity to traditional composite design, where these components contribute to enhanced functionalities such as sensing, actuation, energy storage, and self-healing [6].

9.2.3.1 Types of Functional Components

Functional components in IMMCs can range from smart materials like piezoelectrics, shape memory alloys, and magnetostrictive materials to embedded sensors, actuators, or even microelectronics.

9.2.3.2 Compatibility with Metal Matrix

The chosen functional components must be compatible with the metal matrix in terms of chemical stability, thermal expansion, and mechanical properties. Incompatibility could lead to degradation, delamination, or compromised performance.

9.2.3.3 Manufacturing Techniques

Integrating functional components requires advanced manufacturing methods that ensure precise placement and secure bonding. Techniques like additive manufacturing, co-extrusion, or ultrasonic-assisted casting offer flexibility in embedding functional components within the matrix.

9.2.3.4 Integration Strategies

Sensors can be embedded within the matrix to monitor stress, temperature, or other parameters. These sensors can be piezoelectric, strain gauges, or fibre optics. Coatings on the surface or within specific layers of the composite can introduce additional functionalities like corrosion resistance, thermal management, or electromagnetic shielding [7].

9.3 STRATEGIES FOR SYNTHESIZING INTELLIGENT MMCs

9.3.1 Reinforcement Incorporation Techniques

Synthesizing intelligent metal matrix composites (IMMCs) involves the incorporation of reinforcement materials alongside smart components that offer additional functionalities like sensing, actuation, or self-healing [8].

9.3.1.1 Powder Metallurgy

This technique involves blending metal powders with reinforcement materials and then compacting and sintering them to form a solid composite. Powder metallurgy allows for uniform distribution of reinforcement particles and provides flexibility in incorporating various smart components like sensors or shape memory alloys.

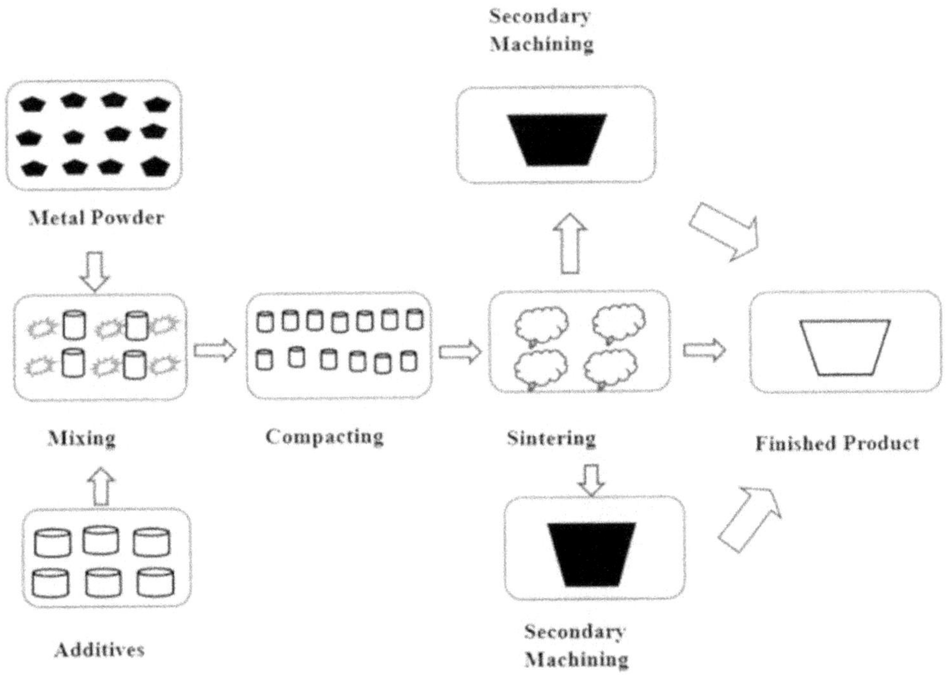

FIGURE 9.3 Basic mixing of powder and sintering process.

9.3.1.2 Stir Casting

In stir casting, reinforcement materials are mixed into molten metal with mechanical stirring to achieve a uniform distribution. This method is often used for particulate reinforcements like ceramics or metal particles.

9.3.1.3 Liquid Infiltration

This technique involves infiltrating a porous structure with molten metal, allowing for the creation of composites with embedded reinforcements like fibres or woven meshes. Liquid infiltration is particularly useful for achieving strong interfacial bonding between the metal matrix and reinforcements.

9.3.1.4 Additive Manufacturing

Additive manufacturing (3D printing) offers a high degree of design flexibility, enabling precise placement of reinforcements and functional components within the metal matrix.

9.3.1.5 Co-Extrusion

This approach allows for the controlled distribution of reinforcements within the metal matrix, facilitating the incorporation of smart components in specific regions of the composite. Co-extrusion is commonly used to create fibre-reinforced MMCs with oriented fibres.

9.3.1.6 Diffusion Bonding

Diffusion bonding involves joining metal layers at high temperatures and pressures, allowing for the incorporation of reinforcements and smart components within specific layers. This technique is useful for creating composites with multi-layered structures, where different layers serve distinct functions, such as reinforcement or intelligence.

9.3.1.7 Advanced Surface Treatments

Surface treatments, such as coatings or claddings, can enhance the properties of MMCs by adding layers of functional materials. This approach is useful for integrating smart components on the surface or near-surface regions of the composite, providing additional functionalities like corrosion resistance, wear resistance, or stress sensing [9].

9.3.2 Functional Coating and Surface Modification

Functional coatings and surface modifications are critical strategies for enhancing the performance of intelligent metal matrix composites [10].

9.3.2.1 Functional Coatings Varieties

Protects the metal matrix from oxidation and corrosion, extending the lifespan of the composite. Materials like ceramic coatings or specialized polymers are used to create a protective barrier. Improves resistance to abrasion and erosion, often achieved with hard ceramics or metal-carbide coatings.

9.3.2.2 Surface Modification Methods

Surface modification techniques alter the surface properties of IMMCs without adding additional layers. These methods can enhance properties like hardness, roughness, or adhesion.

9.3.2.3 Integration with Smart Components

Functional coatings and surface modifications can facilitate the integration of smart components into IMMCs.

9.3.3 Additive Manufacturing Approaches

Additive manufacturing (AM) has revolutionized the way metal matrix composites (MMCs) are designed and produced, offering unique advantages such as design flexibility, material efficiency, and the ability to create complex geometries.

9.3.3.1 Selective Laser Melting (SLM)

SLM uses a high-energy laser to selectively melt metal powder layer by layer, allowing for precise control over the composite's microstructure. This approach is ideal for IMMCs because it supports complex geometries and the integration of smart materials.

9.3.3.2 Electron Beam Melting (EBM)

EBM is similar to SLM but uses an electron beam to melt the metal powder. It offers advantages in terms of faster build times and deeper penetration, which can be beneficial for larger IMMC structures. EBM can integrate a wide range of materials, including those with higher melting points, making it suitable for aerospace and industrial applications.

9.3.3.3 Directed Energy Deposition (DED)

This method allows for material deposition in specific areas, offering flexibility in incorporating reinforcements or smart components. DED is useful for repairing or adding features to existing IMMC structures, as well as for creating complex shapes.

9.3.3.4 Binder Jetting

Binder jetting involves the selective deposition of a binding agent onto a powder bed, followed by post-processing steps like sintering. This technique allows for multi-material composites and can be used to embed smart components or sensors within IMMCs.

9.3.3.5 Material Extrusion

This method, often used in polymer-based 3D printing, has been adapted for metal composites. It extrudes a metal-containing filament to build structures layer by layer. While primarily used for simpler geometries, material extrusion can be useful for creating IMMCs with specific reinforcement patterns or embedded smart components [11].

9.3.4 Hybridization with Nanotechnology

Hybridization with nanotechnology involves incorporating nanoscale materials or structures into metal matrix composites (MMCs) to enhance their mechanical properties, introduce new functionalities, or improve performance characteristics. This approach is particularly valuable for intelligent MMCs (IMMCs), where the integration of nanoscale components can provide additional smart functionalities and improve overall composite behaviour [12].

9.4 INTELLIGENT FUNCTIONALITIES IN MMCs

9.4.1 Sensing and Monitoring Capabilities

Intelligent metal matrix composites (IMMCs) offer advanced sensing and monitoring capabilities, enabling real-time feedback on stress, strain, temperature, and other critical parameters. These capabilities are crucial in applications where structural integrity, safety, and predictive maintenance are paramount. IMMCs achieve sensing and monitoring functions by integrating smart materials, such as piezoelectric sensors, fibre optics, or strain gauges, into the metal matrix [13].

9.4.2 Self-Healing and Damage Detection

Self-healing and damage detection are key functionalities that enhance the longevity and safety of intelligent metal matrix composites (IMMCs). These composites incorporate mechanisms that enable them to detect damage and initiate repair processes autonomously, significantly reducing maintenance costs and preventing catastrophic failures.

9.4.2.1 Self-Healing Mechanisms

IMMCs achieve self-healing through a variety of strategies. One common approach involves embedding microcapsules or vascular networks filled with healing agents within the metal matrix. When the composite sustains damage, such as cracks or fractures, these microcapsules rupture, releasing the healing agents into the damaged area.

9.4.2.2 Damage Detection

IMMCs can detect damage through embedded sensors or smart materials. Piezoelectric sensors, fibre optic sensors, or strain gauges can be incorporated into the metal matrix to monitor stress, strain, or other changes in real time.

9.4.3 Adaptive Thermal Management

Adaptive thermal management is a crucial aspect of intelligent metal matrix composites (IMMCs), especially in applications where temperature control and heat dissipation are critical. By integrating materials and mechanisms that can dynamically respond to thermal changes, IMMCs can maintain optimal operating conditions, protect against thermal damage, and enhance the overall efficiency of systems.

9.4.3.1 Adaptive Thermal Management Approaches

IMMCs utilize various strategies to achieve adaptive thermal management. These approaches often involve integrating materials or devices that can modulate heat transfer or dissipate excess heat [13].

9.4.4 Enhanced Electrical Conductivity

Enhanced electrical conductivity in intelligent metal matrix composites (IMMCs) is achieved through the integration of materials with high electrical conductivity, such as copper, graphene, or carbon nanotubes (CNTs). This property is crucial for applications in electronics, aerospace, and automotive industries, where efficient electrical pathways and signal transmission are essential. By incorporating conductive inclusions, IMMCs can support embedded sensors, electrical circuits, or energy storage devices, enabling intelligent functionalities like real-time monitoring and adaptive responses.

9.4.5 Self-Lubricating and Wear Resistance

Self-lubricating and wear resistance are critical features for intelligent metal matrix composites (IMMCs), especially in applications where moving parts experience significant friction and wear. These features enhance the durability and longevity of IMMCs, reducing the need for external lubrication and frequent maintenance.

9.4.5.1 Self-Lubricating Mechanisms

IMMCs achieve self-lubricating properties by incorporating materials that naturally reduce friction or release lubricants during operation.

9.4.5.2 Wear Resistance

Wear resistance in IMMCs is achieved by selecting materials with high hardness and incorporating reinforcements that enhance resistance to abrasion and erosion [14].

9.5 CHARACTERIZATION TECHNIQUES

9.5.1 Microstructural Analysis

The microstructural analysis is crucial for understanding the properties, performance, and behaviour of metal matrix composites (MMCs). Various characterization techniques are used to examine the microstructure of these composites, providing insights into phase composition, grain structure, reinforcement distribution, and interfacial properties.

9.5.1.1 Optical Microscopy

Optical microscopy involves using visible light to examine the surface and cross-sections of MMCs at relatively low magnifications. This technique is useful for analyzing overall structure, grain boundaries, and large-scale inclusions or defects.

9.5.1.2 Scanning Electron Microscopy (SEM)

SEM provides high-resolution images of MMC surfaces, allowing for detailed analysis of microstructural features. It can magnify samples up to 100,000 times, revealing reinforcement particles, grain structure, and fracture surfaces.

9.5.1.3 Transmission Electron Microscopy (TEM)

TEM allows for high-resolution imaging at the nanoscale, providing insights into fine microstructural details such as dislocations, precipitates, and crystal defects. This technique is crucial for studying the interfaces between the metal matrix and reinforcements, as well as the effects of nano-reinforcements in MMCs.

9.5.1.4 X-Ray Diffraction (XRD)

XRD is used to identify the crystalline phases present in MMCs by analyzing the diffraction patterns of X-rays passing through the sample. This technique helps determine the structure, orientation, and stress within the crystalline lattice.

9.5.1.5 Electron Backscatter Diffraction (EBSD)

EBSD is a technique used to determine the crystallographic orientation of grains in MMCs. It is often used in conjunction with SEM to map grain boundaries and understand the texture and distribution of different phases. EBSD provides valuable information about grain orientation, which is crucial for predicting mechanical properties and behaviour under stress.

9.5.1.6 Atomic Force Microscopy (AFM)

AFM uses a mechanical probe to scan the surface of a sample, providing high-resolution topographical images. This technique can be used to measure surface roughness and characterize surface features, making it useful for analyzing coatings and surface treatments in MMCs.

9.5.1.7 Fourier-Transform Infrared Spectroscopy (FTIR)

FTIR is used to identify chemical bonds and functional groups within a composite. It is particularly useful for characterizing organic components or polymer-based reinforcements in MMCs, as well as assessing the chemical integrity of coatings or self-healing materials.

9.5.1.8 Thermogravimetric Analysis (TGA)

TGA measures changes in weight as a function of temperature, providing insights into the thermal stability and decomposition behaviour of MMCs. This technique is useful for assessing the thermal properties of composites and their reinforcements, particularly when evaluating coatings or organic-based components [15].

9.5.2 Mechanical Testing Methods

Mechanical testing methods are essential for evaluating the performance, strength, and durability of metal matrix composites (MMCs). These tests provide critical data on the composites' mechanical properties, helping engineers and researchers understand how MMCs respond to different loads, stresses, and environmental conditions [16].

9.5.2.1 Tensile Testing

Tensile testing measures the response of MMCs to uniaxial tension, providing data on ultimate tensile strength, yield strength, elongation, and Young's modulus.

9.5.2.2 Compression Testing

Compression testing evaluates the performance of MMCs under compressive loads. It is particularly important for applications where the composite will experience compressive forces, such as in structural components or load-bearing applications.

9.5.2.3 Flexural Testing

Flexural testing (also known as bending testing) examines the response of MMCs to bending stresses. This test is commonly used to determine flexural strength, flexural modulus, and the point of failure under bending loads.

9.5.2.4 Impact Testing

Impact testing assesses the toughness and impact resistance of MMCs by subjecting them to sudden, high-energy impacts. The Charpy and Izod tests are commonly used to measure the energy absorbed during impact and to understand the composite's resistance to fracture.

9.5.2.5 Fatigue Testing

Fatigue testing determines the endurance of MMCs under cyclic loading. It involves applying repeated stress or strain cycles to the composite to simulate long-term use and evaluate its fatigue life.

9.5.2.6 Hardness Testing

Hardness testing measures the resistance of MMCs to indentation, which correlates with their wear resistance and overall durability. Common methods include the Rockwell, Vickers, and Brinell tests [17].

9.5.3 Thermal and Electrical Property Evaluation

Evaluating the thermal and electrical properties of metal matrix composites (MMCs) is crucial for applications that involve temperature management, heat transfer, electrical conduction, or insulation.

9.5.3.1 Thermal Property Evaluation

Evaluating these properties involves various experimental techniques like differential scanning calorimetry (DSC), thermal conductivity measurements using techniques such as laser flash analysis or steady-state heat flow methods, dilatometry for CTE determination, and specific heat capacity measurements through calorimetry.

9.5.3.2 Thermal Conductivity Testing

Thermal conductivity measures a material's ability to conduct heat. It's crucial for applications where heat transfer or dissipation is a concern, such as electronics, aerospace, or automotive components.

9.5.3.3 Thermal Expansion Testing

Thermal expansion evaluates how much a material expands or contracts with temperature changes. Dilatometry is commonly used to measure thermal expansion by monitoring dimensional changes as a function of temperature.

9.5.3.4 Thermal Stability and Degradation

Thermogravimetric analysis (TGA) is used to assess the thermal stability and decomposition behaviour of MMCs. This method measures the change in weight as a function of temperature, indicating the temperatures at which degradation or phase changes occur.

9.5.3.5 Differential Scanning Calorimetry (DSC)

DSC measures the heat flow associated with temperature changes, providing information on phase transitions, melting points, and heat capacity. This technique is useful for evaluating the thermal properties of MMCs and their reinforcements.

9.5.3.6 Electrical Property Evaluation

Evaluating the electrical properties of materials is critical for various applications, from electronics and electrical engineering to material science and industrial applications. This evaluation helps in understanding how materials respond to electric fields, currents, and other electrical stimuli.

9.5.3.7 Electrical Conductivity Testing

Electrical conductivity measures a material's ability to conduct electrical current. This property is critical for applications where MMCs are used in electrical circuits or as conductive pathways. The four-point probe method is commonly used to measure electrical conductivity by passing a current through the material and measuring the voltage drop.

9.5.3.8 Electrical Resistance Testing

Electrical resistance is closely related to conductivity and assesses the opposition to current flow. Resistance testing can help identify defects, impurities, or variations in electrical pathways within MMCs. This information is crucial for applications requiring reliable electrical performance.

9.5.3.9 Dielectric Strength Testing

Dielectric strength evaluates a material's ability to withstand electrical stress without breaking down. This property is vital for MMCs used in insulation or protective applications. Techniques like dielectric breakdown testing apply increasing voltage until breakdown occurs, indicating the composite's resistance to electrical stress.

9.5.3.10 Electromagnetic Shielding Effectiveness

This test measures an MMC's ability to block or attenuate electromagnetic fields. It is important for applications where shielding from electromagnetic interference (EMI) or radiofrequency interference (RFI) is required [18].

9.5.4 In-Situ Monitoring Techniques

In-situ monitoring techniques provide real-time insights into the behaviour and performance of metal matrix composites (MMCs) during manufacturing and in-service conditions. These techniques allow for early detection of defects, stress, or structural changes, enabling proactive maintenance and quality control. Common in-situ monitoring methods include acoustic emission for detecting crack initiation, fibre optic sensors for strain and temperature measurement, and infrared thermography for thermal analysis [19].

9.6 APPLICATIONS OF INTELLIGENT MMCs

9.6.1 Aerospace and Defence

Intelligent metal matrix composites (IMMCs) are revolutionizing aerospace and defence by offering lightweight, high-strength materials with integrated smart functionalities. These advanced composites can withstand the harsh conditions of aerospace environments while providing real-time sensing, monitoring, and adaptive responses. IMMCs with embedded sensors and actuators enable continuous health monitoring, reducing the risk of catastrophic failure and lowering maintenance costs. Self-healing mechanisms in IMMCs can autonomously repair minor damage, enhancing durability and extending the life of critical components [20].

9.6.2 Automotive Industry

Intelligent metal matrix composites (IMMCs) are transforming the automotive industry by providing a combination of strength, lightweight properties, and smart functionalities. In a sector where fuel efficiency, safety, and performance are paramount, IMMCs offer significant benefits. One key application is in vehicle chassis and structural components. IMMCs' high strength-to-weight ratio helps reduce vehicle weight, improving fuel efficiency and reducing emissions. These composites also offer excellent crashworthiness, contributing to enhanced safety in case of collisions [21].

9.6.3 Electronics and Semiconductor Devices

Intelligent metal matrix composites (IMMCs) are playing an increasingly important role in electronics and semiconductor devices. With the growing demand for compact, high-performance electronics, IMMCs offer a unique combination of thermal conductivity, mechanical strength, and electrical properties that make them ideal for this sector. In electronics, IMMCs are used for heat sinks, circuit board substrates, and component housings, where efficient heat dissipation is crucial. Overall, the versatility and performance of IMMCs in electronics and semiconductor applications contribute to the development of more efficient, reliable, and robust devices, addressing the challenges of modern technology and its increasingly demanding requirements [22].

9.6.4 Renewable Energy Systems

Intelligent metal matrix composites (IMMCs) are increasingly vital in renewable energy systems due to their unique properties, which contribute to efficiency, durability, and sustainability. As the demand for renewable energy sources grows, IMMCs offer solutions to address specific challenges in this sector. In wind energy, IMMCs are used in turbine blades, gearboxes, and tower structures, where their high strength-to-weight ratio and fatigue resistance are essential [23].

9.6.5 Biomedical Implants

Intelligent metal matrix composites (IMMCs) are gaining traction in the field of biomedical implants, offering significant benefits in terms of strength, biocompatibility, and multifunctional properties [24].

9.6.6 Structural Engineering

Intelligent metal matrix composites (IMMCs) are transforming the field of structural engineering, offering innovative solutions for constructing more resilient, durable, and adaptable structures. With increasing demands for sustainable and efficient construction, IMMCs provide unique benefits in terms of strength, versatility, and integrated intelligence [25].

9.7 FUTURE PERSPECTIVES AND CHALLENGES

9.7.1 Emerging Trends in Intelligent MMCs

Intelligent metal matrix composites (IMMCs) are experiencing rapid advancements and innovations, driven by the growing demand for smarter, more adaptive materials across various industries [26].

9.7.1.1 Integration of Advanced Nanomaterials

Nanotechnology is playing a pivotal role in enhancing the properties of IMMCs. The incorporation of graphene, carbon nanotubes, and other nanomaterials into metal matrices has resulted in composites with improved electrical conductivity, thermal stability, and mechanical strength.

9.7.1.2 Smart Sensing and Monitoring

The integration of embedded sensors into IMMCs is becoming more sophisticated, allowing for real-time monitoring of structural health, stress, and environmental conditions. Advanced data analytics and IoT connectivity are enabling more intelligent and automated systems.

9.7.1.3 Self-Healing and Adaptive Materials

IMMCs with self-healing capabilities are gaining momentum, with various approaches to autonomous repair of minor damage. This trend is particularly beneficial for high-stress applications, such as aerospace and structural engineering, where self-healing properties can extend the lifespan of components and reduce maintenance costs.

9.7.1.4 Additive Manufacturing and 3D Printing

Additive manufacturing (AM) techniques, including 3D printing, are expanding the possibilities for IMMCs. The ability to create complex geometries and custom designs using AM allows for more efficient use of materials and tailored properties.

9.7.1.5 Sustainability and Eco-Friendly Materials

Sustainability is a growing concern in materials engineering, leading to the development of eco-friendly IMMCs. Researchers are exploring ways to use recycled materials, reduce energy consumption during manufacturing, and create composites that are easier to recycle at the end of their lifecycle.

9.7.1.6 Multifunctional Composites

IMMCs are increasingly designed to perform multiple functions, such as structural support, electrical conductivity, thermal management, and sensing. Multifunctional IMMCs allow for the integration of various components into a single composite, reducing weight and complexity while enhancing performance [27].

9.7.2 Technological and Economic Barriers

Despite the promising advances in intelligent metal matrix composites (IMMCs), several technological and economic barriers need to be addressed to fully realize their potential [28].

9.7.2.1 Complex Manufacturing Processes

The integration of intelligent functionalities into MMCs requires complex manufacturing processes, including precise control over material composition, microstructural design, and embedded sensor integration. Additive manufacturing and other advanced techniques offer solutions, but they can be expensive and require specialized equipment.

9.7.2.2 Quality and Reliability

Ensuring the consistent quality and reliability of IMMCs is a significant challenge. Variations in material properties, reinforcement distribution, and sensor functionality can affect the performance and durability of the composites. Developing robust quality control measures is essential to maintain high standards.

9.7.2.3 Interface and Compatibility Issues

The interfaces between the metal matrix and reinforcements, as well as between the composites and integrated components like sensors, pose potential compatibility issues. Poor bonding or mismatched thermal expansion can lead to delamination or other defects, compromising the structural integrity of IMMCs.

9.7.2.4 Limited Material Options

While certain metals and alloys are commonly used in MMCs, the range of materials suitable for intelligent applications is relatively limited. Finding new metal matrices and reinforcement materials that meet specific requirements for strength, weight, and conductivity is a challenge.

9.7.2.5 High Production Costs

The advanced manufacturing techniques and specialized materials used in IMMCs contribute to higher production costs compared to traditional composites. This factor can be a barrier to widespread adoption, especially in cost-sensitive industries like automotive and construction.

9.7.2.6 Research and Development Expenses

Developing and testing new IMMCs involves significant research and development expenses. The need for specialized facilities, equipment, and skilled personnel increases the costs associated with innovation in this field.

9.7.2.7 Market Acceptance and Standardization

Gaining market acceptance for new technologies like IMMCs requires industry standards and certification processes. The lack of established standards for IMMCs can slow down their adoption, as manufacturers and end-users may hesitate to invest in technology without clear guidelines for safety and performance.

9.7.2.8 Return on Investment

Given the higher initial costs of IMMCs, demonstrating a clear return on investment is crucial for market acceptance. This requires thorough analysis of the long-term benefits, such as reduced maintenance costs, extended lifespan, and improved performance, to justify the investment [29].

9.7.3 Environmental and Regulatory Considerations

As the development and adoption of intelligent metal matrix composites (IMMCs) increase, environmental and regulatory considerations play a critical role in shaping their production and use. Here are some of the key aspects to consider [30].

9.7.3.1 Sustainability and Resource Efficiency

The environmental impact of IMMCs can be significant, depending on the raw materials and manufacturing processes involved. Sustainable practices, such as the use of recycled materials, energy-efficient production methods, and reduced waste, are essential to minimize the environmental footprint.

9.7.3.2 End-of-Life and Recyclability

The recyclability of IMMCs is a crucial consideration, particularly as composite materials often contain a mix of metal and reinforcement. Designing IMMCs with end-of-life recycling in mind can reduce waste and support a circular economy. Efficient separation and recycling processes are needed to recover valuable materials from discarded composites.

9.7.3.3 Toxicity and Safety

Some reinforcements and coatings used in IMMCs may pose health or environmental risks. The use of non-toxic materials and environmentally friendly coatings can mitigate these risks.

9.7.3.4 Compliance with Industry Standards

IMMCs must comply with various industry standards and regulations, which govern the quality, safety, and performance of materials. These standards vary by industry and region, and compliance ensures that IMMCs meet the required benchmarks for use in critical applications such as aerospace, automotive, and construction.

9.7.3.5 Environmental Regulations

Environmental regulations can impact the production and use of IMMCs. Restrictions on hazardous substances, emissions, and waste management require manufacturers to adopt environmentally friendly practices. Compliance with these regulations is essential to avoid penalties and ensure responsible manufacturing processes.

9.7.3.6 Health and Safety Regulations

The safety of IMMCs, particularly in terms of toxicity and exposure risks, is governed by health and safety regulations. This includes proper labelling, safety data sheets, and adherence to workplace safety regulations.

9.7.3.7 Intellectual Property and Patents

As IMMCs continue to evolve, intellectual property and patent considerations become increasingly important. Manufacturers and researchers must navigate the regulatory landscape to protect their innovations while ensuring that their products do not infringe on existing patents [31].

9.8 CONCLUSION

Intelligent metal matrix composites (IMMCs) represent a remarkable advancement in materials science, offering a unique combination of high strength, lightweight design, and intelligent functionalities. As we've explored throughout this discussion, IMMCs have found applications in a wide range of industries, from aerospace and defence to electronics, biomedical implants, renewable energy, and structural engineering. Their ability to integrate sensing, self-healing, adaptive thermal management, and other smart features is driving innovation and enhancing performance in critical applications. Despite these advantages, several challenges must be addressed to fully realize the potential of IMMCs. Technological and economic barriers, including complex manufacturing processes and high production costs, can hinder widespread adoption. Ultimately, the success of IMMCs depends on collaboration among researchers, manufacturers, and industry stakeholders. By working together to develop innovative solutions, establish industry standards, and promote sustainability, we can unlock the full potential of intelligent metal matrix composites, leading to smarter, more resilient, and more sustainable engineering solutions across a variety of sectors.

REFERENCES

[1] Abebe Emiru, A., Sinha, D.K., Kumar, A. and Yadav, A., 2023. Fabrication and characterization of hybrid aluminum (Al6061) metal matrix composite reinforced with SiC, B_4C, and MoS_2 via stir casting. *International Journal of Metalcasting*, *17*(2), pp. 801–812.

[2] Ananth, G., Muthu Krishnan, T., Thirugnanam, S. and Tariku Olkeba, T., 2023. Optimization on stir casting process parameters of Al7050/nano-B_4C metal matrix composites. *Journal of Nanomaterials*, *2023*, pp. 1–7.

[3] Angadi, S.B., Nagaral, M., Pilankar, V., Yarnalkar, P., Purohit, B. and Shewale, S., 2023. Assessment of mechanical behaviour and microstructure of micro boron carbide particles reinforced Al2011 alloy metal matrix composite for aerospace applications. *International Journal of Vehicle Structures & Systems*, *15*(2).

[4] Arunachalam, J., Sathish, T., Makki, E. and Giri, J., 2024. Influence of machining process of MoS2/B4C/Az31 Mg alloy composite and its tribological characteristics. *AIP Advances*, *14*, p. 035342.

[5] Asif, M., Chandra, K. and Misra, P.S., 2011. Development of aluminium based hybrid metal matrix composites for heavy duty applications. *Journal of Minerals & Materials Characterization & Engineering*, *10*(14), pp. 1337–1344.

[6] Bharathi, P. and Kumar, T.S., 2023. Mechanical characteristics and wear behaviour of Al/SiC and Al/SiC/B4C hybrid metal matrix composites fabricated through powder metallurgy route. *Silicon*, *15*(10), pp. 4259–4275.

[7] Chaudhary, B., Patel, M., Jain, N.K., Murugesan, J. and Patel, V., 2023. Friction stir powder additive manufacturing of Al 6061/FeCoNi and Al 6061/Ni metal matrix composites: Reinforcement distribution, microstructure, residual stresses, and mechanical properties. *Journal of Materials Processing Technology*, *319*, p. 118061.

[8] Chauhan, A., Vates, U.K., Kanu, N.J., Gupta, E., Singh, G.K., Sharma, B.P. and Gorrepati, S.R., 2021. Fabrication and characterization of novel nitinol particulate reinforced aluminium alloy metal matrix composites (NiTip/AA6061 MMCs). *Materials Today: Proceedings*, *38*, pp. 3027–3034.

[9] Cramer, C.L., Cakmak, E. and Unocic, K.A., 2023. Hardness measurements and interface behavior of SiC-B_4C-Si multiple phase particulate composites made with melt infiltration and additive manufacturing. *Journal of Composites Science*, *7*(4), p. 172.

[10] Dewangan, S., Bohra, P.M. and Sharma, P., 2024. Mechanical properties and microstructure analysis of Al-6061 metal matrix prepared with alumina and silicon carbide powders. *Journal of the Institution of Engineers (India): Series D*, pp. 1–11.

[11] Gafur, M.A., Abrar, R. and Soshi, S.S., 2023. Development and characterization of aluminium-based metal matrix composites. *Materials Sciences and Applications*, *14*(1), pp. 1–19.

[12] Gaylan, Y., Avar, B., Panigrahi, M., Aygün, B. and Karabulut, A., 2023. Effect of the B_4C content on microstructure, microhardness, corrosion, and neutron shielding properties of Al–B_4C composites. *Ceramics International*, *49*(3), pp. 5479–5488.

[13] Göçer, A. and Karamis, M.B., 2024. Production and mechanical characterization of steel/Al-B_4C layered circular hybrid composite materials. *Arabian Journal for Science and Engineering*, pp. 1–19.

[14] Hima Gireesh, C., Durga Prasad, K.G. and Ramji, K., 2018. Experimental investigation on mechanical properties of an Al6061 hybrid metal matrix composite. *Journal of Composites Science*, *2*(3), p. 49.

[15] Jayappa, K., Kumar, V. and Purushotham, G.G., 2021. Effect of reinforcements on mechanical properties of nickel alloy hybrid metal matrix composites processed by sand mold technique. *Applied Science and Engineering Progress*, *14*(1), pp. 44–51.

[16] Karabulut, Ş., Karakoç, H., Bilgin, M., Canpolat, H., Krolczyk, G.M. and Sarıkaya, M., 2023. A comparative study on mechanical and ballistic performance of functionally graded Al6061 composites reinforced with B_4C, SiC, and Al_2O_3. *Journal of Materials Research and Technology*, *23*, pp. 5050–5065.

[17] Kumar, C., Sarkar, S., Mukhopadhyay, G., Chakraborti, P.C., Sen, I. and Roy, S., 2023. Systematic study of the effect of K2TiF6 flux content on the microstructure and mechanical properties of Al–B_4C composites. *Materials Science and Engineering: A*, *871*, p. 144913.

[18] Lal, C., Dhiman, B., Tejyan, S. and Singh, V., 2024. Aluminum 6061 alloy and metal matrix composites: A bibliometric analysis (1973–2023). *Materials Today: Proceedings*. https://doi.org/10.1016/j.matpr.2024.04.018

[19] Lee, T., Jeong, W., Chung, S., Kim, J.J., Lee, D., Kim, J. and Ryu, H.J., 2023. Role of B_4C in the directed energy deposition of aluminum alloy 6061/B_4C composite using core-shell powder for crack annihilation, pore mitigation, and grain refinement. *Additive Manufacturing*, *72*, p. 103622.

[20] Mydur, K., Kumar, M., Nagaral, M., Auradi, V., Bharath, V. and Sudarshan, T.A., 2023. Microstructure, physical, tensile and wear behaviour of B_4C particles reinforced Al7010 alloy composites. *Manufacturing Review*, *10*, p. 3.

[21] Parikh, V.K., Badheka, V.J., Badgujar, A.D. and Ghetiya, N.D., 2021. Fabrication and processing of aluminum alloy metal matrix composites. *Materials and Manufacturing Processes*, *36*(14), pp. 1604–1617.

[22] Raj, P.P., Sridhar, R., Pugazhenthi, R., Anbuchezhiyan, G. and Ganesh, M., 2023. Investigating the mechanical properties of tungsten carbide metal matrix composites with Al 6061. *Materials Today: Proceedings*. https://doi.org/10.1016/j.matpr.2023.03.605

[23] Raksha, M.S., Nagaral, M., Boppana, S.B., Anjinappa, C., Khan, M.S., Wahab, M.O.A., Islam, S., Bhardwaj, V., Palavalasa, R.K., Khan, M.A. and Razak, A., 2023. Impact of boron carbide particles and weight percentage on the mechanical and wear characterization of Al2011 alloy metal composites. *ACS Omega*, *8*(26), pp. 23763–23771.

[24] Ramadoss, N., Pazhanivel, K., Ganeshkumar, A. and Arivanandhan, M., 2023. Microstructural, mechanical and corrosion behaviour of B4C/BN-reinforced Al7075 matrix hybrid composites. *International Journal of Metalcasting*, *17*(1), pp. 499–514.

[25] Ranjitha, P., Bavan, D.S., Ajaykumar, B.S., Raju, T.H. and Udayashankar, S., 2024. Investigation of mechanical properties of Al6061–SiC–B_4C composites produced by using stir casting method. *Journal of the Institution of Engineers (India): Series D*, pp. 1–13.

[26] Senthilnathan, T. and Balachandar, K., 2023. Mechanical and microstructure evaluation of novel hybrid TiB2/B_4C aluminium metal matrix composite. *Materials Research Express*, *10*(6), p. 066509.

[27] Singh, B., Kumar, I., Saxena, K.K., Mohammed, K.A., Khan, M.I., Moussa, S.B. and Abdullaev, S.S., 2023. A future prospects and current scenario of aluminium metal matrix composites characteristics. *Alexandria Engineering Journal*, *76*, pp. 1–17.

[28] Sivalingam, S., Kumaresan, G., Ganesh, P. and Azhagan, M.T., 2023. Superplastic formability and cavitation analysis of AA7075 matrix composite reinforced with B_4C particles produced by stir casting. *International Journal of Metalcasting*, *17*(4), pp. 2917–2927.

[29] Tasci, U., Yilmaz, T.A. and Bostan, B., 2023. Investigation of microstructure, wear and transverse rupture strength of WE43/nano B_4C composites produced by powder metallurgy method. *Tribology International*, *180*, p. 108231.

[30] Yi, J., Zhang, X., Liu, G., Rao, J.H. and Liu, H., 2023. Microstructure and dynamic microhardness of additively manufactured (TiB2+TiC)/AlSi10Mg composites with AlSi10Mg and B_4C coated Ti powder. *Journal of Alloys and Compounds*, *939*, p. 168718.

[31] Yunus, M. and Alfattani, R., 2023. Assessment of mechanical and tribological behavior of AA6061 reinforced with B_4C and Gr hybrid metal matrix composites. *Coatings*, *13*(9), p. 1653.

10 Investigation of the Effects of Process Parameters on Metal Matrix Composites of A356 with SiC on Mechanical and Tribological Characteristics

Padmanabhan Sambandam, Giridharan Krishnan, Deepak James Raj, Vinod Kumar Thangaraj, and Chakravarthi Gurijala

10.1 INTRODUCTION

The assessment of emerging trends and the promotion of innovation is an ongoing endeavour aimed at creating cutting-edge materials that satisfy the needs of various strategic applications. Functionally graded materials (FGMs) and composites are exemplary instances of advanced materials. Engineers are increasingly prioritizing the study and development of FGMs and composites. These materials are being recognized as a novel category of advanced materials that provide dependable and high-performance solutions. In recent years, the transportation and aerospace industries have increasingly utilized these composite materials due to their excellent mechanical and thermophysical qualities, both in structural and non-structural applications [1, 2].

Many commercially used particle or discontinuous reinforced metal matrix composites are mass-produced using the well-respected stir casting process. The ease, adaptability, and low cost of this method make it a promising contender for the fabrication of large-scale components. The high production cost of even components with minimally complicated geometries has prevented particle metal matrix composites from being used widely in engineering. Although casting technique is currently coupled with a number of technological hurdles, it does have the potential to offer a solution to this difficulty [3].

Because of its versatile mechanical properties, metal matrix composites are gaining interest for use in various structural applications. These include the aerospace, transportation, military, and sports industries, among others [4]. Using high-strength aluminium alloys or aluminium metal composites, fabricating components from sheet metal, coming up with new designs for parts, using additive manufacturing, and reducing friction on moving parts are all ways to make engines lighter. A metal matrix with scattered non-metal reinforcement is the building block of a metal composite. Metal composites combine the best qualities of metal and reinforcing material to create a material with superior qualities than either component could deliver on its own. Transforming aluminium and other lightweight technical materials into reasonably strong ones makes them ideal for use in the manufacture of engine parts and other automotive components. Some examples of common reinforcing materials are silicon carbide (SiC), titanium dioxide (TiO_2), and aluminium oxide (Al_2O_3) [5].

 DOI: 10.1201/9781003499718-10

An investigation was conducted to examine the effect of stirring duration and speed on the strengthening of silicon carbides in A384 aluminium alloy during the stir casting process. An alloy was infused with a solution containing 10 weight percent silicon carbide, with an average particle size of 60 micrometres. The results indicate that the addition of hard SiC particles to the aluminium alloy increased the hardness of the composite. Micrographs demonstrated the occurrence of particle clustering at lower velocities and shorter stirring durations. The study found that higher stirring speed and longer stirring time resulted in a more uniform dispersion of SiC particles [6]. The casting technique was used to augment the aluminium 6061 alloy with silicon particles ranging from 0 to 4 wt% in a related work on metal matrix composites. The findings demonstrated a positive correlation between an increased weight percentages of SiC elements and enhanced mechanical properties [7].

Because of their increased hardness, resistance to oxidation at high temperatures, and better wear resistance, composites that contain hard oxides, such silicon carbide, are preferred. Taking into account the smallest friction coefficient, the results show that the matrix alloy wears out faster than the composite materials. It also improves micro hardness and tensile strength, which is a nice bonus. Wear is minimized and density is nearly three times lower than grey cast iron thanks to the addition of basalt and fly ash [8]. The integration of SiC lead to an increase in the density of the composites, making them denser than pure Al 6061. The incorporation of SiC in the Al6061 composite resulted in a 35% increase in hardness and this indicates that the addition of additive has enhanced the hardness. The ultimate tensile strength of the 2.5% SiC composite exhibited a 7% improvement, reaching 157 MPa. In contrast, the 5% SiC composite had a 47% loss in performance, measuring 76.5 MPa. The energy absorption of a 5% SiC composite decreases by 47% and fails at a lower load as the percentage of reinforcement increases [9].

This study examines the effects of slight variations in the SiC particle reinforcement on the AZ91. The study found that SiC reinforcement led to a substantial increase in resistance, along with a slight improvement in tensile and yield strength. As the wt% of SiC particles rose, the composites based on magnesium alloy showed an increase in density and porosity [10]. The study establish that the use of SiC led to enhanced wear resistance and hardness of the aluminium MMCs. The stir casting method was engaged to obtain effective dispersion of SiC particles within the aluminium matrix. The findings suggest that the produced composites have the ability to be used in situations that need enhanced wear resistance and hardness properties [11].

SiC reinforcement and the stiff Al_3Ni phase enhanced the wear characteristics of the aluminium nickel composite and increase linearly with the concentration of reinforcing particles. The composite material containing 16 wt. % SiC has a compressive strength of 732.48 MPa and a hardness of 49 HV. An instance of elevated wear rate can be observed in a composite material containing 4% SiC and subjected to a 20 N strain at each frequency. By examining a damaged surface, one can determine whether an adhesive, abrasive, or oxidative wear mechanism is at work [12].

It has been demonstrated through the results of the experiments that the incorporation of SiC into the composites that were produced led to an enhancement in their vital properties. In addition, silicon carbide matrix demonstrated improved wear property in comparison to the alloy, which led to improvements in tribological parameters [13]. The addition of SiC particles, at weight percentages of 5% and 10%, to the Al2014 alloy led to improvements in its yield, compression, and ultimate strength. An inclusion of ten weight percentage of SiC to the Al2014 alloy resulted in significant improvements in mechanical properties. Specifically, there were sequential increases of 15.96%, 35.2%, 43.8%, and 38.7% respectively [14]. The study centres on producing A356 alloy and SiC composites by various processing procedures, while also altering the volume fractions of SiC particles by increments of 0.25%, 0.50%, and 0.75%. The efficiency of SiC reinforcement is assessed by systematically evaluating vital metal properties such as tensile strength (ASTM E8) and hardness (ASTM E10). In addition, the composites' resistance to wear is examined under various loading situations and sliding velocities.

10.2 MATERIALS AND METHODS

10.2.1 Silicon Carbide

Composite materials made from aluminium alloys reinforced with silicon carbide (SiC) have several advantages. The yield, tensile strength of the aluminium matrix were greatly improved with the presence of SiC particles, which function as powerful strengthening agents within the matrix. Reinforcing with SiC makes the composite stiffer than with aluminium because of SiC's higher modulus of elasticity; this makes it ideal for uses where structural integrity and dimensional stability are critical [15]. Furthermore, SiC's remarkable wear resistance makes it perfect for components that are exposed to abrasive environments, such parts in automobiles and aircraft. In addition, the addition of SiC reinforcement can improve the composite's thermal conductivity, leading to better heat dissipation. This is particularly important for heat sinks and electronic packaging, which both rely on efficient thermal management. Despite these benefits, optimizing the performance of Al-SiC composites and mitigating potential downsides requires addressing issues such as regulating processing parameters and establishing uniform dispersion of SiC particles [16].

Analyzing a scanning electron microscopy (SEM) image of silicon carbide yields useful information about its surface structure and microscopic characteristics in Figure 10.1. The scanning electron microscope image provides a visual representation of the surface features, including grain boundaries, surface roughness, and any deviations from a regular pattern. Silicon carbide commonly exhibits a polycrystalline structure, which is clearly discernible from the arrangement, size, and orientation of the grains depicted in the image. This research provides significant insights into the distribution of grain sizes, configurations of grain boundaries, and the overall texture of the material's surface. In addition, SEM can be used to identify particle agglomerates, analyze pore distribution, and detect surface defects such as cracks or contaminants.

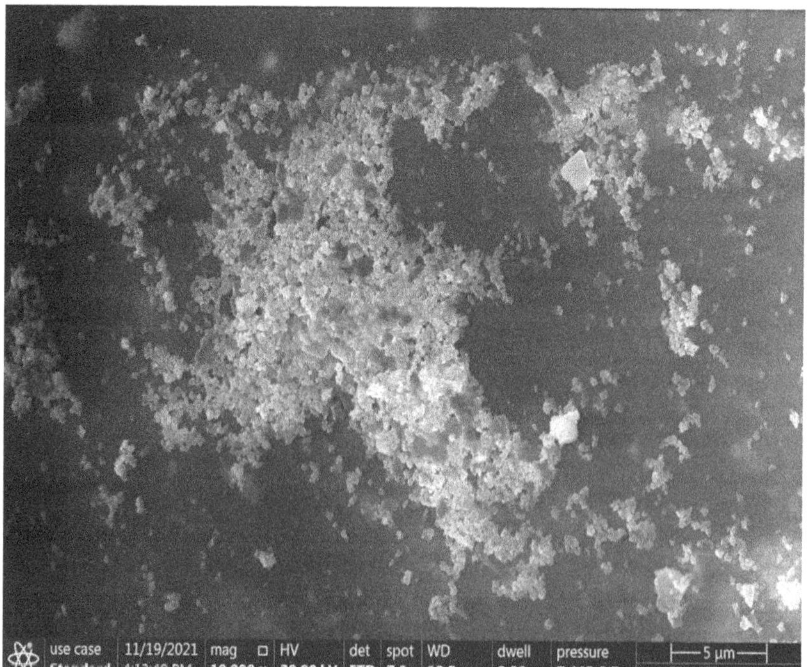

FIGURE 10.1 SEM image of SiC.

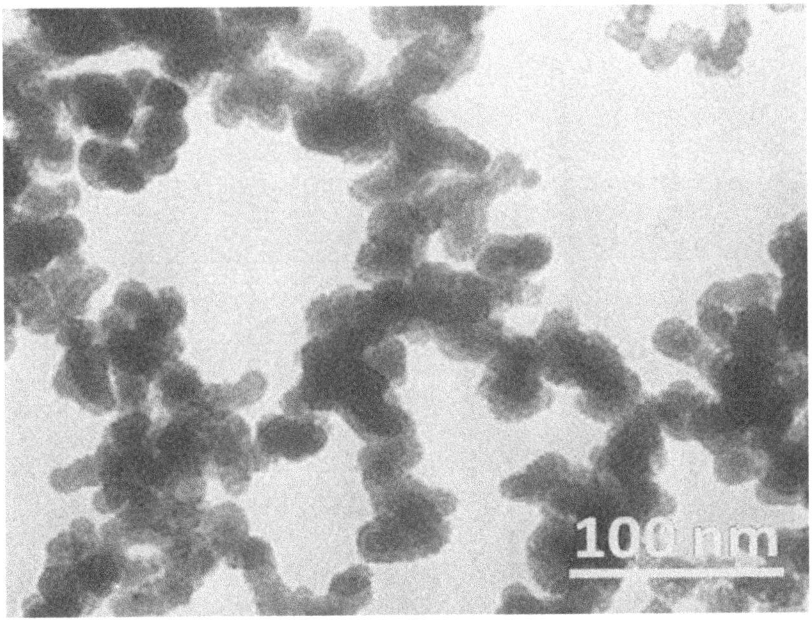

FIGURE 10.2 TEM image of SiC.

Analyzing a transmission electron microscopy (TEM) image of silicon carbide yields vital information about its microstructure and imperfections in Figure 10.2. The image commonly displays the hexagonal crystal structure of SiC, which is distinguished by alternating layers of silicon and carbon atoms. These layers often manifest as recurring patterns across the image. Grain boundaries, which occur when there is a shift in crystal orientations between neighbouring grains, might appear as narrow lines or areas with differing levels of contrast. Dislocations, which can be observed as lines or loops, indicate aberrations in the crystal lattice structure. Stacking faults, which are aberrations from standard stacking sequences, are observable as differences in contrast. Additional imperfections such as point defects, vacancies, or impurity atoms can also be observed, affecting the characteristics of the material.

The crystal structure of silicon carbide has tetrahedra of silicon and carbon atoms bound together strongly. The result is a substance that is both tough and durable. Up to 800°C, no acid, alkali, or molten salt can break down silicon carbide. Because of its great strength, moderate thermal expansion, and strong thermal conductivity, this material is very resistant to thermal shock. The material's capacity to hold strength even at higher temperatures, resistance to thermal chemical assault, and chemical purity have made it an incredibly attractive reinforced material, according to researchers [17].

10.2.2 Silicon Carbide Reinforcement

The process of mechanical stir casting involves mixing a dispersed phase with a molten matrix metal through mechanical stirring. This liquid state approach is used to fabricate composite materials. The aluminium metal matrix is reinforced with three different percentages of SiC, 0.25%, 0.5%, and 0.75% in volume fractions (Figure 10.3). Reinforcing the metal matrix are silicon carbide with a usual size of 30 micrometres. It takes one hour to warm silicon carbide to 500 K before adding it to the melt. In order to keep the compounds from reacting too much, the furnace is kept at a constant temperature of around 700°C. A connection between the furnace's relay and the thermocouple

FIGURE 10.3 Samples of A356-SiC.

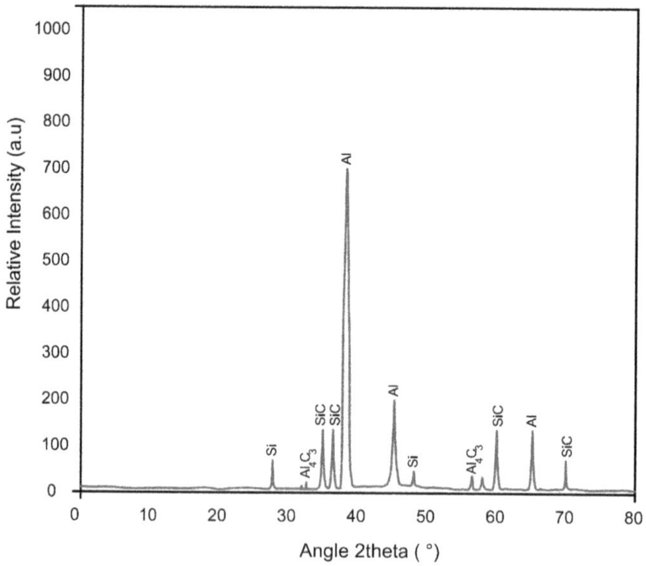

FIGURE 10.4 XRD reinforcement A356–0.5%SiC.

allows for precise temperature regulation. The relay's job is to turn off the electricity when the temperature rises over 700°C. To ensure that the silicon carbide particles in the matrix alloy are evenly distributed, mechanical stirring is employed. Stir the mixture at 650 rpm for 10 minutes after adding the preheated reinforcing particles. The mixing process is prolonged until the composite enters the mushy zone [18].

The current study employed X-ray diffraction (XRD) to determine the phases contained in the metal matrix composite. Figure 10.4 displays XRD graphs illustrating the relationship amongst the

diffraction angle (measured in 2-theta degrees) and relative intensity for a sample produced using mechanical stir casting. The figure clearly depicts the distinct peaks that correspond to SiC and Al, respectively. The data demonstrates that the peak values associated with SiC are very low, suggesting a low percentage of SiC in the spreading inside the metal matrix alloy. A limited number of Al_4C_3 peaks were detected in the stir casting material.

10.3 RESULTS AND DISCUSSION

10.3.1 Study on Hardness

The incorporation of silicon carbide reinforcement into A356 aluminium matrix composites leads to a notable improvement in hardness owing to several underlying mechanisms. The higher percentage of Al MMC could be influencing the increased hardness level. In this case, A356 + 0.75% (Al MMC) has a greater quantity, suggesting that the percentage of Al MMC may directly correlate with the hardness of the material as shown in Figure 10.5. The hardness of A356 with 0.75% aluminium metal matrix composite (MMC) is significantly higher than that of A356 with 0.5% aluminium MMC and other materials. First, SiC particles are significantly harder than the aluminium matrix, acting as effective obstacles to dislocation movement within the material. This hindrance to dislocation motion increases the material's resistance to plastic deformation, resulting in enhanced hardness. Additionally, the presence of SiC particles creates a complex microstructure within the composite, which contributes to increased hardness through mechanisms such as solid solution strengthening and grain refinement. The arrangement of SiC particles within the aluminium matrix is also a critical factor in determining hardness. A more even distribution results in more consistent mechanical characteristics across the material. Moreover, the bonding at the interface between the silicon carbide and the aluminium matrix enables the transmission of loads, resulting in a more effective distribution of stresses and further improving hardness [9–11].

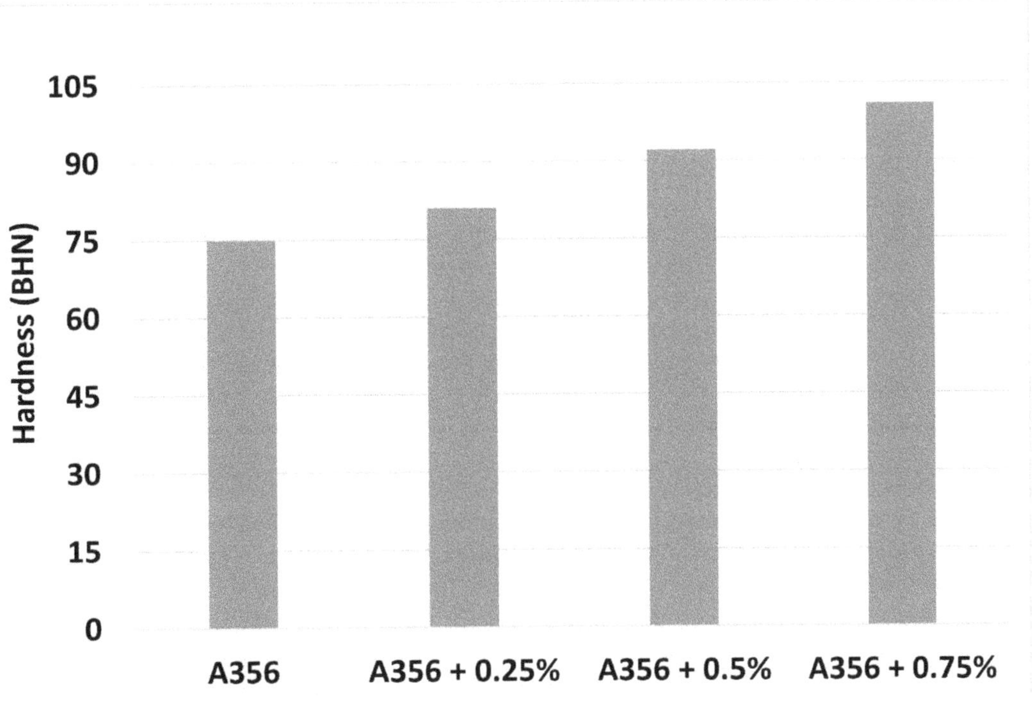

FIGURE 10.5 Study on hardness A356-SiC.

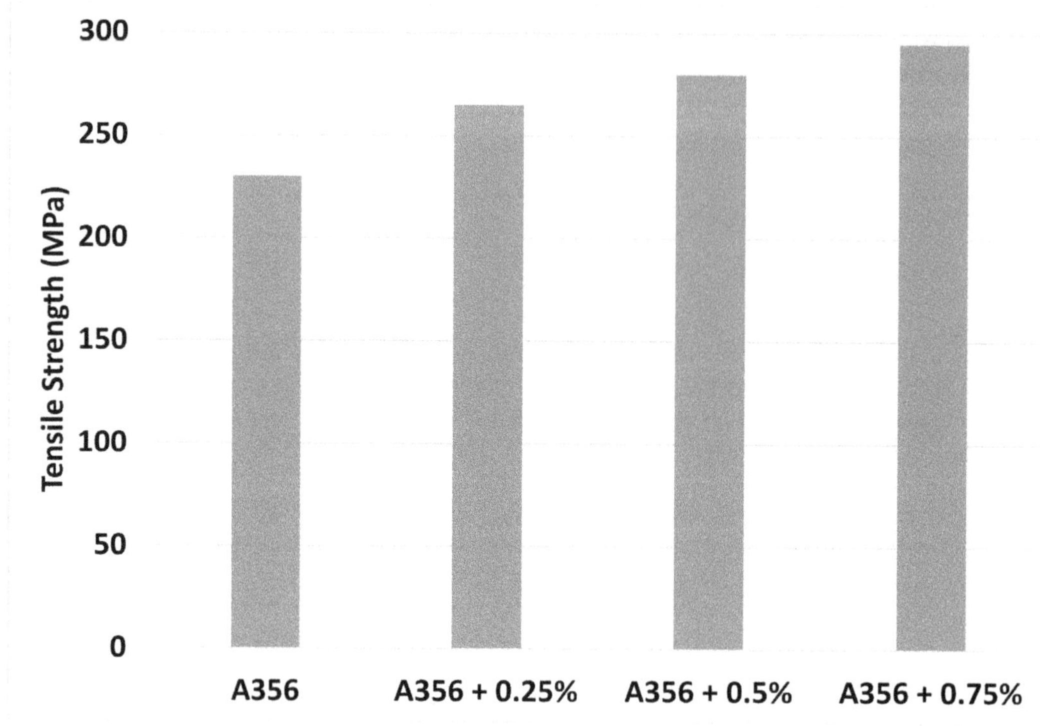

FIGURE 10.6 Study on tensile strength A356-SiC.

10.3.2 Study on Tensile Strength

Adding silicon carbide reinforcement to A356 aluminium alloy significantly enhances its tensile strength in metal matrix composites for various reasons as shown in Figure 10.6. First, the incorporation of SiC particles into the aluminium matrix results in high intrinsic strength and stiffness. These particles effectively hinder the movement of dislocations during deformation. This hinders the spread of plastic deformation, thus improving the material's ability to bear loads. Additionally, the interaction amongst the SiC and the metal matrix enhances load transfer, resulting in a more effective distribution of stresses throughout the composite structure. By ensuring that the applied load is effectively distributed between the matrix and the reinforcement, the overall tensile strength is enhanced. Furthermore, the even dissemination of SiC particles in the aluminium composites, achieved through precise processing methods, guarantees uniform mechanical properties across the composite material [15, 19].

10.3.3 Study of Density and Porosity

The qualities of materials are significantly impacted by porosity and the types of pores. The relationship between porosity and the number of linked pores in a material is often negative. This means that thermal conductivity is going to be higher, but strength, water absorption, permeability, and frost resistance are all going to be better. The samples' porosity levels were ascertained by the density measurements. This was accomplished by contrasting the theoretical and experimental densities of the various SiC reinforced composites with respect to volume percent. Weighing the test samples allowed us to assess their experimental density. In each example, the volume of the samples was divided by the weight that was measured.

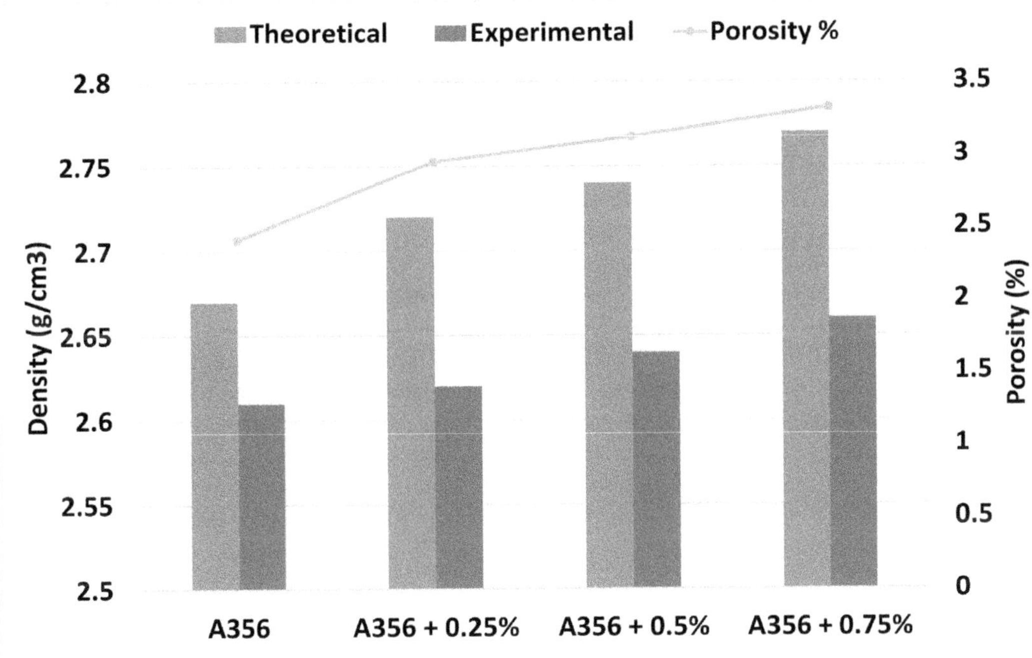

FIGURE 10.7 Study on density and porosity A356-SiC.

When silicon carbide reinforcement is added to A356 aluminium matrix composites, the percentage porosity can be reduced through different mechanisms. The porosity percentage ranges from 2.41% to 3.32%, with a mean of 0.6% and a median of 0.7%. The standard deviation is 0.4%, suggesting moderate variability in porosity across the samples. Notably, porosity increases with the addition of material to the Al MMC, as seen in the sample A356 + 0.25%, A356 + 0.5%, etc., indicating a potential correlation between the added percentage and porosity. The theoretical density of Al MMCs increases with the addition of alloying elements, as seen with A356 + 0.75% having the highest theoretical density at 2.77 g/cm^3. However, the experimental density is consistently lower than the theoretical values, indicating the presence of porosity. The experimental densities for A356, A356 + 0.25%, A356 + 0.5%, and A356 + 0.75% are 2.61, 2.62, 2.64, and 2.66 g/cm^3, respectively as shown in Figure 10.7. The porosity percentage increases with the addition of alloying elements, which could explain the discrepancy between theoretical and experimental densities. The dataset shows a positive correlation between the amount of alloying element and porosity, which suggests that alloying may contribute to increased porosity in these materials. Silicon carbide particles are essential for promoting the solidification of aluminium during the casting procedure [10, 20]. By employing techniques akin to those used by a physical chemist, the composite undergoes a transformation that results in a more refined and consistent microstructure. This, in turn, decreases the chances of porosity formation during the solidification process.

Furthermore, the incorporation of SiC particles enhances the viscosity and feeding characteristics of the molten aluminium alloy, facilitating a more seamless flow and improved filling of the mould cavity in the casting procedure. By employing certain techniques, the occurrence of gas entrapment and void formation within the material can be significantly reduced. In addition, the inclusion of SiC reinforcement enhances the wetting behaviour between the aluminium matrix and the mould surface, resulting in a decrease in air entrapment and porosity formation. In addition, SiC particles have the ability to effectively filter and trap impurities and gas bubbles in molten aluminium alloy during casting. This results in a reduction of porosity in the final material. Furthermore, implementing appropriate processing techniques like degassing and precise control of casting parameters can effectively reduce the formation of porosity in A356/SiC MMCs. In A356 MMCs, the inclusion of SiC reinforcement leads to a decrease in percentage porosity. This is achieved by improving solidification behaviour, enhancing fluidity, improving wetting characteristics, and filtering impurities during casting [10, 15, 17].

10.3.4 Study on Wear

A major improvement in the material's resistance to wear is brought about by the inclusion of silicon carbide reinforcement into the A356 alloy as shown n Figure 10.8a–8c as per the ASTM G99 test methods. In the A356 matrix, SiC, which is well-known for its remarkable hardness and resistance to abrasion, functions as an efficient reinforcement material when it is spread throughout the matrix. The microstructure of the alloy is changed as a result of this reinforcement, which results in the creation of a composite material that has improved mechanical properties. By acting as barriers against abrasive forces, the existence of SiC slows down the progression of wear mechanisms associated with the A356 alloy. This grades in a decrease in the alloys' wear rate [12].

10.3.5 Study on Co-Efficient of Friction

The addition of silicon carbide additives into the A356 metal matrix induces modifications in the surface properties and interfacial dynamics of the material, ensuing in notable alterations

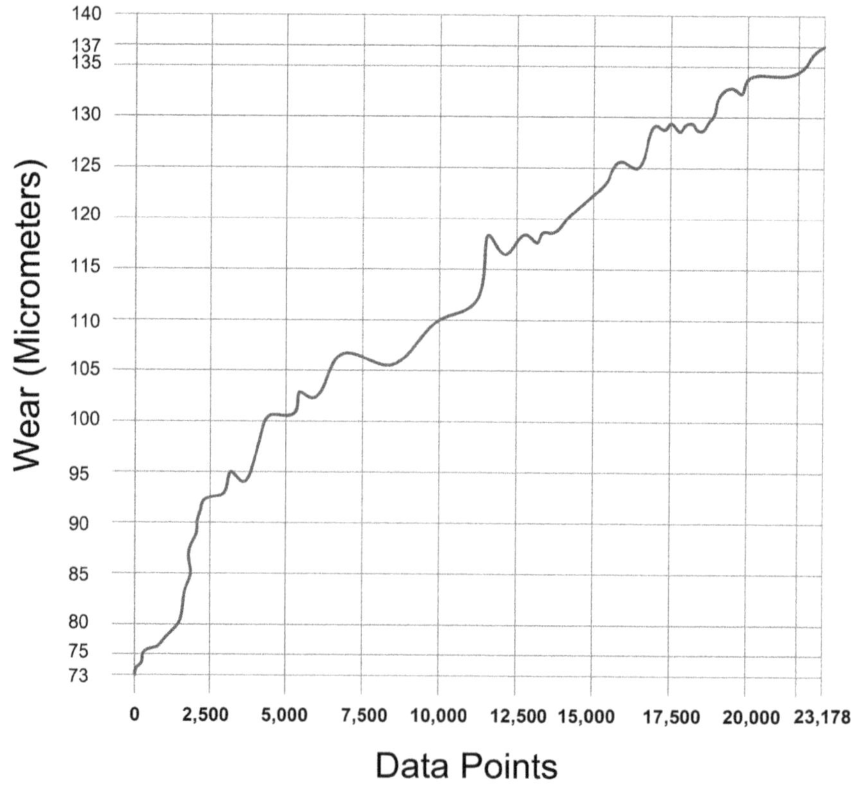

FIGURE 10.8A Study on wear A356+0.25%SiC.

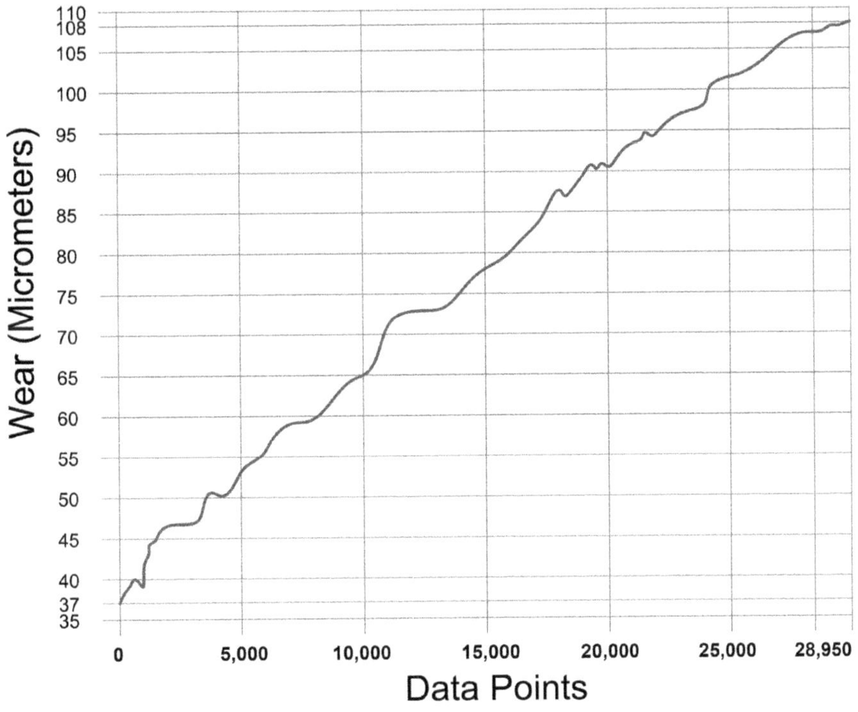

FIGURE 10.8B Study on wear A356+0.5%SiC.

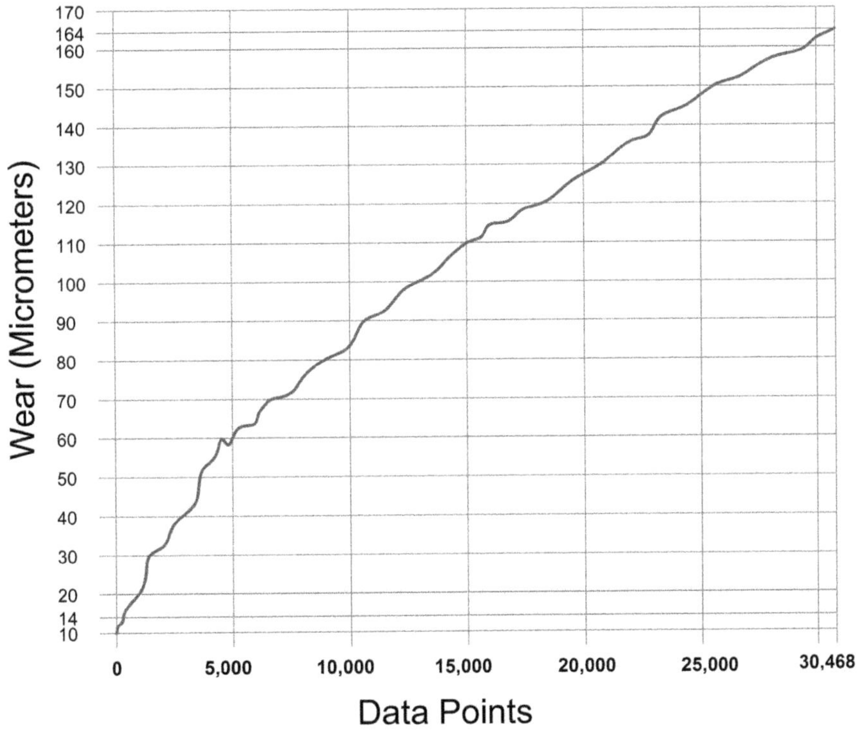

FIGURE 10.8C Study on wear A356+0.75%SiC.

in friction and wear characteristics as shown in Figure 10.9a–c as per ASTM D1894. To begin with, the incorporation of SiC reinforcement in the composite material serves to augment its hardness and abrasion resistance, hence leading to a decrease in contact area and surface deformation during sliding contact. The reduction in contact area signifies a decline in the coefficient of friction caused by the existence of SiC particles, which act as solid lubricants. This leads to

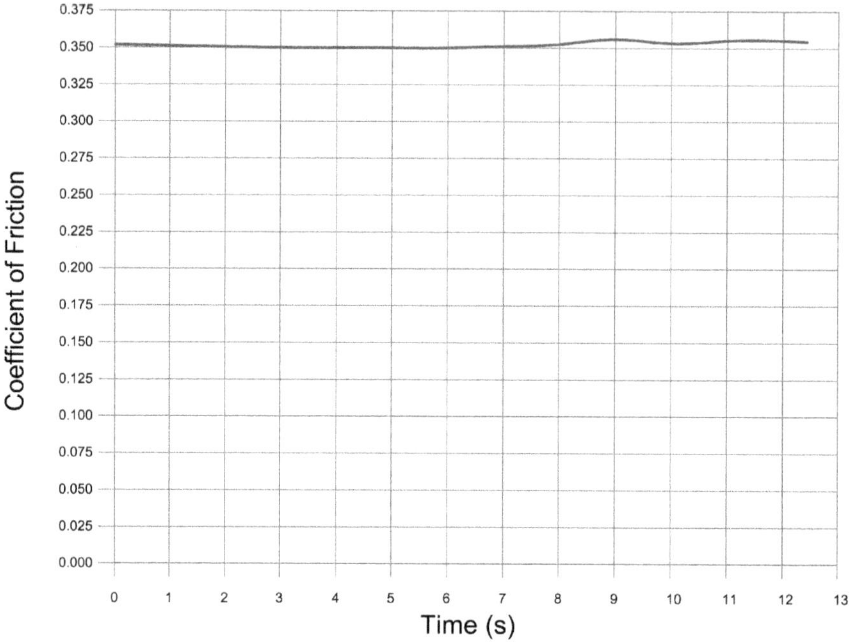

FIGURE 10.9A Study on co-efficient of friction A356+0.25%SiC.

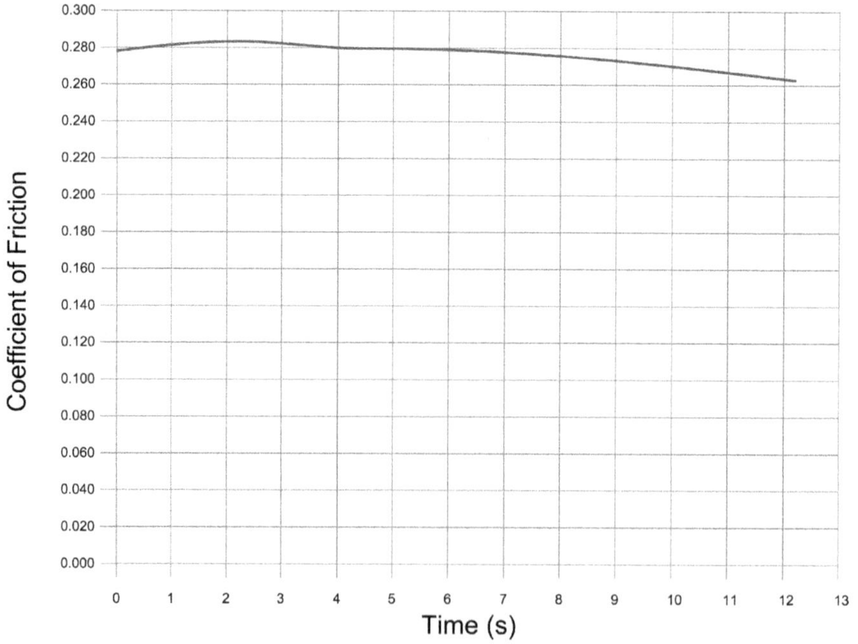

FIGURE 10.9B Study on co-efficient of friction A356+0.50%SiC.

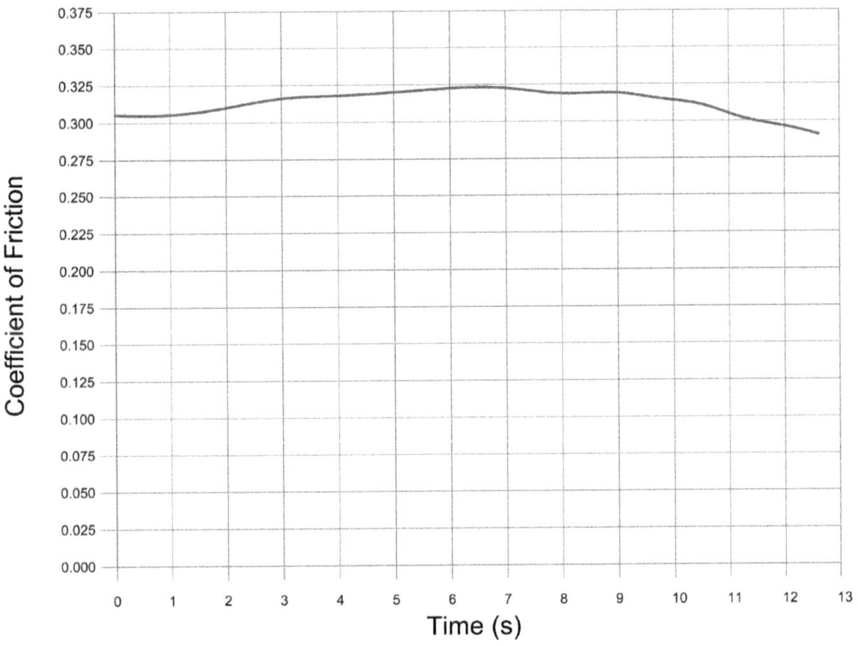

FIGURE 10.9C Study on co-efficient of friction A356+0.75%SiC.

a decrease in both adhesion and frictional forces between the surfaces that are being compared. Furthermore, the inclusion of SiC reinforcement in the composite material modifies the wear mechanisms, effectively impeding abrasive wear by acting as barriers to the spread of micro cracks and the creation of wear debris. Hence, the coefficient of friction stays minimal, and the wear rate significantly diminishes as an effect of the combined impact of SiC addition on the A356 alloy matrix [8].

10.4 CONCLUSIONS

Aluminium alloys, such as A356, have gained significant popularity as a result of their advantageous characteristics, including their lightweight composition, remarkable mechanical qualities, and resistance to corrosion. The integration of silicon carbide as a reinforcing phase is intended to augment these characteristics, rendering it a promising pathway for enhancing material performance. The study reveals the synergistic benefits between A356 aluminium and SiC reinforcements by systematically evaluating mechanical parameters such as tensile strength, hardness, and wear resistance under different situations. An analysis of the microstructure offers vital evidence about the bonding and spreading of SiC particles inside the aluminium matrix. This facilitates a more comprehensive comprehension of the factors that lead to the reported enhancements in mechanical and wear performance. These findings have the prospective to facilitate the creation of advanced composite materials that are specifically designed for challenging engineering applications that demand exceptional mechanical and wear resistance.

ACKNOWLEDGMENTS

We thank Vel Tech Rangarajan, Dr Sagunthala, and the R&D Institute of Science and Technology, Chennai, for the providing the resources to carry out this research work.

FUNDING

This research work is not funded by any organization.

CONFLICT OF INTEREST

The authors declare that there is no conflict of interest regarding the publication of this article.

AUTHOR CONTRIBUTIONS

All authors contributed equally to bringing out this research article.

REFERENCES

[1] Elzayady, N.E., Rashad, R.M., Elenany, A. and Elhabak, A., 2014. Mechanical and microstructure characterization of A356 (Al/SiC) MMC. *Advanced Materials Research*, *875–877*, pp. 135–144.

[2] Hashim, J., Looney, L. and Hashmi, M.S.J., 2002. Particle distribution in cast metal matrix composites—Part I. *Journal of Materials Processing Technology*, *123*(2), pp. 251–257.

[3] Hashim, J., Looney, L. and Hashmi, M.S.J., 1999. Metal matrix composites: Production by the stir casting method. *Journal of Materials Processing Technology*, *92–93*, pp. 1–7.

[4] Pech-Canul, M.I., 2000. Aluminum alloys for Al/SiC composites. In *Recent Trends in Processing and Degradation of Aluminium Alloys*, IntechOpen, vol. 18, pp. 299–317.

[5] Ibhadode, A.O.A., 2022. Engine lightweighting: Use of green materials as reinforcement in aluminum metal matrix composites. In *Renewable Energy—Recent Advances*, IntechOpen, pp. 1–12.

[6] Prabu, S.B., Karunamoorthy, L., Kathiresan, S. and Mohan, B., 2006. Influence of stirring speed and stirring time on distribution of particles in cast metal matrix composite. *Journal of Materials Processing Technology*, *171*(2), pp. 268–273.

[7] Sivananthan, S., Ravi, K. and Samson Jerold Samuel, C., 2020. Effect of SiC particles reinforcement on mechanical properties of aluminium 6061 alloy processed using stir casting route. *Materials Today: Proceedings*, *21*, pp. 968–970.

[8] Venkatachalam, G. and Kumaravel, A., 2017. Mechanical behaviour of aluminium alloy reinforced with SiC/Fly ash/basalt composite for brake rotor. *Polymers and Polymer Composites*, *25*(3), pp. 203–208.

[9] Sachin, D., Uday, K.N., Rajamurugan, G. and Krishnasamy, P., 2021. Effect of SiC reinforcement on the mechanical and tribological behaviour of Al6061 metal matrix composites. *Materials Today: Proceedings*, *46*, pp. 7082–7089.

[10] Kumar, A., Kumar, S., Mukhopadhyay, N.K., Yadav, A. and Winczek, J., 2020. Effect of SiC reinforcement and its variation on the mechanical characteristics of AZ91 composites. *Materials*, *13*(21), pp. 4913–4921.

[11] Wahyuni, N., Nur, R., Renreng, I. and Adnan, M., 2019. Effect of adding SiC on resistance wear and hardness through stir casting of aluminum matrix composites. *AIP Conference Proceedings*, *2114*, pp. 1–11, 050020.

[12] Mahali, M., Sinha, N.K., Choudhary, I.N., Singh, J.K. and Mohan, S., 2023. Effect of Al_3Ni and SiC on mechanical and wear behaviour of Al-Ni-SiC composite. *Silicon*, *15*(9), pp. 3897–3908.

[13] Dey, D., Bhowmik, A. and Biswas, A., 2022. Effect of SiC content on mechanical and tribological properties of Al2024-SiC composites. *Silicon*, *14*(1), pp. 1–11.

[14] Vasanth Kumar, H.S., Revanna, K., Kumar, N., Sathyanarayana, N., Madeva, N., Manjunath, G.A. and Adisu, H., 2022. Impact of silicon carbide particles weight percentage on the microstructure, mechanical behaviour, and fractography of Al2014 alloy composites. *Advances in Materials Science and Engineering*, *2839150*, pp. 1–10.

[15] Ajagol, P., Anjan, B.N., Marigoudar, R.N. and Preetham Kumar, G.V., 2018. Effect of SiC reinforcement on microstructure and mechanical properties of aluminum metal matrix composite. *IOP Conference Series: Materials Science and Engineering*, *376*(1), pp. 1–9, 12057.

[16] Chechi, P., Maurya, S.K., Prasad, R. and Manna, A., 2022. Microstructural and mechanical characterization of stir cast Al-SiC/Flyash/Graphite hybrid metal matrix composite. *Materials Today: Proceedings*, *64*, pp. 637–642.

[17] Bharathi, P. and Kumar, T.S., 2023. Mechanical characteristics and wear behaviour of Al/SiC and Al/SiC/B_4C hybrid metal matrix composites fabricated through powder metallurgy route. *Silicon*, *15*(10), pp. 4259–4275.
[18] Dwivedi, S.P., Sharma, S. and Mishra, R.K., 2015. Microstructure and mechanical behavior of A356/SiC/Flyash hybrid composites produced by electromagnetic stir casting. *Journal of the Brazilian Society of Mechanical Sciences and Engineering*, *37*(1), pp. 57–67.
[19] Gurusamy, P., Raj, S.H.K., Bhattacharjee, B. and Bhowmik, A., 2023. Assessment of microstructure and investigation into the mechanical characteristics and machinability of A356 aluminum hybrid composite reinforced with SiC and MWCNTs fabricated through rotary centrifugal and squeeze casting processes. *Silicon*, *16*(1), pp. 367–382.
[20] Karvanis, K., Fasnakis, D., Maropoulos, A. and Papanikolaou, S., 2016. Production and mechanical properties of Al-SiC metal matrix composites. *IOP Conference Series: Materials Science and Engineering*, *161*(1), pp. 1–8, 12070.

11 Mechanical Characteristics of Hybrid Metal Matrix Composites Fabricated through Liquid State Technique

Alphonse Bovas Herbert Bejaxhin, Selvaraj Hari Krishna Raj, Pathinettampadian Gurusamy, and Nandagopan Ramanan

11.1 INTRODUCTION TO COMPOSITES IN MANUFACTURING

Automobile manufacturers are adapting to the demand for enhanced fuel efficiency by turning to alternative materials like metal matrix composites (MMCs). Components fashioned from MMCs not only offer significant weight reductions but also uphold or even enhance performance when compared to traditional materials. Over the past decade, MMCs have steadily progressed toward mainstream adoption in the automotive industry. To meet the criteria for reduced fuel consumption, minimized pollution, and heightened efficiency standards, automotive producers are earnestly exploring alternative materials for their product designs. Advanced materials such as MMCs present a promising avenue for achieving substantial performance improvements.

11.1.1 Literature Reviews Based on Advancements in Aluminium Hybrid Metal Matrix Composites

Abhay et al. [1] conducted a comprehensive study on Al-SiC spur gears under static loading conditions, illustrating the structural advantages and stress distribution improvements achieved through the incorporation of SiC particulates. Similarly, Ajith Arul Daniel [2] investigated the dry sliding wear behaviour of Al 5059/SiC/Al_2O_3 hybrid composites, highlighting the role of SiC and Al_2O_3 in enhancing wear resistance and mechanical properties. These findings are corroborated by the work of Ajmal Hussain et al. [3], who performed experimental analysis on Al 6063 with SiC, demonstrating significant improvements in hardness and tensile strength. In a related study, Ravi et al. [4] have studied that the AA6061 aluminium alloy composites reinforced with boron carbide (B4C) using stir casting. The research revealed enhanced hardness, tensile strength, and wear resistance with increased B_4C content. The uniform dispersion of B_4C particles significantly improved the composite's mechanical properties, highlighting its potential for advanced engineering applications. Antony Arul Prakash [5] provided a detailed microstructural analysis of aluminium hybrid metal matrix composites developed using the stir casting process, underscoring the uniform distribution of reinforcements and resultant mechanical benefits.

11.1.2 Enhancing Performance and Properties of Hybrid Aluminium Metal Matrix Composites

Ghasali [6] explored the production of Al-SiC-TiC hybrid composites using both microwave and conventional heating methods, finding that microwave-assisted synthesis offered better particle

DOI: 10.1201/9781003499718-11

distribution and mechanical performance. Additionally, Kanchan [7] investigated the tribological properties of Al 7075/TiC/Al_2O_3 hybrid composites, demonstrating superior wear resistance and reduced friction coefficients. Kawin et al. [8], employed the Taguchi-GRA method for multi-criteria optimization of turning parameters in Al/Bagasse ash/Gr hybrid composites, revealing optimal processing conditions for improved tribological behaviour. Nagesh and Shashikanth [9] optimized the tribological properties of Al 6061 metal matrix composites reinforced with boron and graphite using the stir casting method. Their study revealed that incorporating boron and graphite significantly enhanced the wear resistance and reduced the friction coefficient of the composites. This improvement is attributed to the solid lubrication effect of graphite and the hardening effect of boron, leading to better performance under tribological conditions. Sudalai and Udaya Prakash [10] conducted a multi-objective optimization of tribological parameters of hybrid composites using grey relational analysis, focusing on the complex interactions between various tribological factors to enhance the performance of hybrid composites. Their systematic evaluation and improvement of the wear and friction characteristics of these materials demonstrated the effectiveness of this method in optimizing multiple tribological parameters simultaneously.

11.1.3 Advancements in Characterization and Optimization

Further investigations by Rajesh and Kaleemulla [11] focused on the wear behaviour of SiC and Al_2O_3-reinforced hybrid aluminium composites, documenting substantial improvements in wear resistance and mechanical strength. Rajesh [12] studied the influence of TiC particles in Al 6061 MMCs, reporting enhanced hardness and wear properties. Rajesh Prabha [13] analyzed the effects of TiC and Al_2O_3 reinforcements on the microstructure and thermogravimetric properties of aluminium composites, noting significant improvements in thermal stability. Ram Kumar et al. [14], characterized the mechanical and physical properties of Al 6061 MMCs reinforced with SiC and Al_2O_3, highlighting the benefits of the stir casting method in achieving uniform reinforcement distribution. Senthil Murugan [15] developed hybrid composites for automotive applications, using ANSYS to analyze structural stability and predict performance under various loading conditions. Yashpal et al. [16], reviewed the fabrication methods and particulate reinforcement effects in aluminium MMCs, providing a comprehensive overview of current advancements and challenges. Brucely et al. [17] have measured tensile strength and wear rate in AA8011-TiC-ZrB2 hybrid composites using online acoustic emission techniques, demonstrating innovative approaches to real-time monitoring of composite performance. Rajkumar et al. [18] have applied fuzzy logic for the parametric optimization of powder-mixed EDM in AA2014/Si3N4/Mg/cenosphere hybrid composites, offering insights into machining characteristics and mechanical properties.

11.1.4 Parameters Used with Various Ratios

This study investigates the effects of varying weight percentages (5%, 10%, 15%) and particle sizes (10 μm, 20 μm, 40 μm) of silicon carbide (SiC) reinforcements within an aluminium matrix, while maintaining a constant 2% content of molybdenum disulfide (MoS2). The pin-on-disc technique is employed to assess the wear resistance and friction coefficient of the composite to determine its wear performance. Under dry sliding conditions, experiments are conducted by varying parameters such as sliding speed (1.5 m/s, 2.5 m/s, 3.5 m/s), loads (30 N, 50 N, 70 N), and sliding distances (500 m, 1,000 m, 1,500 m). To understand the impact of variables such as ceramic particle reinforcement, sliding distance, sliding speed, and applied load on the friction coefficient and wear rate, the Taguchi experimental design and ANOVA analysis are utilized. The findings reveal that sliding distance and applied load have the greatest effect on the friction coefficient, while load and SiC percentage significantly influence the wear rate. Gwyddion software and an optical microscope are used to examine the worn surfaces of the composites.

The findings indicate that for aluminium hybrid metal matrix composites (AHMMCs), a 15% SiC weight percentage with a particle size of 10 μm provides the best wear resistance and friction coefficient. In different research, turning parameter optimization is done to reduce temperature and surface roughness during machining the Al/Bagasse Ash/Gr hybrid composite. Squeeze casting is used to create this composite, and turning tests are conducted using the Taguchi L9 orthogonal array as a basis. Surface roughness and temperature are the response parameters, while variables like feed, speed, and bagasse ash % are taken into account. Grey relational analysis (GRA) based on Taguchi is used to determine the ideal combination of parameters that results in a better surface finish and reduced temperature during turning. When machining the Al/Bagasse Ash/Gr hybrid composite, it is advised to use a minimal proportion of reinforcement, a lowered feed rate, and an enhanced working speed. Because of their superior strength, low density, durability, machinability, availability, and affordability as compared to other materials, aluminium alloys are preferred in advanced applications. However, by using aluminium matrix composite (AMC) materials, the characteristics of aluminium alloys may be further enhanced. AMC materials are synthetically created blends of a minimum of two chemically different materials, one of which is aluminium, with unique interfaces dividing the components. Compared to current monolithic materials, these materials provide better property combinations. The benefits of AMCs in terms of performance, economy, and ecology are what motivate their use in the automobile industry. Reduced airborne pollutants, noise pollution, and fuel usage are among the main advantages.

The use of AMCs in the automobile industry is expected to become both necessary and desirable in the coming years due to stricter environmental regulations and an increased focus on fuel efficiency. AMCs are anticipated to replace monolithic materials such as polymer-based composites, ferrous alloys, titanium alloys, and traditional aluminium alloys in various applications. Additionally, through selective reinforcement approaches, AMCs can offer commercially viable solutions for a range of uses. Recent successes in military and commercial applications of AMCs have been partly attributed to innovative modifications in component design.

11.2 MATERIALS AND METHODS

11.2.1 Details of Aluminium-LM-25

The LM25 alloy complies with the requirements specified in BS 1490:1988. Precipitation treated (TE), solution treated, artificially aged and stabilized (TF7), and completely heat treated (TF) are the three typical states under which castings are offered. LM25 is a well-known aluminium casting alloy that finds extensive use in several sectors due to its adaptability. With silicon and magnesium added for improved mechanical qualities, LM25 is largely composed of aluminium and provides high corrosion resistance, machinability, and thermal stability. By adhering to guidelines like BS 1490:1988, LM25 guarantees constant casting quality. Its robust strength and durability make it suitable for components exposed to mechanical stresses, while its ability to withstand high temperatures renders it ideal for applications requiring thermal resistance. Commonly found in automotive, marine, and aerospace sectors, LM25 alloy stands as a reliable choice for manufacturers seeking a material that balances strength, corrosion resistance, and ease of processing. Table 11.1 lists the elements and their respective composition ranges in percentage for the aluminium alloy LM-25, highlighting the major and trace elements present.

11.2.2 Chemical Composition of Aluminium LM-25

TABLE 11.1
Typical Chemical Composition for Aluminium Alloy LM-25

Element	Percentage
Copper	0.2 max
Magnesium	0.2–0.6 max
Silicon	6.5–7.5 max
Iron	0.5 max
Manganese	0.3 max
Nickel	0.1 max
Zinc	0.1 max
Lead	0.1 max.
Tin	0.05 max
Titanium	0.2 max.
Aluminium	Remainder
Others: each	0.05 max.
Others: total	0.15 max.

11.2.3 LM-13 Aluminium Mechanical Properties

TABLE 11.2
Aluminium Alloy LM-25 Properties

0.2% Proof Stress (N/mm^2)	80–100
Tensile Stress (N/mm^2)	130–150
Elongation (%)	2
Impact Resistance Izod (Nm)	-
Brinell Hardness	55–65
Endurance Limit (5 × 10^7 cycles N/mm^2)	70–100
Modulus of Elasticity (×10^3 N/mm^2)	71
Shear Strength (MPa)	80–150

11.2.4 Applications of LM 25

Because of its adaptable qualities, LM25 alloy is widely used in a variety of sectors mentioned with their properties in the Tables 11.1 and 11.2. It is used for braking systems, transmission housings, and engine parts in the automobile industry, where its machinability, corrosion resistance, and thermal stability are sought. In marine applications, LM25 alloy is favoured for marine engine parts, propellers, and hull fittings, owing to its ability to withstand corrosive seawater environments. Additionally, because of its strength-to-weight ratio and resilience to high temperatures, LM25 alloy is used in the aerospace sector to make fuselage structures, wing assemblies, and landing gear elements. Moreover, LM25 alloy is used in the manufacturing of industrial equipment, such as pumps, valves, and machinery components, where its durability and ease of fabrication are advantageous. Overall, LM25 alloy's combination of properties makes it a versatile material choice across diverse sectors, contributing to its widespread adoption in numerous applications. Table 11.2 provides various physical and mechanical properties of aluminium alloy LM-25, including density, melting range, tensile strength, proof stress, elongation, hardness, thermal and electrical conductivity, modulus of elasticity, and coefficient of thermal expansion.

11.2.5 Machinability

When machining LM25 alloy, there is reduced drag compared to high silicon alloys like LM6, which lack additional alloying elements. Ordinary steel tools wear out rapidly and are not recommended. Optimal results are achieved with carbide tipped tools featuring large rake angles and lower cutting speeds, while the use of cutting lubricants and coolants is advisable. Diamond tools are often preferred for finishing operations.

11.2.6 Silicon Carbide (SiC)

In the chemical compounds, the substance SiC stands for silicon carbide, often referred to as carborundum. It is a product of Si and C in the ratio of 1:1. And it manifests itself naturally and is found as a pure mineral moissanite in a very rare form. The consolidation of silicon carbide particles into stable ceramics, which are invaluable for automobile parts like brakes, clutches, heavy-duty machine parts, and armours, is another notable aspect of energy-efficient manufacturing. Besides, therefore, the silicon carbide can replace quartz crystal and other devices for electronic purposes, including development of LED lights. In metallic matrix composites, SiC was additionally used as reinforcement with 10% of volume-weighted approach.

11.2.7 Aluminium Oxide (Al2O3)

As one of the elements that is more valuable and widely used in AI, aluminium oxide (alumina in its common name) is such a cost effective and diverse engineering ceramic. Utilizing general purpose raw (materials) for the fabrication of technical grade (high performance) alumina and fuelling procurement are factors that contribute to (providing) a favourable cost for (meeting) the value (ratio). Equipped with an aluminium-aluminium weighted interatomic interaction, aluminium oxide exerts advantageous material nature. It can appear in various crystalline forms and, upon heating, restores to the most energetically stable phase, i.e., hexagonal alpha form. This form possesses high density and strength, and as a result the material is highly suitable for use in construction. Alpha alumina phase, which is among the most dominant anion-based ceramics, is characterized by high hardness, perfection in thermal, refractory, and dielectric properties; and these properties, therefore, make it the favourite for a diverse range of applications.

In addition, super-purity alumina shows that it is one of the few components resistant to corrosion and capable of withholding attack by most gases and regular acids, that is, except wet fluorine and a few others. Meanwhile, anodizing enhances the thickness and characteristics of the oxide layer on aluminium alloys metallically, thus resulting to the better corrosion product of the said metal. It was last mentioned on the United States EPA's list of harmful agents in 1988, but it remains on the OTC's list if it is in a fibre form.

11.2.8 Casting Characteristics

Because of its high fluidity, this alloy may be cast into thin portions. It is appropriate for leak-tight castings due to its acceptable pressure tightness rating. It functions wonderfully in hot conditions and has good resilience to heat tearing which is poured normally at 700°C.

Depending on how the mould is set up, the actual pouring temperatures may range from 670 to 780°C. This alloy may develop internal shrinkage pipes in heavier parts. Because of their many advantages including low density, higher surface polish, high tensile strength, and exceptional wear resistance, composite materials have attracted more attention recently. Among the cost-effective and low-density reinforcements, aluminium oxide stands out. The consideration of tensile strength and wear is paramount. To address these requirements, an experimental setup has been devised where all necessary parameters will be controlled. The study aims to develop a composite by incorporating SiC and Al_2O_3 with aluminium metal at different weight ratios. The composite will be fabricated using the stir casting technique and subjected to comprehensive analysis for various mechanical properties. The key constituents utilized in this investigation are aluminium-LM-25, silicon Carbide (SiC), and aluminium oxide (Al_2O_3), commonly known as alumina. The stir casting method is employed for the casting process in this experiment.

11.2.9 Aluminium-LM-25

This alloy meets LM-25 specifications specified in BS 1490:1988. Precipitation treated (TE), solution treated, artificially aged and stabilized (TF7), and completely heat treated (TF) are the three standard treatments for castings. Pulleys (sheaves), food, chemical, maritime, electrical, and engine components working at high temperatures are just a few industries that use LM 25 alloy. Its low coefficient of thermal expansion, strong resistance to wear, and advantageous bearing qualities are its main features. A 10% volume-based approach is used to integrate silicon carbide in the context of metal matrix reinforcement. One of the most affordable and often used materials in the field of engineering ceramics is alumina. Its inexpensive and easily available raw ingredients provide excellent value for formed alumina forms. Fine grain technical grade alumina is widely used in many different sectors due to its desirable price point and remarkable mix of characteristics. Figure 11.1 illustrates a sample of the aluminium alloy LM-25 material, showcasing its typical structure and surface characteristics. Figure 11.2 depicts samples of SiC (silicon carbide) and Al2O3 (aluminium oxide) powders, highlighting their granular forms and textures.

FIGURE 11.1 Aluminium alloy LM-25 sample material.

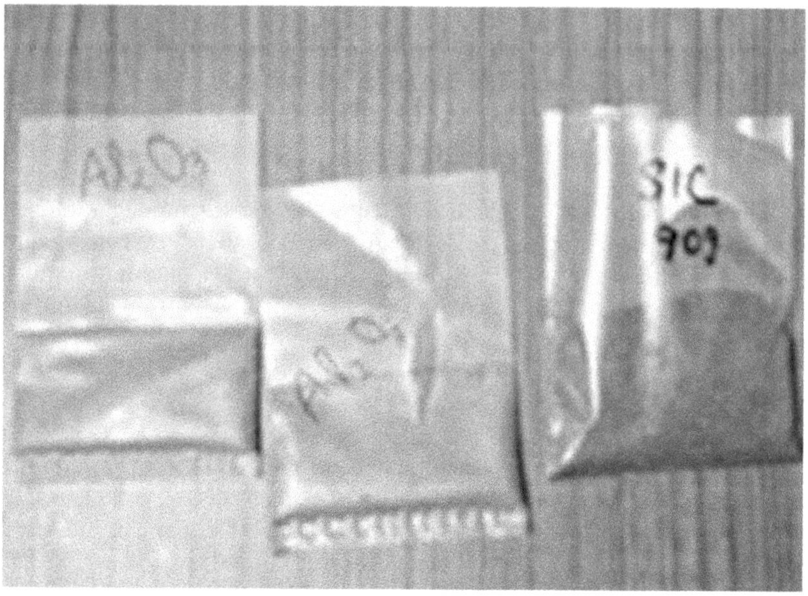

FIGURE 11.2 SiC and Al_2O_3 powder sample.

11.3 FABRICATION THROUGH STIR CASTING PROCESS

11.3.1 Material Composition

The Al_2O_3 and SiC powder, with respective particle sizes of 90 μm and 120 μm, were selected as reinforcing materials, with the aluminium LM25 alloy acting as the basis material. Three cast samples 0% SiC and 0% Al_2O_3, 2.5% SiC and 5% Al_2O_3, and 5% SiC and 10% Al_2O_3 were made with different weight percentages of Al_2O_3 and SiC. The aluminium LM25 alloy was first heated to 850°C in a blower furnace for 15 minutes, which resulted in the alloy melting in a pot. Al_2O_3 and SiC powder were warmed in a separate muffle furnace at 575°C in the interim. The LM25 alloy was gradually added to the Al_2O_3 and SiC powder that had been heated before it was fully melted. For about 15 minutes, a drilling machine was used to enable stirring at a rate of around 950 revolutions per minute. After that, the resultant slurry was put into the mould cavity and given time to cool to ambient temperature. To create three specimens with various Al_2O_3 and SiC compositions, this process was repeated.

11.3.2 Mechanism of Casting

A stirrer is used in stir casting to stir the molten metal matrix. Usually, a cylindrical rod with connections to an impeller on one end and a motor shaft on the other is employed as a graphite stirrer as depicted in Figure 11.3. Usually mounted vertically, the stirrer is turned at different speeds by a motor. After that, a die is filled with the molten metal combination to be cast. The possible segregation of reinforcement particles owing to different process parameters and material qualities, which might result in non-uniform metal distribution, is a major problem with stir casting. Variations in stirrer velocity, stirrer angle, and vortices cone may affect how the particles are distributed in the molten metal matrix. To guarantee full melting, the matrix metal is first heated above its liquid temperature in this procedure. It becomes semi-solid when cooled to a temperature in between those of the liquid and solidus phases. The molten matrix is then mixed thoroughly by adding preheated reinforcement particles and heating the mixture to a completely liquid state. Figure 11.3 shows the stir casting setup, which is used for mixing and producing metal matrix composites.

FIGURE 11.3 Stir casting setup.

11.4 CALCULATIONS

First, the volume of a cylindrical object with a diameter of 2 cm and a length of 40 cm is calculated using the equations 11.1 and 11.2, resulting in 125.66 cm^3. Adding an extra volume of 22.5 cm^3 gives a total volume of 148.16 cm^3. The mass of this volume of aluminium, with a density of 2.7 g/cm^3, is calculated to be 400 grams. Including a 30% extra allowance, the mass becomes 520 grams.

$$\begin{aligned} \text{Volume} = \text{V} &= \frac{\pi}{4}\text{D}^2\text{L} \\ = \text{V} &= \frac{\pi}{4}(2)^2(40) \\ &= 125.66\ \text{cm}^3 \end{aligned} \tag{11.1}$$

$$\begin{aligned} \text{Plate area} = \text{A} &= \text{LBH} \\ \text{A} &= 10(1.5)(1.5) \\ &= 22.5\ \text{cm} \end{aligned} \tag{11.2}$$

Additional volume = 125.66 + 22.5 = 148.16 cm^3
Mass of Al = 148.16 cm * 2.7
Mass of Aluminium = 400 g
Including 30% extra for additional material
Al = 148.16 * 2.7 = 400 g + 30% Extra – 120 (Given the density of aluminium is = 2.7 g/cm^3)
Total mass LM-25 = 520 gram

Mixing ratio of samples:
Sample 1: LM-25 = 100%
Sample 2: SiC-2.5% + Al_2O_3-5% and LM-25
Sample 3: SiC-5% + Al_2O_3-10 % and LM-25

TABLE 11.3
Calculation for All Ratios

Ratio	LM-25 (grams)	SiC (grams)	Al_2O_3 (grams)
R1	520	0	0
R2	520	2.5% = 13	5% = 26
R3	520	5% = 26	10% = 52

The attached data indicates three ratios of mixing including the primary sample, the secondary sample, and the tertiary sample in the Table 11.3. This displays the amounts of LM-25 alloy, silicon carbide (SiC), and aluminium oxide (Al_2O_3) in grams for three different ratios. Sample 1 consists solely of the LM-25 alloy as its only constituent. Sample 2 comprises a mixture of the LM-25 alloy with 2.5% silicon carbide (SiC) and 5% aluminum oxide (Al_2O_3). Finally, Sample 3 is composed of the LM-25 alloy, containing 10% porous aluminum oxide (Al_2O_3) and 5% silicon carbide (SiC). Every analysis by sample is a kind of raw material, which has its own system of composition with different proportions of all the elements for creating specific or predetermined properties and special features of composites.

11.5 RESULT AND DISCUSSION

11.5.1 Effect of Hybrid Composition on Hardness

Table 11.4 lists the hardness values of hybrid composites with different ratios, demonstrating how varying compositions affect material hardness. R1 represents 100% LM-25 with a hardness value of 35 HRB. R2 includes 2.5% silicon carbide (SiC) and 5% aluminium oxide (A12O3) with LM-25, having a hardness value of 34 HRB. R3 consists of 5% SiC and 10% A12O3 with LM-25, showing an increased hardness value of 45 HRB.

The provided data outlines the mixing ratios of different materials and their corresponding Rockwell hardness (HRB) values in Figure 11.4. In Ratio R1, where LM-25 alloy constitutes 100% of the material composition, the resulting hardness is measured at 35 HRB. In contrast, Ratio R2 combines LM-25 alloy with 2.5% silicon carbide (SiC) and 0.5% aluminium oxide (A12O3), resulting in a slightly lower hardness of 34 HRB. Finally, Ratio R3 features a blend of LM-25 alloy with 5% silicon carbide (SiC) and 10% aluminium oxide (A12O3), yielding a notably higher hardness of 45 HRB. These findings suggest that variations in material composition as mentioned in Figure 11.4, particularly the presence and concentration of silicon carbide and aluminium oxide alongside LM-25 alloy, significantly impact the resulting hardness properties. Such insights into the relationship between material ratios and hardness can be pivotal in optimizing material compositions for specific applications, where hardness and structural integrity are critical factors.

11.5.2 Effect of Hybrid Composition on Tensile Strength Values

The given data includes several ratios, together with readings of diameter (mm), cross-sectional area (CSA) in square millimetres (mm^2), tensile load (TL) in kilonewtons (kN), and tensile strength in newtons per square millimetre (N/mm^2). In ratio R1 that has the sectional constancy of 16 mm, the cross-sectional area is counted to be 201.06 mm^2, thus providing the tensile load that reached 20.84 kN and the tensile resistance that equals 103.65 N/mm^2. As the R1 is replaced by R2 with the same diameter and cross-sectional area, there is both an increase in tangent loads and tangent strength.

TABLE 11.4
Hardness value of different ratio based hybrid composites

Ratio	Material	HRB
R1	LM-25 – 100%	35
R2	SiC-2.5% + Al_2O_3-5% and LM-25	34
R3	SiC-5% + Al_2O_3-10 % and LM-25	45

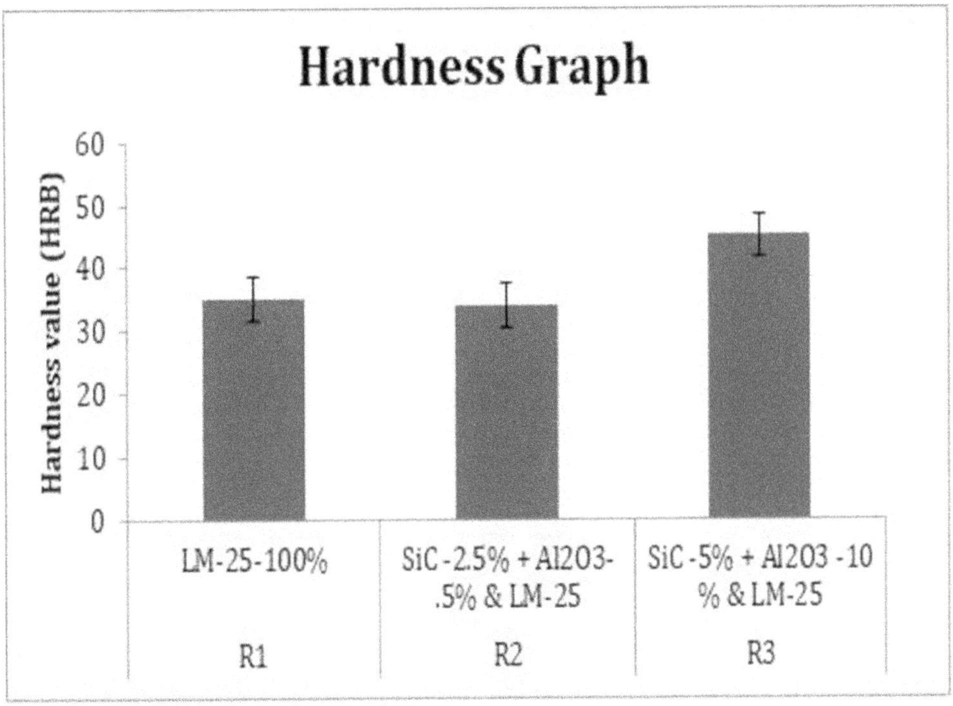

FIGURE 11.4 Hardness strength graph.

They encompass 22.86 kN and 113.69 N/mm², correspondingly. Likewise, when we look at the ratio metric profile R3, the constant diameter and the cross-sectional area still do not give us any opportunity to reduce tension and tensile strength because the further we progress, the higher they are, namely 24.42 kN and 121.46 N/mm², respectively. This data shows clear relationships between different ratios, for example, as we progress from R1 to R3 the load and strength are increasing, indicating a better stability or improved material properties with composition going from R1 to R3. Identifying the relationships in ratio changes to mechanical properties is a thing that needs to be known as this is the basis of an effective material composition for the engineering applications where the tensile strength and the load-bearing capacity are seemingly paramount considerations. Table 11.5 presents the tensile strength values for three different composite ratios, showcasing their comparative strength characteristics. Figure 11.5 depicts the tensile strength levels for different ratios of the hybrid composite, highlighting the variations in strength with changing compositions.

TABLE 11.5
Tensile Strength of Three Ratios

Ratio	DIA (mm)	CSA (mm²)	TL (kN)	TS (N/mm²)	IGL (mm)	FGL (mm)	%E
R1	16	201.06	20.84	103.65	73.2	73.2	0.0
R2	16	201.06	22.86	113.69	73.2	74.0	1.09
R3	16	201.06	24.42	121.46	73.2	74.0	1.09

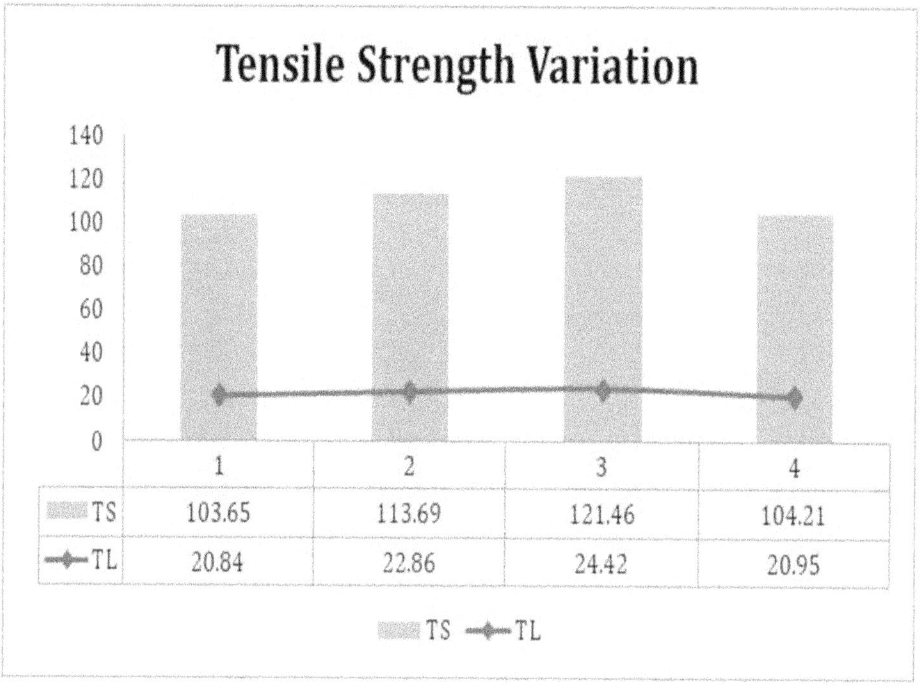

FIGURE 11.5 Tensile strength levels for different ratios of hybrid composite.

11.5.3 Effect of Hybrid Composition on Compressive Strength

It is also worth mentioning that the hybrid compositions, which are based on LM25 alloy combined with other materials like SiC and Al_2O_3, can extensively affect the compressive strength of LM25 alloy. Compressive strength measures how much pressure a material can tolerate before it collapses or crushes under it, therefore it is of critical importance in many engineering projects. Considering LM25 alloy combined with materials for instance SiC and Al_2O_3, these factors are capable of having a strong influence on the compressive strength of both of them. The utilization of fibrous material like SiC and Al_2O_3 will serve as the main strengthening components of the composite comprised of the fibre with higher quality. In LM25 Al-SiC epoxy composites, both SiC and Al_2O_3 are the main reinforcement material. Incredibly, they both offer rigid and hard properties that are capable of increasing the LM25 alloy matrix compression resistance.

Second, the distribution and dispersion of re-enforcing phases are alloy matrix parameters that are directly related to LM25 alloy mechanical properties. By a comprehensive dispersion and a more or less homogeneous distribution of the phase of SiC and Al_2O_3, the matrix can better spread the load and resist compression, hence higher compressive strength is obtained. Nevertheless, the combination of the compressive strength of hybrid composites is likely to be also controlled by additional factors like the volume fraction, the particle size, and the processing conditions of the reinforcing agents as mentioned in the Table 11.6. To obtain such results, the processing methods and the amount of each type of material that would be suitable remains to be established. Table 11.6 provides an evaluation of the compressive strength, outlining the performance of the hybrid composite materials under compressive loading conditions. Figure 11.6 illustrates the compression strength graph, displaying the relationship between compressive strength values and various parameters or conditions of the hybrid composite material.

The provided compression strength values, measured in N/mm², are 308.4, 345.2, and 358.6 MPa. To provide a comprehensive explanation, additional context regarding the corresponding materials or ratios associated with these compression strength values would be necessary.

TABLE 11.6
Evaluation of Compressive Strength

S.No	Composition	Compression strength (N/mm²)
R1	LM-25 – 100%	308.4
R2	SiC-2.5% + Al_2O_3-5% & LM-25	345.2
R3	SiC-5% + Al_2O_3-10 % & LM-25	358.6

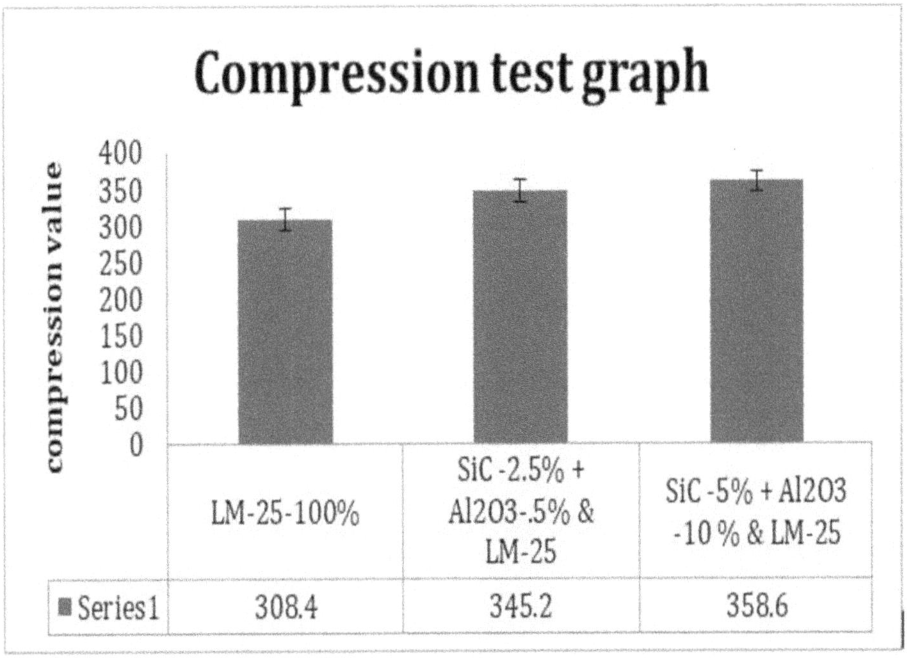

FIGURE 11.6 Compression strength graph.

11.5.4 Effect on Hybrid Metal Matrix Composite on Impact Strength

Here, the given data is showing the comparative strength impacts on the material, in joules, through various compositions. In the paper where hammering LM-25 alloy is equal to 100% of the material (LM-25–100%), the strength of impact is equal to 4 joules. This figure, however, increases again when LM-25 alloy is combined with 2.5% silicon carbide (SiC) and 0.5% aluminium oxide (Al_2O_3), resulting in a harmonic resonance with a value of 3.5 joules. Also, note that the impact strength is drastically reduced to 2J of the SiC-10Al_2O_3/LM-25 hybrid composite. Such a result proves that the type and proportion of additives in the compound illustrated as LM-25 in this example, in particular the incorporation of silicon carbide and aluminium oxide to LM-25 substance, affects the eventual mechanical strength of the finished product significantly. Such disclosures are instrumental for revealing the material's performance under the influence of dynamic loadings, which, in turn, are the basis of the engineering design of applications in mechanical engineering, where impact resistance is a key factor. This Table 11.7 provides the impact strength in Joules for different compositions of aluminium alloy LM-25 combined with silicon carbide (SiC) and aluminium oxide (Al_2O_3). Figure 11.7 displays the impact strength values of three different samples, providing insight into their ability to withstand sudden forces or impacts.

TABLE 11.7
Impact Strength of Different Ratios of HMMC

S.NO	Composition	Impact strength (Joules)
R1	LM-25–100%	4
R2	SiC-2.5% + Al_2O_3-5% & LM-25	3.5
R3	SiC-5% + Al_2O_3-10 % & LM-25	2

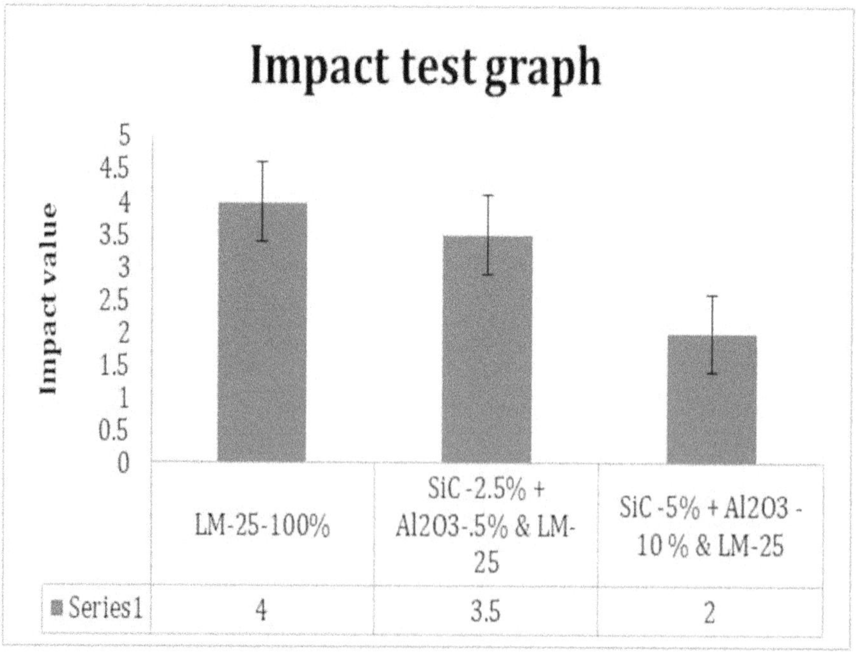

FIGURE 11.7 Impact strength values of three samples.

11.5.5 Effect of Corrosion Rates on HMMC

11.5.5.1 Salt Spray Testing Method Testing

Assured Testing Services in Ridgway, Pennsylvania uses salt spray testing, which is carried out in accordance with ASTM B117 standards and produces a controlled corrosive environment to assess the relative corrosion resistance of metal specimens. While this practice is commonly used to provide corrosion resistance data, its correlation with performance in natural environments is limited when used in isolation. The salt spray system is made up of a reservoir for salt solution, atomizing nozzles, specimen supports, compressed air supply, fog chamber, chamber heating, and control mechanisms. It's essential to prevent solution droplets from accumulating on the chamber's ceiling or cover and falling onto the exposed specimens. Additionally, solution droplets falling from the specimens should not be recycled into the solution reservoir for re-spraying. A corrosion testing tub, commonly used for salt spray testing, is depicted in Figure 11.8.

Table 11.8 shows the initial and final weights of different composite samples after undergoing corrosion testing, along with the corresponding weight loss

The provided data presents the initial weight, final weight, and weight loss in grams for three different ratios or compositions (R1, R2, and R3). To calculate the corrosion rate of the various

FIGURE 11.8 Corrosion (salt spray testing tub).

TABLE 11.8
Corrosion Rates of the Different Various Composites

Ratio	Initial weight in grams	Final weight in grams	Weight loss in grams
R1	7.873	7.768	0.105
R2	7.181	7.175	0.006
R3	6.268	6.252	0.016

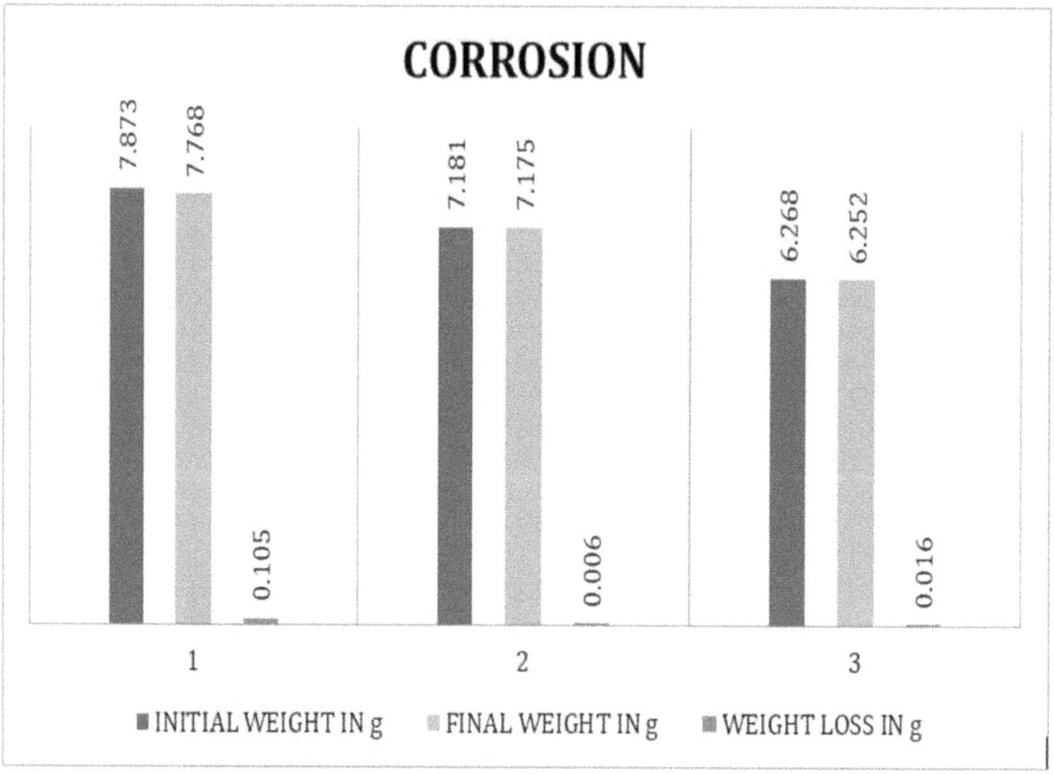

FIGURE 11.9 Corrosion deviations of R1, R2, and R3 ratio samples.

composites, we can utilize the weight loss data over a specified time period as mentioned in Table 11.8 and Figure 11.9. Corrosion rate typically indicates the amount of material loss per unit time due to corrosion processes. It is commonly expressed in units such as millimetres per year (mm/year) or grams per square meter per day (g/m²/day). Using the weight loss data provided, we can calculate the corrosion rate using the formula

$$\text{Corrosion rate} = 30 \text{ days} * 1\text{m}^2 * 0.105 \text{ grams}. \quad (11.3)$$

After performing the calculations, we obtain the corrosion rates for each ratio. These values represent the rate at which the respective composites are corroding over the specified time period. Higher corrosion rates indicate more rapid material loss due to corrosion processes, whereas lower rates suggest slower corrosion. Interpreting these corrosion rates in the context of specific applications is crucial. For instance, a lower corrosion rate may indicate better corrosion resistance and durability, making the material more suitable for environments prone to corrosion, such as marine or chemical processing industries. Conversely, higher corrosion rates may necessitate protective coatings or material modifications to mitigate corrosion effects and prolong the lifespan of the components.

11.6 CONCLUSION

Composite materials, particularly those incorporating LM-25 alloy with reinforcements like molybdenum disulfide, silicon carbide, and graphite, exhibit superior mechanical properties compared to LM aluminium alloy without reinforcements. The investigation revealed that hybrid LM-25 metal matrix composites experienced enhancements in mechanical and tribological properties.

Notably, the addition of silicon carbide and alumina oxide reinforcements, particularly in a ratio of 3 (SiC-5% + Al_2O_3-10% & LM-25), resulted in improved tensile and compressive strength while maintaining low corrosion rates. Although maximum hardness was achieved with varying reinforcement compositions, impact strength was adversely affected in Ratio-3 due to reinforcement agglomeration. Moreover, during salt spray corrosion analysis, specimens with the composition Ratio—LM-25SiC-5% + Al_2O_3-10% exhibited minimal corrosion deposition compared to other samples. Overall, the study underscores the effectiveness of silicon carbide and alumina oxide reinforcements in enhancing the mechanical and corrosion-resistant properties of hybrid LM-25 metal matrix composites.

The developed composite materials discussed in the referenced studies have diverse applications across industries such as automotive, aerospace, marine, manufacturing, renewable energy, and electronics. With their exceptional properties including high strength, wear resistance, corrosion resistance, and lightweight nature, these composites find utility in automotive components, aircraft structures, marine vessels, machinery parts, renewable energy infrastructure, and electronic devices. Their versatility makes them valuable for enhancing performance, efficiency, and durability in a wide range of applications, contributing to technological advancements across various sectors.

REFERENCES

[1] Abhay, A., Utpata, P.B. and Pawar, P.B., 2018. Modeling and analysis of Al-SiC spur gear under static loading condition. In *Proceedings of 3rd International Conference on Recent Trends in Engineering & Technology*, Elsevier, pp. 123–128.

[2] Ajith, A.D., Sakthivel, M., Manojkumar and Sudhagar, S., 2018. Dry sliding wear behaviour of aluminium 5059/SiC/Al_2O_3 hybrid metal matrix composites. *Materials Research*, *20*(6), pp. 1697–1706.

[3] Ajmal Hussain, S., Rajaneesh, R., Hashim, N. and Jithin, K., 2020. Experimental analysis on aluminum alloy (6063) with silicon carbide: An experimental investigation. *International Journal of Research in Engineering and Technology*, *6*(10), pp. 88–94.

[4] Ravi, B., Balu Naik, B. and Udaya Prakash, J., 2015. Characterization of aluminium matrix composites (AA6061/B4C) fabricated by stir casting technique. *Materials Today: Proceedings*, *2*(4–5), pp. 2984–2990.

[5] Antony, M.D., Prakash, A. and Jaswin, M.A., 2019. Microstructural analysis of aluminium hybrid metal matrix composites developed using stir casting process. *International Journal of Advances in Engineering*, *1*(3), pp. 333–339.

[6] Ghasali, E., Yazdani-rad, R., Asadian, K. and Ebadzadeh, T., 2017. Production of Al-SiC-TiC hybrid composites using pure and 1056 aluminum powders prepared through microwave and conventional heating methods. *Journal of Alloys and Compounds*, *690*, pp. 512–518.

[7] Kanchan, A. and Suyash Pawar, Y., 2019. Tribological investigation of Al7075/Tic/Al_2O_3 hybrid composite material. *International Research Journal of Engineering and Technology*, *5*(7), pp. 2052–2058.

[8] Kawin, N., Jagadeesh, D., Saravanan, G. and Periasamy, K., 2019. Optimization of turning parameters in sugarcane bagasse ash reinforced with Al-Si10-Mg alloy composites by taguchi method. *Materials Today: Proceedings*, *21*, pp. 474–476.

[9] Nagesh, D. and Shashikanth, G.S., 2021. Optimization of tribological properties of Al6061, boron and graphite MMCs using stir casting method. *IOSR Journal of Mechanical and Civil Engineering*, pp. 55–72.

[10] Perumal, S. and Udaya Prakash, J., 2016. Multi-objective optimization of tribological parameters of hybrid composites using grey relational analysis. *International Journal of Chemical Sciences*, *14*(2), pp. 1172–1182.

[11] Rajesh, A.M. and Kaleemulla, M., 2019. Material characterization of SiC and Al_2O_3 reinforced hybrid aluminum metal matrix composites on wear behavior. *Advanced Composites Letters*, *28*, pp. 1–10.

[12] Rajesh, S., Suresh, S., Nithyananth, S.B., Sivakumar, S. and Prabhusubramaniam, S., 2019. Influence and wear characteristics of TiC particles in Al6061 metal matrix composites. *International Journal of Applied Engineering Research*, *13*(9), pp. 6514–6517.

[13] Rajesh Prabha, N. and Edwin Raja Dhas, J., 2017. Effect of TiC and Al_2O_3 reinforced aluminium metal matrix composites on microstructure and thermogravimetric analysis. *Research Journal of Chemistry and Environment*, *10*(3), pp. 729–737.

[14] Ram Kumar, S., Ramakrishnan, S., Risvak, M., Thauffeek, S. and Yuvaraj, T., 2021. Experimental analysis and characterization of mechanical, physical properties of aluminium (Al6061) metal matrix composite reinforced with SiC and Al_2O_3 using stir casting. *IOP Conference Series: Materials Science and Engineering*, *1145*(1), p. 012109.

[15] Senthil Murugan, S. and Jegan, V., 2017. Development of hybrid composite for automobile application and its structural stability analysis using ANSYS. *International Journal of Modern Studies in Mechanical Engineering*, *3*(1), pp. 23–34.

[16] Yashpal, S., Jawalkar, C.S., Ajay Singh, V. and Suri, N.M. 2017. Fabrication of aluminium metal matrix composites with particulate reinforcement: A review. *Materials Today: Proceedings*, *4*(2), Part A, pp. 2927–2936.

[17] Brucely, Y., Shaji, Y.C., Bejaxhin, A.B.H. and Abeens, M., 2022. Online acoustic emission measurement of tensile strength and wear rate for AA8011-TiC-ZrB_2 hybrid composite. *Surface Topography: Metrology and Properties*, *10*(4), p. 045009.

[18] Rajkumar, G., Saravanan, M., Bejaxhin, A.B.H., Sharma, S.P. and Dwivedi, S., 2023. Parametric optimization of powder-mixed EDM of AA2014/Si_3N_4/Mg/Cenosphere hybrid composites using fuzzy logic: Analysis of mechanical, machining, microstructural, and morphological characterizations. *Journal of Composites Science*, *7*(9), pp. 1–25, 380.

12 Investigation on Properties of Friction Stir Welded Metal Matrix Composites

Ganapathy Srinivasan Ramachandran, Rajaravi Chinnathambi, Selvam Murugan, and Palani Subbiah Kone

12.1 INTRODUCTION

Due to its advantages over traditional fusion welding techniques, including low deflection, a small heat-affected zone, and better mechanical properties, friction stir welding (FSW) has become a viable approach for incorporating sophisticated materials, especially metal matrix composites (MMC). FSW's reliability and effectiveness in joining MMCs have been the subject of extensive research efforts in recent years, focusing on a number of factors, such as composites, process variables, and tool design. This review of the literature intends to provide a broad overview of the current state of research on the fibre shear strength of MMC, focusing specifically on the mechanical and metallurgical characterization of aluminium matrix composites reinforced with silicon carbide particles (Al/CICP MMC). In a study on the mechanical and metallurgical evaluation of friction stir welded AA6061-ZrO_2-C hybrid MMCs, Pandiarajan et al. [1] emphasized the importance of understanding the evolution of microscopic structure and how it relates to mechanical properties. A review of FSW of Al 6061 MMC using cylindrical tapered and cylindrical taper threaded tools was published by Kumar et al. [2], emphasizing the impact of tool geometry on mechanical parameters and weld quality. A similar an analysis of FSW of several metal matrix compounds by Kumar et al. [3] highlights the need for thorough investigations to improve process parameters for various material systems. The impact of FSW on the mechanical properties of hybrid Mg/(SiCp+Al_2O_3p+Grp)-MMC was investigated by Hira et al. [4] providing insight into microstructural changes and how they affect machine behaviour. By conducting experimental analysis of the mechanical properties of in-situ manufactured MMCs, Rajavi et al. [5] were able to shed light on facial binding features and strengthening mechanisms. The impact of tool material on friction stirring butt welding of AA2124-T4 alloy matrix MMC was investigated by Bozkurt and Boumerzoug [6], emphasizing the importance of tool material selection in the production of defect-free welds. A thorough analysis of the field's opportunities and difficulties in incorporating aluminium matrix alloys using FSW was provided by Mir et al. [7]. The performance of NI-WC-based MMC overlays deposited by plasma transferred arc (PDA) welding was studied by Sundaramurthy et al. [8] underscoring the importance of matrix-reinforcement interactions in defining the characteristics of MMCs [9]. The friction ruminated clarified the relationship between processing parameters, microstructure, and mechanical efficiency by analyzing the microstructure and mechanical properties of the AA6061/AA6061+40 vol% SiC plates. Ash-related analysis was used by Gosavi and Jaybai [10] to improve the FSW process parameters of Al 7075/SiC composites, highlighting the importance of process optimization in improving mechanical properties and weld quality. The study by Chenniangiri et al. [11] examined the effect of B_4C content on metal matrix compounds based on AA5052 in the mechanical properties of welded joints, emphasizing the importance of reinforcing content that affects MMC properties. Using the friction stir spot welding of Al6061-T6, Chaudhary and Singh [12] developed in-situ MMC junctions, providing a unique method for connecting MMCs with certain defects.

DOI: 10.1201/9781003499718-12

Using a taper threaded tool, Ravi Kumar and others [13] investigated the mechanical properties of friction-stirred ignited aluminium 6061 MMC, highlighting the impact of tool design on mechani cal performance and weld integrity. The friction stir of the Al Alloy-Al-Cc metal matrix is a process of utilizing a solid-state joining method to enhance the weld's mechanical properties and structural integrity. The process causes grain refinement, homogenous distribution of the reinforcing particles, and reduced porosity, leading to higher tensile strength and hardness. In the comparison of tool wear during welding, Pist et al. [14] drew attention to the difficulties caused by tool deformation and how it affects the quality of the weld. Through the use of nano-SiC, Subramani et al. [15] evaluated the mechanical properties and microstructure of stir zone, providing insights into the evolution of microscopic structural features and how they affect mechanical behaviour. The study of the literature, in a nutshell, emphasizes how important it is to understand the complex interactions between process variables, material composition, tool design, and microstructural evolution in order to determine the mechanical and metallurgical properties of frictional syllable welded MMC. While the basic mechanisms governing the FSW of MMCs have been largely clarified, there are still many unanswered questions and areas where additional investigation could be used. These include process parameter optimization, developing state-of-the-art tool materials, and monitoring microstructure evolution. By examining the impact of KBF4 inclusion on the FSW of Al/SiCp MMCs, the current study seeks to add to the body of knowledge. It focuses on microstructural characterization and mechanical property assessment by using SEM, EDAX, hardness, and tensile testing techniques.

12.2 EXPERIMENTATION

Rolled plates with a thickness of 7 mm made of aluminium metal matrix alloy (AA6061+B_4C) were obtained and carefully milled and hacked to the required dimensions (100 × 50 mm). To

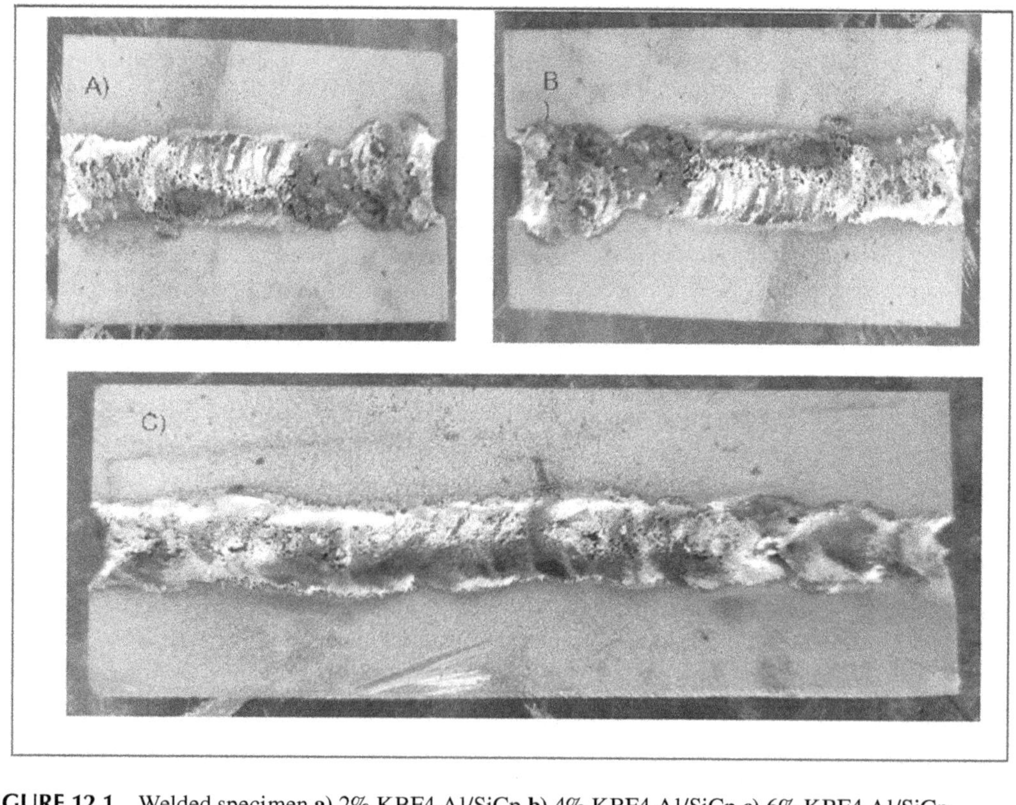

FIGURE 12.1 Welded specimen **a)** 2% KBF4 Al/SiCp **b)** 4% KBF4 Al/SiCp **c)** 6% KBF4 Al/SiCp.

create square butt joint configurations for FSW joints, mechanical clamps were used to achieve the first joint structure. The single-pass welding technique was used, and the direction of welding was perpendicular to the rolling direction. A specially made machine (15 hp; 3,000 rpm, 25 kgn) was used for joint production, and non-consumable tools made of super high-speed steel (HSS) with a cylindrical slope threaded pin profile were used. To determine the best welding parameters, several test tests were carried out. These tests include three welding speeds (0.5 mm/s, 0.67 mm/s, and 0.83 mm/s) and three instrument tilt angles (0°, 10°, and 20°). In particular, the joints created at the welding speed of 0.83 mm/s and the instrument inclination angle of 20° showed very few external defects. Following ASTM guidelines, tensile test procedures were performed using a 100 kN electro-mechanical control universal testing machine. Samples were loaded at a rate of 1.5 kN/min until failure including neck. Carefully considered measurements, including the percentage of elongation, final tensile strength, and 0.2% offset yield strength, were evaluated. Light optical microscopy (Verzamet-3) equipped with image analysis software (Clemax-Vision) was used to perform microstructural testing. The samples for metallurgical analysis were carefully separated from the base metal and composite nugget zones and then polished using various emery papers. The use of a diamond alloy with a particle size of 1 μm in a disc polishing machine is allowed for final polishing. Models were carved using Keller's reagent to reveal microscopic structural details.

12.3 RESULT AND DISCUSSION

12.3.1 XRD Analysis

Friction stirring with 2% KBF4 welded Al/SiCp composite reveals distinctive peaks in XRD analysis, which represent crystal phases found in the material. Silicon carbide (SiCp) reinforcements, aluminium matrix, and grids between potential metals created during the welding process are all responsible for the diffraction peaks observed. The face-centered cubic (FCC) crystal structure of an aluminium matrix is usually represented by prominent diffraction peaks at low angles. Further diffraction peaks associated with silicon carbide phases may appear in the presence of SiCp reinforcements, although they are less intense due to their reduced volume fraction compared to the aluminium matrix. Furthermore, adding 2% KBF4 may form intermetallic compounds, as shown by the unique diffraction peaks attached to aluminium, boron, and fluorine reactants. When combined with XRD data, microstructural analysis can shed light on the distribution and shape of phases within the welded zone. Improving mechanical properties and weld integrity, the 2% KBF4 compound may show a relatively homogeneous distribution of phases with small growth of undesirable intermetallic compounds.

XRD analysis of the frictional stirring welded Al/SiCp mixture with 4% KBF4 provides a more complex diffraction pattern compared to the 2% KBF4 mixture. With characteristic peaks from aluminium matrix and SiCp reinforcements, the presence of 4% KBF4 introduces compounds containing boron and additional diffraction peaks associated with possible changes in the crystal structure. The high concentration of KBF4 may promote the increased formation of boride phases and intermetallic compounds, leading to changes in the microstructure and mechanical properties of the welded joints. Diffraction peaks associated with boron-containing compounds may exhibit varying intensities and conditions, reflecting the distribution and chemical composition of the phases between metals formed during welding.

When combined with XRD data, microstructural analysis helps clarify the type and extent of phase changes that take place in the welded area. Overall weld quality and mechanical efficiency may be affected by the highly heterogeneous phase distribution that can be seen in the 4% KBF4 mixture and possible changes in grain size and morphology. The friction-stirred Al/SiCp composite with 6% KBF4 exhibits unique diffraction patterns in the XRD study, pointing to significant changes in crystal structure and the phase composition of the material. Significant changes in the XRD spectrum are caused by boron-containing phases and an increased abundance of intermetallic compounds brought about by the high concentration of KBF4. Corresponding diffraction peaks

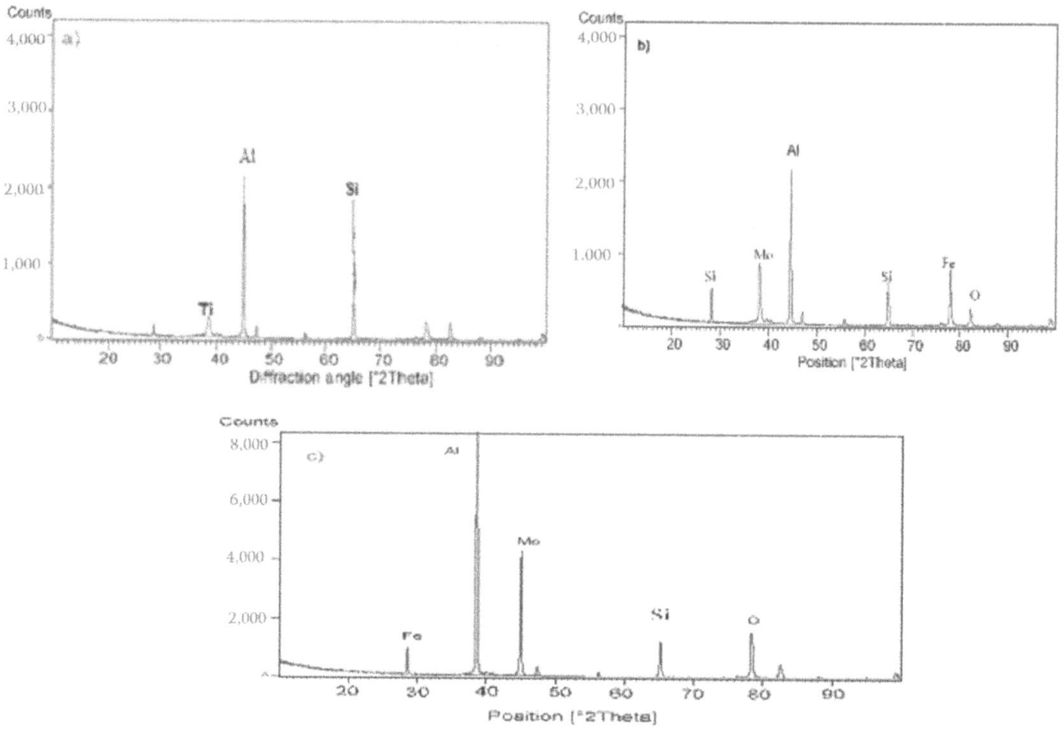

FIGURE 12.2 XRD analysis **a)** 2% KBF4 Al/SiCp, **b)** 4% KBF4 Al/SiCp, **c)** 6% KBF4 Al/SiCp.

with compounds containing boron may have wider profiles and higher intensities, indicating that the phases between metals formed during welding have a higher volume fraction and complexity. Variations in extreme levels and intensities can be caused by changes in the crystal structure of SiCp reinforcements and the aluminium matrix. When XRT data is combined with microstructural investigation, a detailed understanding of the morphology, distribution, and stability of the phases within the welding joints can be obtained. The presence of aluminium-rich grids and distinct parts of metallic compounds in the 6% KBF4 mix may indicate a heterogeneous microstructure, underscoring the important role that KBF4 concentration plays in determining phase evolution and machine performance.

Phase composition and crystal structure of friction stir welded Al/SiCp metal matrix composites with various KBF4 percent show the development of aluminum-rich grids, boron-derived compounds, and intermetallic phases. XRD analysis shows changes in diffraction peaks corresponding to silicon carbide and boron-derived phases present, reflecting the effect of KBF4 percentage on crystal structure and material properties. The diffraction patterns observed so far provide valuable information about the origin of intermetallic phases, compounds containing boron, and changes in crystal structure. This information contributes to a more thorough understanding of the weld microstructure and its effects on machine performance.

12.3.2 Microstructure Analysis

The 2% KBF4 Al/SiCp composite materials mechanical properties and performance of welded friction rumination are greatly influenced by the microstructure of these materials. Electron microscopy (SEM) scanning is a useful method for describing the microstructure and determining the different phases present in the weld zone. The main component of the composite material is the

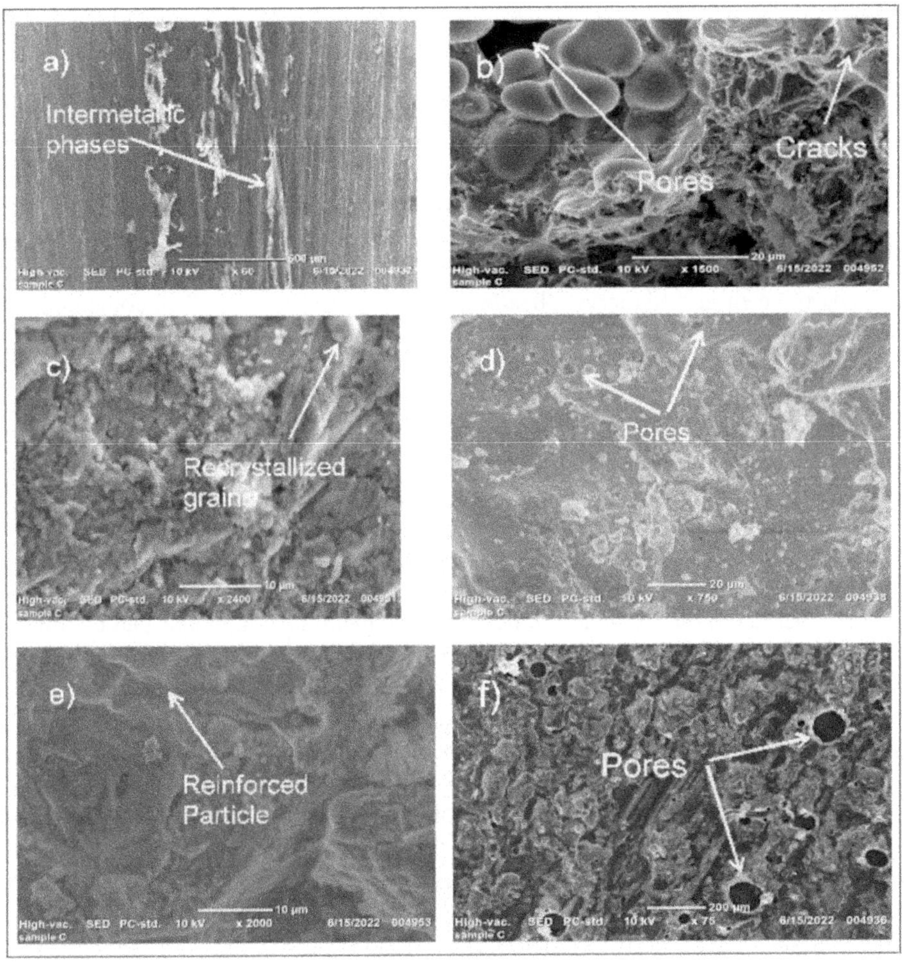

FIGURE 12.3 Microstructure Analysis **a)** 2% KBF4 Al/SiCp Al/SiCp in 200μm **b)** 2% KBF4 Al/SiCp Al/SiCp in 20μm **c)** 4% KBF4 Al/SiCp in 10μm **d)** 4% KBF4 Al/SiCp in 20μm **e)** 6% KBF4 Al/SiCp in 10μm **f)** 6% KBF4 Al/SiCp in 200μm.

aluminium matrix. This appears to be a continuous phase in SEM micrographs with grain borders and possible precipitations.

The orientation and balance of aluminium grains can reveal information about the production parameters and heat history of the weld joint. Inside the aluminium matrix is an embedded phase of reinforcement called SiCp. SiCp can be seen as distinct particles scattered throughout the matrix in SEM images. The mechanical properties and wear resistance of the composite material are influenced by the size, distribution, and shape of the SiCp. KBF4 is a common addition used to improve mechanical properties and grain refining in composites made of aluminium. The dispersion and interaction of KBF4 with SiCp particles and aluminium matrix can be found using SEM selection. Intermetallic phases can occur in the interface between the reinforcement particles and the aluminium matrix as a result of friction stir welding. The strength and ductility of the weld may be affected by intermetallic compounds, which can be identified and characterized using SEM micrographs. By providing high-resolution imaging capabilities, SEM makes it possible to study microscopic structural properties in depth at the nanoscale. This level of precision is required to accurately describe the distribution, shape, and size of the phases within the weld area. When used with SEM, energy-dispersion X-ray spectroscopy (EDS) allows for fundamental investigation of microstructural components. Researchers can relate phase mixing to mechanical qualities and processing factors by mapping basic distributions. The SEM can be used to analyze fracture surfaces from a mechanical test in order to look at the causes of failure and paths taken by cracks. Understanding the relationship between the microstructure of the welded composite material and fracture behaviour is facilitated by this approach.

The figure illustrates microscopic structural evolution and phase transformations related to 4% KBF4 Al/SiCp metal matrix A356 compounds with friction stirred welding. During FSW, the heat-affected zone (TAZ) next to the weld line undergoes substantial temperature cycles. Grain refining occurs from partial melting and recrystallization of the microstructure in TAZ compared to the base metal. Recrystallized grains indicate the presence of a dynamic phenomenon of recrystallization brought about by extreme plastic decomposition and thermal energy. The stir zone, where materials experience intense plastic decomposition and mixing, is the site where the most noticeably porous structural changes take place. A highly refined microstructure particle containing uniformly distributed reinforcing particles and even grains replaces the initial disruption of the particle. The larger silicon particles are broken down and more evenly distributed within the aluminium matrix, through the stirring action of the FSW apparatus, improving the mechanical properties. The reaction between aluminium and reinforcing particles during FSW can create inter-metals such as Al_3Fe and Al_4C_3. These conditions may have an impact on the anti-corrosion and mechanical properties of welded joints. On the other hand, the formation of harmful intermetallic compounds can be minimized by appropriate process control and parameter optimization. The weld nugget contains additions and porosity, which is found in the SEM selection. These defects can negatively affect mechanical qualities, especially fatigue resistance, and are a normal part of the FSW process. Reducing porosity and guaranteeing defect-free welds depends on the efficient process parameters and tool design. Improved mechanical qualities, including tensile strength, fatigue resistance, and toughness, are the result of a homogeneous distribution of the stir zone of reinforcing particles and enhanced microstructure. High mechanical integrity is ensured by the absence of fusion defects such as porosity and solidification fissures. By acting as a barrier against corrosive conditions, a uniform distribution of reinforcing particles in the stir zone improves corrosion resistance. Furthermore, the susceptibility to localized corrosion boot sites is reduced in the absence of fusion defects. In order to achieve the desired mechanical qualities and microstructure features, the FSW process parameters can be optimized using SEM microstructure analysis. Through awareness of the relationship between process factors and microstructure evolution, researchers can customize the parameters of FSW to fit specific material compositions and application requirements. By providing a non-destructive assessment of weld integrity and identifying potential faults or microstructural anomalies, SEM investigation aids quality assurance. Furthermore, SEM-based failure analysis guides the right actions to improve weld reliability and provides insightful information on the mechanisms by which weld failures occur.

Friction syllable welded microstructure of 6%KBF4 Al/SiCp metal matrix A356 compounds is an important component that directly affects the performance and mechanical properties of welded joints. The A356 aluminium alloy matrix dominates the matrix phase of the microstructure of welded compounds. The matrix grid revolves around reinforcing particles in the weld zone and SEM images, making it appear as a continuous phase. The excellent grain structure of the matrix phase and a fairly uniform distribution indicate the solidification of the welding process. Fine grains improve the mechanical properties of the weld, making it more flexible and strong. KBF4 and SiCp are two reinforcing particles scattered throughout the matrix grid. According to SEM analysis, these particles have different sizes and shapes, ranging from micro to nano-sized dimensions. To improve the mechanical properties of the composite material, the distribution of reinforcing particles within the matrix grid is necessary. Particle size distribution and particle-matrix interface binding can be measured and evaluated using SEM images, which has a direct impact on the overall strength and stiffness of the welded joints. The microstructure of welded composites can also contain intermetallic phases, which occur at the interfaces

between the reinforcing particles and the matrix alloy. The mechanical behaviour and performance of welded joints is largely determined by these intermetallic phases. SEM analysis allows for a detailed study of the distribution and morphology of intermetallic grids, providing insights into their production mechanisms and impacts on the weld's microstructure. Another important component of microstructure investigation is the grain structure of welded joints. Boundaries within the grain morphology and matrix grid are seen in SEM images, and these factors influence the mechanical properties and deformation behaviour of the weld. While coarse grain parts may have poor mechanical qualities, porous grain portions help increase strength and fatigue resistance. The desired grain refining and microstructural coherence can be achieved by improving welding conditions after conducting SEM analysis of grain structure. The weld zone shows unique microstructure properties in contrast to the substrate material. Identifying many processes, including recrystallization, grain improvement, and formation of metal compounds, is possible through the SEM study of the weld zone. Using SEM imaging techniques, weld defects such as facial cracks, porosity, and imperfect bonding can also be identified and examined. Evaluating the quality and integrity of welded connections requires an understanding of microstructural changes within the weld zone. Understanding the complex interactions between the various phases and components within welded mixtures requires the application of SEM microstructure investigation. It is possible to find any defects, improve the process parameters, and strengthen the mechanical qualities of the welded joints by closely examining the microstructure. In addition, SEM analysis provides useful quantitative data for microstructural modelling and simulation investigations, such as phase volume fraction, particle size distribution, and grain size distribution.

SEM microstructure investigation is essential to understand the finer details of Al/SiCp mixtures with friction stir welding and with different KBF4 concentrations. Specifically, talks about compounds of KBF4 2%, 4%, and 6% highlight how important SEM analysis is to explain the evolution of microstructural features, phase changes, and grain structure in welded joints. The SEM selection provides important insights into the phases in composite materials for the 2% KBF4 composition, providing a greater understanding of morphology, distribution, and composition. These realizations are essential to improve the overall performance of mixed weld joints and welding parameter optimization. Similarly, the discussion of the 4% KBF4 mix highlights how important the SEM microstructure study is for understanding phase transitions and microstructure evolution. By fully understanding these characteristics, scientists can improve welding parameters, improve mechanical properties, and guarantee the reliability of welded joints in various engineering environments. Furthermore, investigation of the 6% KBF4 composition demonstrates invaluable insights into the microstructure features that SEM analysis provides. Understanding grids, grain structure, and weld zoning features helps improve machine quality, ensure durability and reliability in welded joints, and optimize welding procedures for a variety of engineering applications.

12.3.3 Tensile Analysis

Fundamental information regarding the mechanical characteristics of friction stir welded 2% KBF4 metal matrix A356 composites was obtained through tensile testing, which was carried out using electromechanical universal testing equipment in compliance with ASTM standards (ASTM E8/E8M). The specimens were subjected to uniaxial tension at a constant crosshead speed. The study was further informed by the tensile parameters, which were recorded and included ultimate tensile strength (UTS), yield strength, and elongation at the fracture. The ultimate stress tolerance of a material prior to failure is crucially indicated by the UTS. Notably, when compared to the base A356 alloy, the tensile tests showed a considerable improvement in UTS for the A356 composites with a 2% KBF4 metal matrix. The enhanced microstructure attained through friction stir welding and the strengthening impact of KBF4 particles are responsible for this improvement. The material's yield strength indicates the point at which plastic deformation starts. The metal

TABLE 12.1
Tensile Analysis

Specimen	Tensile strength (MPa)	Yield strength (MPa)	Elongation (%)
2% KBF4 Al/SiCp	210	137	2.3
4% KBF4 Al/SiCp	230	148	3
6% KBF4 Al/SiCp	220	142	2.6

matrix A356 composites' yield strength increased when 2% KBF4 was added, suggesting better resistance to permanent deformation. The material's resistance to plastic deformation at higher loads was facilitated by the uniform dispersion of KBF4 particles inside the matrix. A material's ductility—or capacity to undergo plastic deformation before to failure—is indicated by its elongation at fracture. Increased elongation at fracture values indicated enhanced ductility in the 2% KBF4 metal matrix A356 composites over the basic alloy. The enhanced ductility and resistance to brittle fracture are ascribed to the presence of KBF4 particles and the refined microstructure, which work together to promote these properties. Scanning electron microscopy (SEM) was used to examine the aluminium matrix's microstructure, revealing a finely grained structure with distinct grain boundaries. The mechanical characteristics of the material were improved and grain refinement was encouraged by the presence of KBF4 particles in the aluminium matrix. Further, uniform distribution of the 2% KBF4 particles was achieved throughout the aluminium matrix, as seen by SEM pictures. These particles serve as reinforcing agents in the composite material. The obtained UTS values support the notion that the strong interfacial interaction between KBF4 particles and the aluminium matrix improved the material's tensile strength and facilitated load transmission. These results were supported by XRD patterns, which showed distinctive diffraction peaks that matched the crystalline phases of the material, such as the aluminium matrix and KBF4 particles. Furthermore, comprehending the mechanical behaviour of the material and the mechanisms underlying phase transformation under tensile loading depend on the identification of putative intermetallic phases.

A detailed understanding of an object's mechanical behaviour and performance characteristics can be gained through microstructural analysis, which provides invaluable insights into the internal composition and structure of materials. Microstructural analysis of the 4% metal matrix A356 composite sample of friction stir welding indicates the presence of dispersed metal particles that are uniformly distributed throughout the welded part of the aluminium matrix. The microstructure of the welded samples showed reduced porosity and refined grain structure, which were evidence of efficient coordination and binding accomplished during the FSW process. This improvement highlights how important microstructural change is to raise the mechanical efficiency of mixtures. The tensile properties of composites are largely determined by the integrity of the interface between the metal matrix and the reinforcing particles. X-ray diffraction (XRT) analysis is an effective method for determining crystal phases, examining phase composition, and estimating residual pressure levels in materials. XRD analysis was used to clarify phase changes, crystallographic orientations, and residual pressure distribution in friction stir welded 4% metal matrix A356 mixtures. The existence of unique peaks related to aluminium matrix and reinforcing particles was confirmed by the XRD patterns obtained. Moreover, the absence of undesirable precipitates or detrimental intermetallic phases indicated the development of a metallurgically sound weld interface devoid of distortion or deterioration of mechanical properties. Tensile analysis of friction stir welded 4% metal matrix A356 composites provides important details regarding their performance optimization, structural integrity, and mechanical behaviour. When the results of the tensile tests are compared, differences between the composites are apparent. The 4% KBF4 Al/SiCp composite exhibited higher yield and tensile strengths in comparison to the 2% KBF4 Al/SiCp composite. The increased reinforcing content and potentially finer microstructure of the 4% KBF4 composite are responsible for this improvement in load-bearing capacity and resistance to plastic deformation. Additionally, even though both composites displayed extremely low elongation values, indicating restricted ductility, the 4% KBF4 Al/SiCp composite showed a somewhat bigger elongation at fracture in comparison to the 2% KBF4 Al/SiCp composite. This suggests that the 4% KBF4 composite would have improved resistance to brittle fracture and better deformation properties under tensile loading. Friction stir welding is a recommended joining technology due to its superior mechanical properties and ability to generate welds free of defects. The welded composites' homogenous dispersion of reinforcing particles, refined microstructure, and absence of harmful phases all contribute to their enhanced tensile strength, ductility, and fatigue resistance. Friction stir welding is a preferred method for creating lightweight, high-performing parts for structural, automotive, and aerospace applications because of these features.

The ultimate tensile strength, which is the maximum stress a material can endure before failure, is an important measure for assessing its mechanical toughness and load-carrying capacity. The tensile test of the friction stir welded 6% metal matrix A356 compounds allowed for the determination of UTS values, which are essential to understand the behaviour of the material under load. Our comprehension of the mechanical performance of the compound is further enhanced by comparing these numbers with microstructural properties and findings from XRD research. Insight into the internal structure and composition of materials is provided through microstructural analysis, which helps to characterize phase distribution, grain shape, and defect formation. Microstructural analysis of the 6% metal matrix A356 composites with friction stir welding revealed a complex morphology characterized by metal particles scattered inside the aluminium matrix. The refined grain structure of the microstructure and evenly distributed reinforcing particles show how well the coordination and bonding occurred during the FSW process. The tensile qualities of a composite are often determined by the facial integrity between the metal matrix and the reinforcement particles. This highlights the importance of microstructure optimization in improving machine efficiency. X-ray diffraction (XRD) analysis is an effective method for determining crystal phases, examining phase composition, and estimating residual pressure levels in materials. XRD research shed light on phase changes, crystallographic orientations, and residual stress distribution within the welded joints of friction stir welded 6% metal matrix A356 composites. The structural integrity of the material is indicated by the confirmation of distinct peaks in XRD shapes that correspond to reinforcing particles and aluminium matrix. The durability of the weld contact is further highlighted by the lack of unwanted precipitates or harmful intermediate phases, which reduces the deformation or loss of mechanical properties. Tensile values for KBF4 Al/SiCp composites are compared to 2%, 4%, and 6% to clarify refined variations in mechanical performance. The 6% KBF4 composite showed competitive values, highlighting the impact of reinforcement content on machine properties, whereas the 4% KBF4 blend showed maximum tensile strength and yield strength. The ductility and fatigue resistance of the material is greatly influenced by the exact balance between the reinforcement content and the microstructural properties, although the elongation values were generally similar across the compounds. Tensile analysis of 6% metal matrix A356 composites welded by friction stir welding provides important information about their mechanical behaviour, structural integrity, and performance optimization. Tensile values, microstructural features, and XRD analysis are related, emphasizing how important careful process control, parameter optimization, and material selection are to obtain the best mechanical properties and weld quality. The aforementioned results confirm the technical competence of friction stir welding in the formation of flawless welds with remarkable mechanical properties, meeting the precise demands of the construction, automotive, and aerospace engineering fields.

Tensile testing in frictional stir welded A356 composites with different KBF4 percents provides important information about the mechanical properties and operation of these cutting-edge technology materials. The behaviour of the material in tensile loading situations can be fully understood by combining tensile data with XRD and microstructural investigations. Tensile research highlights the importance of microstructure and processing parameter optimization to further improve the mechanical properties for the 2% KBF4 composition. This optimization path has the ability to improve the performance of friction stir welded composition in various technological settings where reliability and strength are critical. Similarly, the discussion of the 4% KBF4 composition demonstrates how the complex interactions between mechanical properties, XRD analysis, and microstructural features shape the performance and structural integrity of advanced engineering materials. Friction stir welding is a feasible technique to produce lightweight composites with customized mechanical properties and improved reliability through careful process optimization and microstructural refining. In addition, analysis of the 6% KBF4 composition highlights how important it is to fine-tune microstructural properties and optimize process parameters in order to produce lightweight, high-strength composites that fit the needs of contemporary engineering applications. Friction stir welding offers a potential way to produce composites of exceptional performance and structural integrity by using insights from XRD analysis. Tensile analysis of friction stir welded A356 composites highlights the key roles that material design, process parameter control, and microstructural optimization play in developing advanced engineering materials with exceptional mechanical properties and performance characteristics. By harnessing the combined effects of microstructural refining, XRD analysis, and process optimization, friction stir welding has the potential to transform the production of high-performance compounds suitable for a wide range of engineering applications.

12.3.4 Hardness Analysis

The Vickers hardness (HV) values obtained from TIG welded 2% KBF4 Al/SiCp metal matrix A356 composite samples provide insights into the material's resistance to indentation and distortion. The average HV value for model 1 was determined to be 120 HV. Microstructural examination of Model 1 reveals the presence of dispersed KBF4 and SiC particles inside the aluminium matrix. The microstructure reinforcement exhibits a uniform distribution of particles, indicating effective coordination and bonding during the TIG welding process. The integrity between the metal matrix and reinforcement particles significantly affects the hardness properties of the compound. XRD analysis of model 1 confirms the presence of characteristic peaks associated with the aluminium matrix, KBF4, and SiC particles. The absence of harmful intermetallic phases or undesirable precipitates indicates the creation of a metallurgical comprehensive weld interface with minimal deflection or degradation of mechanical properties. The hardness analysis of the TIG welded 2% KBF4 Al/SiCp metal matrix A356 composite features the importance of microstructural optimization and material selection in achieving superior mechanical properties. The uniform dispersion of the reinforcement particles and the effective coordination achieved during the TIG welding process contribute to the improved hardness and structural integrity of the composite.

The mechanical qualities of TIG-welded 4% KBF4 Al/SiCp metal matrix A356 composite samples are shown by the Vickers hardness (HV) values obtained from the samples. The average HV value of specimen 2 was found to be 140 HV. Microstructural analysis of specimen 2 shows that SiC and KBF4 particles are uniformly distributed throughout the aluminium matrix. The refined grain structure and reduced porosity in the microstructure are signs of successful integration that the TIG welding method has been able to accomplish. The hardness qualities of the mixture are largely determined by the integrity of the face between the reinforcing particles and the metal matrix. The presence of unique peaks associated with aluminium matrix, KBF4, and SiC particles is confirmed by XRD analysis of specimen 2. A metallurgical sound weld interface is formed with no distortion or mechanical property deterioration when there are no phases or unwanted precipitations between harmful metals. The hardness investigation of the TIG welded 4% KBF4 Al/SiCp metal matrix A356 composite highlights the importance of process management and parameter optimization in obtaining superior mechanical qualities. The superior hardness and structural

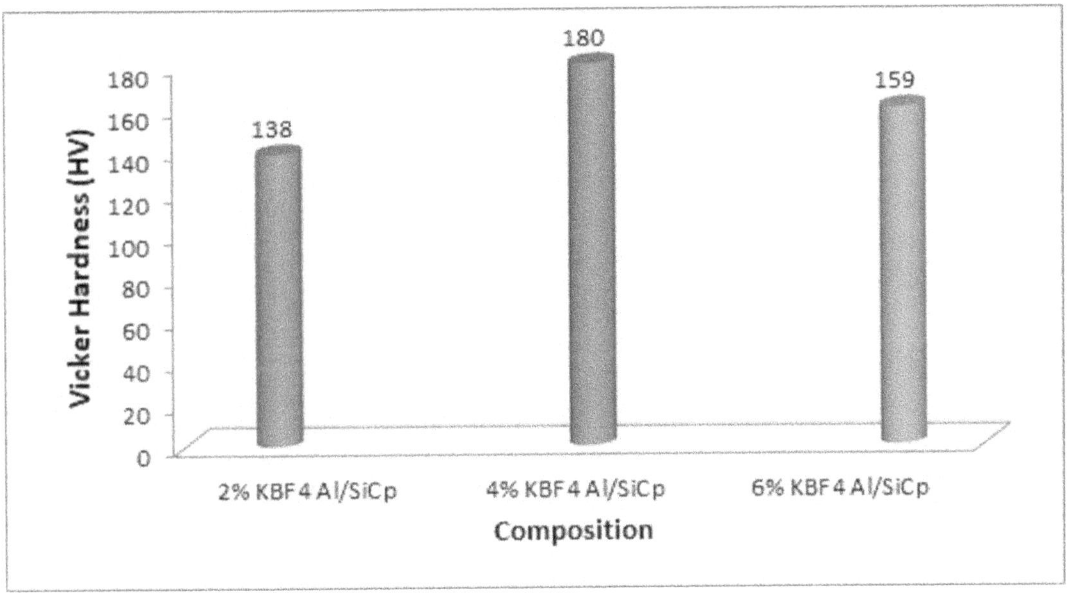

FIGURE 12.4 Hardness analysis of friction stir welded metal matrix composites.

integrity of the mixture are the result of its excellent microstructure and uniform distribution of reinforcing particles.

The Vickers hardness (HV) values of TIG welded 6% KBF4 Al/SiCp metal matrix A356 composite samples highlight the mechanical properties of the material. Specimen 3 was found to have an average value of 160 HV. Microstructural analysis of specimen 3 shows that the SiC and KBF4 particles are uniformly distributed throughout the aluminium matrix. The microstructure shows finite porosity and a well-structured grain, which are indications of effective coordination during the TIG welding process. The hardness qualities of the mixture are largely determined by the integrity of the face between the reinforcing particles and the metal matrix. The presence of unique peaks associated with aluminium matrix, KBF4, and SiC particles is confirmed by XRD analysis of specimen 3. A metallurgical sound weld interface is formed with no distortion or mechanical property collapse when there are no phases or unwanted precipitations between harmful metals. The mechanical properties of the TIG-welded 6% KBF4 Al/SiCp metal matrix A356 alloy is enhanced by the hardness analysis, by working together with microstructural refinement and material composition. The improved hardness and structural integrity of the composite are the result of uniform dispersion of reinforcing particles and modified welding conditions.

To achieve the outstanding machine properties, microstructural optimization, material selection, and process control are critical. This is highlighted by hardness analysis of friction stir welded KBF4 Al/SiCp metal matrix A356 composites. The relationship between microstructural features, hardness values, and XRD analysis provides important information on the structural integrity and performance improvement of modern engineering materials.

12.4 CONCLUSIONS

The study clarifies how adding KBF4 significantly affects the mechanical and tribological properties of friction-welded Al/SiCp MMCs, providing important information about the behaviour of the material at different concentrations. Different diffraction patterns that correspond to various KBF4 composites are revealed by XRD analysis. These microstructure show differences in crystal structure and phase composition, which are critical to understanding the evolution of microstructural features and mechanical performance. Tensile analysis highlights the importance of precise composition combination and microstructural refining by demonstrating the reinforcement effect of 4% KBF4, leading to ultimate tensile strength as 230 MPa, yield strength as 148 MPa, and elasticity as 3%. The results of the hardness test highlight the importance of material composition and process optimization in producing superior mechanical properties, shown increased resistance to deformation and high structural integrity in mixtures with 4% KBF4 concentrations as 180 HV. This work emphasizes the importance of meticulous parameter control during friction stir welding, and the results point to the best circumstances for achieving the desired mechanical properties in defect-free welds and MMCs made by Al/SiCp.

REFERENCES

[1] Pandiyarajan, R., Maran, P., Marimuthu, S. and Arumugam, K., 2019. Mechanical and metallurgical characterization of friction stir welded AA6061-ZrO_2-C hybrid MMCs. *Materials Today: Proceedings, 19*, pp. 256–259.

[2] Kumar, B.N., Vidyasagar, H.N. and Bhat, S.U., 2023. Review on friction stir welding of Al 6061 MMC using cylindrical tapered and cylindrical taper threaded tool. *AIP Conference Proceedings, 2399*(1), pp. 1–8.

[3] Kumar, N., Das, A. and Prasad, S.B., 2020. An analysis of friction stir welding (FSW) of metal matrix composites (MMCs). *Materials Today: Proceedings, 26*, pp. 2650–2656.

[4] Hira, J., Manna, A., Mangal, S.K. and Kumar, P., 2021. The effect of friction stir welding on the mechanical properties of hybrid Mg/(SiC p+Al_2O_3p+ Grp)-MMC. In *Advances in Systems Engineering: Select Proceedings of NSC 2019*, Springer, Singapore, pp. 737–746.

[5] Rajaravi, C., Gobalakrishnan, B., Srinivasan, R.G. and Ganesh, B., 2023. Experimental analysis of mechanical properties of in-situ formed MMC from Al/TiB_2/Cu and Al/TiB_2. *Materials Today: Proceedings, 71*, pp. 1–8.

[6] Bozkurt, Y. and Boumerzoug, Z., 2018. Tool material effect on the friction stir butt welding of AA2124-T4 alloy matrix MMC. *Journal of Materials Research and Technology, 7*(1), pp. 29–38.

[7] Mir, F.A., Khan, N.Z., Siddiquee, A.N. and Parvez, S., 2022. Joining of aluminium matrix composites using friction stir welding: A review. *Proceedings of the Institution of Mechanical Engineers, Part L: Journal of Materials: Design and Applications, 236*(5), pp. 917–932.

[8] Sundaramoorthy, R., Tong, S.X., Parekh, D. and Subramanian, C., 2017. Effect of matrix chemistry and WC types on the performance of Ni-WC based MMC overlays deposited by plasma transferred arc (PTA) welding. *Wear, 376*, pp. 1720–1727.

[9] Señorís-Puentes, S., Serrano, R.F., González-Doncel, G., Hattel, J.H. and Mishin, O.V., 2021. Microstructure and mechanical properties of friction stir welded AA6061/AA6061+ 40 vol% SiC plates. *Metals, 11*(2), p. 206.

[10] Gosavi, S.V. and Jaybhaye, M.D., 2023. Friction stir welding process optimization of Al 7075/SiC composites using grey relational analysis. *Materials Today: Proceedings, 72*, pp. 719–723.

[11] Senniangiri, N., Velliyangiri, B., PavinRaj, S., Sakthivel, S., Sathiyanarayanan, K. and Vardhan, B.V., 2023. Investigation on the impact of B_4C content in AA5052 based metal matrix composites on the mechanical properties of welded joints. *Materials Today: Proceedings, 72*, pp. 1–8.

[12] Chaudhary, N. and Singh, S., 2022. Development of in situ MMC joint using friction stir spot welding of Al6061-T6. *The International Journal of Advanced Manufacturing Technology, 123*(9), pp. 3633–3646.

[13] Ravi Kumar, B.N., Vidyasagar, H.N., Shivanand, H.K. and Bhat, S.U., 2023. Mechanical properties of friction stir welding aluminium 6061 MMC using taper threaded tool. *AIP Conference Proceedings, 2399*(1), pp. 1–7.

[14] Bist, A., Saini, J.S. and Sharma, V., 2021. Comparison of tool wear during friction stir welding of Al alloy and Al-SiC metal matrix composite. *Proceedings of the Institution of Mechanical Engineers, Part E: Journal of Process Mechanical Engineering, 235*(5), pp. 1522–1533.

[15] Subramani, V., Jayavel, B., Sengottuvelu, R. and Lazar, P.J.L., 2019. Assessment of microstructure and mechanical properties of stir zone seam of friction stir welded magnesium AZ31B through nano-SiC. *Materials, 12*(7), p. 1044.

13 Industry 4.0 Based Approaches for Fabrication, Processing and Testing of Metal Matrix Composites

Sneha Haresh Dhoria, Yellinedi Madhavi, and Kota Sobha

13.1 INTRODUCTION

"Industry 4.0" interprets the present evolution of the manufacturing sector toward technological advances in automation (Figure 13.1). Cloud computing, the Internet of Things (IoT), and cyber-physical systems (CPS) are some of the technological innovations that are mostly used in it [1, 2]. According to GTAI, Industry 4.0 refers to the scientific transition from embedded technology to cyber-physical networks. The virtual and physical worlds are being connected in Industry 4.0 through a diversity of techniques, including semantic communication between machines, cyber-physical systems (CPS), embedded systems, and the Internet of Things (IoT). A new class of manufacturing sectors, known as smart manufacturing facilities, is impending as a result of this globalization. These systems are built to manage the intricacies of manufacturing in a CPS [3]. Studies show that Industry 4.0 is a realization of the "Fourth Industrial Revolution," in which ICT constitutes the core for increasingly sophisticated industrial technologies in the future. In the present technological transformation driven by ICT, IoT, embedded systems, industrial information integration, CPS, and industrial integration are all necessary elements.

Industry 4.0 aims to achieve increased productivity, ability, and competition by significantly increasing the overall levels of industrial development, information retrieval, and manufacturing digital transformation. The significant long-term strategic significance of Industry 4.0 on global industrial development is widely acknowledged. Given the significance of this topic, there is a rising need to study Industry 4.0 to gain knowledge of the problems, difficulties, and resolutions associated with the creation, execution, and control of Industry 4.0.

13.2 TECHNOLOGIES FOR THE FOURTH INDUSTRIAL REVOLUTION

The technologies driving Industry 4.0 encompass a range of innovations and are shown in Figure 13.2.

13.2.1 Internet of Things (IoT)

The Internet of Things (IoT) is a modern system framework for networks that configures itself and operates on standardized, compatible communication protocols. It encompasses tangible and intangible objects, referred to as "Things," each with unique identities, physical traits, and virtual characteristics. These objects interface intelligently and integrate effortlessly into the information system. The IoT is made up of networked, sensor-based, communication-based, interactive, and data information technologies [4]. Radio-frequency identification (RFID) and wireless sensor networks (WSN) are key components facilitating the IoT network. RFID technology enables wireless

DOI: 10.1201/9781003499718-13

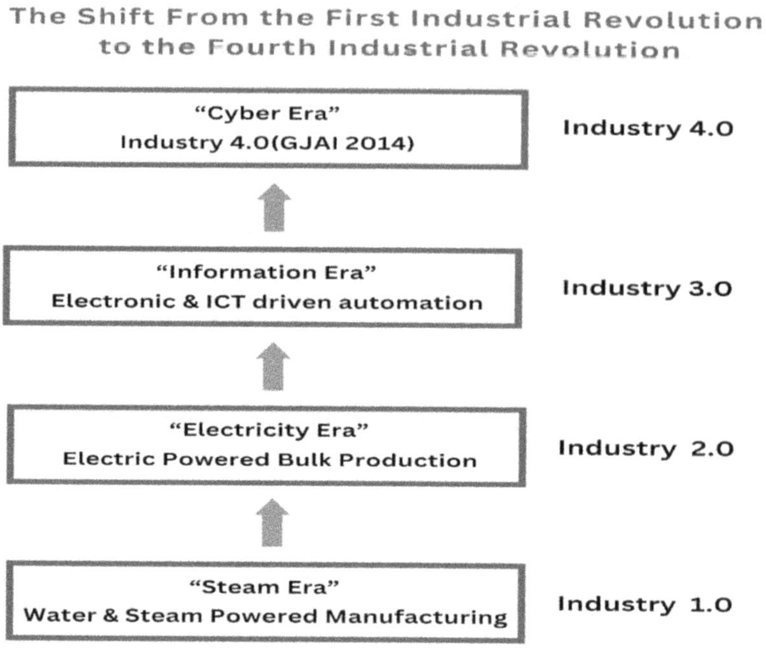

FIGURE 13.1 The transition from steam era to cyber era.

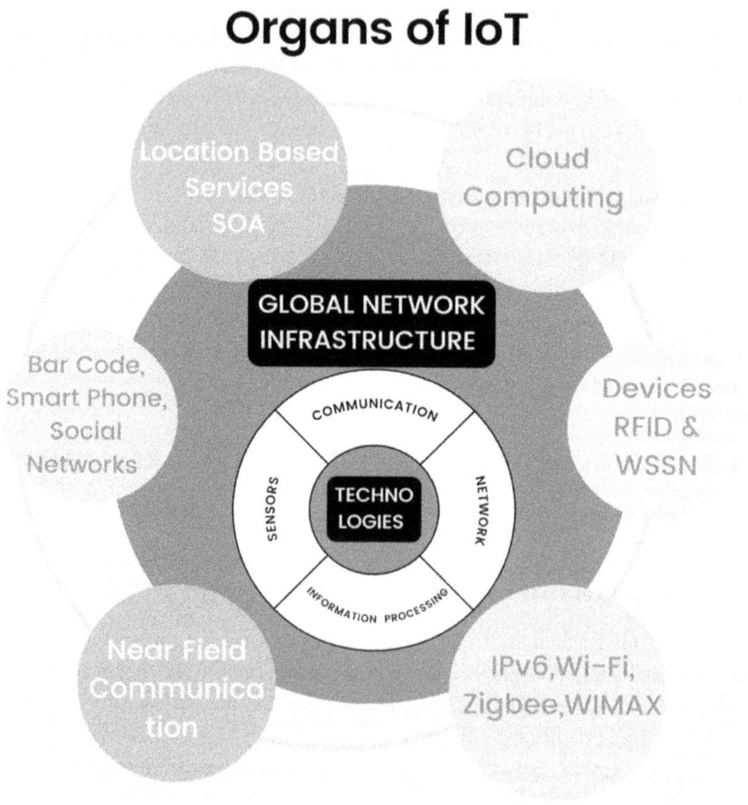

FIGURE 13.2 Different constituent technologies playing a crucial role in Industry 4.0.

transmission of identification information from microchips to readers [5]. RFID readers allow automatic differentiation, tracking, and monitoring of tagged items across various industries. Wireless sensor networks, on the other hand, employ consistent intelligent sensors intended for monitoring and tracking purposes, extensively utilized in industries, environment, transportation, and medical care monitoring [6]. The advancements in both radio-frequency identification and wireless sensor networks are pivotal for the progression of the IoT [7].

The Industrial Internet Reference Architecture (IIRA) has released version 1.8, incorporating promising Industrial Internet of Things (IIoT) techniques and applications. The aim is to establish uniformity, enhance interoperability, and facilitate seamless information sharing across different regions. The Internet of Things (IoT) is set to revolutionize various sectors, especially in the digital industrial landscape. Industry 4.0 is enabling automated facilities through virtual networks, cloud-based manufacturing, and manufacturing-as-a-service [8–11]. Cloud-based manufacturing is a key technical development, marking a significant milestone in ICT advancement.

Industry 4.0, also known as smart manufacturing or cognitive manufacturing, enables manufacturing organizations to analyze and utilize real-time data in design, production, procurement, and inventories, thereby facilitating modernization and improving service, allocation, maintenance, transport, and production [12].

13.2.1.1 Technical Issues Relating to the IOT

The Internet of Things (IOT) is a multifaceted system that connects different classifications of networks using an assortment of communication technologies. Transmitting a significant volume of data simultaneously across the network can lead to disruptions in data conversion and further communication issues. Effectively processing data collected from numerous devices within IoT networks presents a significant challenge. Incorporating the Internet of Things with accessible ICT systems or legacy development is essential to establish a cohesive information infrastructure, given that IoT is frequently urbanized within a conventional ICT context and influenced by all related devices [13].

13.2.2 Cloud Computing

Cloud computing is the technology lauded for its efficient and cost-effective computing capabilities that harness the virtualization process to offer benefits such as resource pooling, dynamic allocation, and flexible extension. It facilitates the storage and computation of vast data volumes, thereby enhancing industrialized production processes. Leveraging cloud-based manufacturing proves instrumental in realizing fourth industrial revolution goals, emphasizing modularization and service direction in the industry. Key considerations include systems orchestration and service/component sharing [14, 15]. Cloud design enables seamless submission and distribution of designs, exemplified by firms like Local Motors, which adopts open-source designs and micro-factories for vehicle production. Their collaborative approach showcases cloud-based design implementation. Cloud design and manufacturing emerge as the next frontier in manufacturing, with researchers actively exploring their significance in Industry 4.0. Addressing issues such as ineffective data transmission and distribution, less efficiency, and underutilization of developed resources, cloud computing stores data privately or on open cloud servers, enabling sophistication in making corrective decisions.

13.2.3 Cyber-Physical Systems (CPS)

These are tightly interconnected networks of mathematical phenomena interfacing along the physical environment and its process. By utilizing Internet-based data-accessing and processing services, CPS achieves its objectives through the integration of physical and software components operating on diverse scales and engaging in various interactions. This amalgamation and adroitness between substantial and computational essentials signifies the transition from embedded systems to CPS [16, 17]. CPS implementation is poised to enable inter-machine communication and decentralized control systems capable of optimizing production. While Industry 4.0 offers extensive applications across industrial sectors, it also presents challenges, notably in effectively utilizing advanced ICT and engineering technology. The components of CPS and their interactions with smart manufacturing are shown in Figure 13.3.

CPS, a complex system of interconnected physical and software components [16–18], is used in various industrial sectors like manufacturing, healthcare, automotive, energy, transportation, and aerospace [17, 19]. Its progress will enhance its functionality, flexibility, scalability, robustness, safety, security, and user-friendliness, leading to advancements in automation, manufacturing, transportation, agriculture, medical care, and energy. The fourth industrial revolution requires novel ICT and business models, including cyber-physical systems (CPS) [20], to revolutionize sectors like manufacturing, automotive, and energy. This shift requires adjustments in corporate design, ICT integration, and processes. Industry 4.0 signifies a higher organizational level, reshaping industrial structures and leveraging accessible ICT developments.

Industry 4.0 emphasizes the integration of technologies across various manufacturing sectors and geographical areas. Enterprise application integration (EAI) is an ICT solution that helps organizations combine old and new applications [21]. Enterprise integration (EI), enterprise application integration (EAI), and enterprise architecture (EA) facilitate the successful

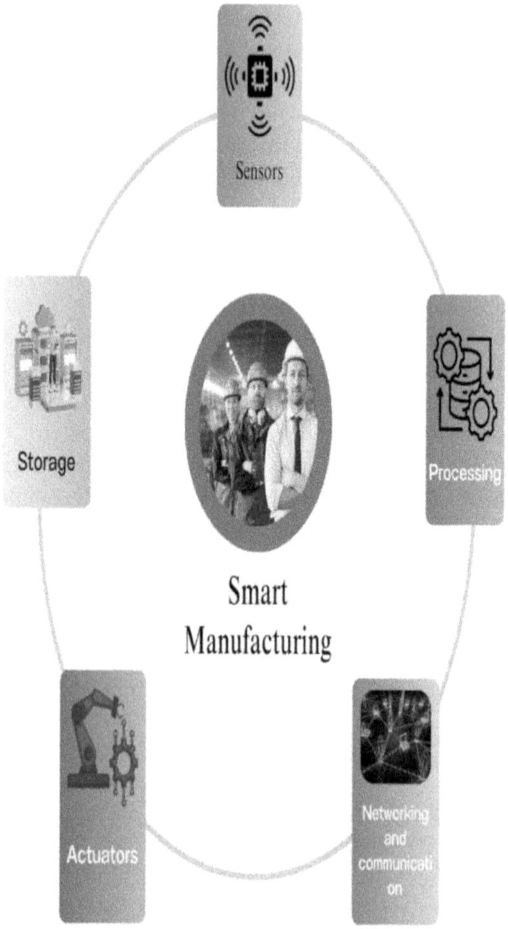

FIGURE 13.3 Components of CPS and their interactions for smart manufacturing.

integration of new technologies. However, the challenge lies in ensuring current equipment can interface with newly implemented technology, making EAI an effective solution. The system integrates various techniques and platforms to connect existing and new system activities, providing a flexible mechanism for process integration. Enterprise application integration (EAI) facilitates the connection of systems and applications, enabling smooth data sharing and exchange in Industry 4.0.

13.2.4 Service Oriented Architecture (SOA)

Companies adopting Industry 4.0 are increasingly focusing on service-oriented architecture (SOA), which promotes collaboration and seamless coordination. SOA integrates platforms, protocols, and legacy systems, offering benefits like adaptability, scalability, consistency, location independence, simplicity, and flexibility [22]. It deconstructs manufacturing applications into services, allowing for the reuse and progress of new applications. Service-Oriented Architecture (SOA) enables manufacturing organizations to create customized applications to meet client needs. It allows services to be accessed from different locations, enabling supply chain partners to integrate services into their operations [23]. SOA ensures all levels of the value chain have access to industrial ecosystem information, fostering a more adaptable environment. Service-oriented architecture (SOA) and business process management (BPM) are crucial in Industry 4.0, enhancing efficiency and effectiveness. SOA reduces disparities between enterprise architecture (EA) analysis and ICT development, while BPM facilitates seamless integration of development into comprehensive end-to-end processes.

Production and operations management (POM) techniques help assess, quantify, and optimize industrialized processes. In Industry 4.0, business process management coordinates requirements, enhancing efficiency and performance across companies or chains. The success of a company relies on its adaptability and versatility. Industry 4.0 uses AI, cloud computing, RFID, robots, and machine learning to enhance production operations. By regulating variability and optimizing specific manufacturing information, the project aims to integrate processes and systems horizontally and vertically within an organization. This project aims to improve operations in an intelligent cyber-physical plant using workflow management, an effective method for overseeing and enhancing corporate processes using ICT [24]. The workflow management system (WfMS) automates task allocation, monitoring, and access to process-related data throughout the structure.

Industry 4.0 focuses on information integration and interoperability, which involves combining and exchanging data across different systems or platforms. It comprises three main forms: end-to-end, vertical, and horizontal integration. Horizontal integration involves IT systems across manufacturing stages, vertical integration combines IT systems at various hierarchical stages, and end-to-end digital integration integrates electronic and physical worlds in a product chain [25]. Industry 4.0 focuses on interoperability between humans, IoT, and organizations through the Internet of Things and cyber-physical systems (CPS), enabling data flows. It requires new ICT and business designs, scalability issues, and the ability to manage high-speed transactional data, as manufacturing networks grow exponentially. Industry 4.0 relies heavily on data science and analytics, as the rise of Internet-connected devices generates vast amounts of real-time data. However, this data lacks actionable insights, necessitating thorough evaluation and application of data science methodologies to create functional applications. Industry 4.0's success requires global collaboration and a comprehensive system-level perspective [26]. The Reference Architecture Model for Industry 4.0 (RAMI 4.0) was introduced by the German Electrical and Electronic Manufacturers' Association, provide a 3D coordinate system for addressing standardization challenges and enable the breakdown of intricate interconnections into subsystems, clusters, or modules [27]. The Industrial Internet Reference Architecture (IIRA), launched in 2017, aims to standardize the fourth industrial revolution by promoting interoperability management, identifying relevant technologies, and guiding skill and standard growth across various system types.

13.3 INDUSTRY 4.0 IN MODERN MANUFACTURING COMPONENT (MMC)

13.3.1 Fabrication and Processing

In the paradigm of Industry 4.0, the fabrication and processing of modern manufacturing components are undergoing a transformative evolution. Key technologies such as additive manufacturing optimization, robotics and automation, collaborative robotics (Cobots), predictive maintenance, and smart sensors are revolutionizing traditional manufacturing processes.

Table 13.1 outlines various areas and problem statements in current research relevant to Industry 4.0, along with the materials and technologies applied for practical applications within manufacturing processes.

Outlines various areas and problem statements in current research relevant to Industry 4.0. It also lists the materials and technologies applied for practical applications within manufacturing processes.

TABLE 13.1
A List of a Few Selected Research Reports Relevant to Industry 4.0 Implementation

S. No.	Some interesting areas/problem statements of current research	Materials/Technologies for Industry 4.0	Application
1.	Utilizing data-driven modelling to forecast load versus displacement curves of specific composite materials aligns with the objectives of Industry 4.0 and smart manufacturing	Composites, ANN	Production/manufacturing
2.	Current state and future trajectories of automation levels (LoA) in aerospace composite manufacturing heading toward Industry 4.0	Composites	Aerospace composite manufacturing
3.	In the context of Industry 4.0, does metal additive manufacturing strengthen the significance of subtractive machining?	Polymeric/ceramic/metal materials	Automotive/electronic/medical manufacturing
4.	Exploring the relationship between Industry 4.0 and lean manufacturing: Charting present study and setting a research agenda	Examining the interplay of lean manufacturing and Industry 4.0 alongside the influence of environmental factors on the performance of integrated industry and lean manufacturing	Technical solutions of lean manufacturing
5.	A comprehensive review of Industry 4.0 technologies' applications in the manufacturing context	Combined use of Internet of Things, big data analytics, and cloud	Production scheduling in the smart factory domain
6.	Critical factors for implementation and enhancement of Industry 4.0 technologies in manufacturing enterprises	Continuous improvement and lean management as crucial drivers for successful implementation (other key factors encompass quality and flexibility in competition, leadership from top management, formation of inter-functional teams, execution of preparatory activities, project planning, and training)	Technology implementations in manufacturing

TABLE 13.1 (Continued)
A List of a Few Selected Research Reports Relevant to Industry 4.0 Implementation

S. No.	Some interesting areas/problem statements of current research	Materials/Technologies for Industry 4.0	Application
7.	Advanced robotics and additive manufacturing of composites: pioneering the next phase of Industry 4.0	Future of fibre-manufacturing reinforced polymer composites: robotic and additive processes for advanced fibre-reinforced composites	Manufacturing/production of polymer composites
8.	Implementing Industry 4.0 principles in 3D printing of composite materials	Utilizing smart machines, storage systems, and production facilities, autonomously exchange data, initiate actions, and synchronize tasks through software platforms like Siemens PLM, including Mind Sphere	Production processes integrate equipment from diverse manufacturers, enabling seamless communication facilitated by sensors, feedback loops, and feedforward mechanisms
9.	Exploring the prospects of additive manufacturing in the context of Industry 4.0: A comprehensive review	Innovative production approaches: Design, manufacturing, and distribution strategies for complex component creation, product customization, and optimal waste reduction	Technology, applications, and materials aspects in AM
10	Assessment of friction stir processing for composite fabrication within the framework of Industry 4.0: a bibliometric analysis	Friction stir processing emerges as a promising solid-state technique capable of integrating reinforcements uniformly into the matrix, thereby minimizing defects and enhancing surface properties	Monitoring and assessing friction stir processing for the fabrication of surface hybrid composites
11.	Utilizing data science for predictive maintenance and materials science within the context of Industry 4.0	Methods, algorithms, and processes for extracting insights from vast industrial data sets to enhance the efficiency and reliability of manufacturing systems	Anticipate material quality to reduce production line costs and time
12.	Framework for implementing Industry 4.0 in the composite manufacturing sector	Automation and horizontally and vertically integrated business models are seen as enablers for addressing the intersection of challenges within the composites industry	A conceptual implementation framework, grounded in the systems engineering V model, is suggested for the composite manufacturing industry
13.	Milling of graphene-reinforced Ti6Al4V nanocomposites: an Industry 4.0 approach utilizing artificial intelligence	An advanced hybrid adaptive neuro-fuzzy inference system (ANFIS) coupled with a multi-objective particle swarm optimization method has been devised	This system aims to determine the optimal combination of milling parameters and reinforcement ratio to minimize feed force, depth force, and surface roughness
14.	A ranking model for non-conformity detection methods in casting within the framework of Industry 4.0	The X-ray method is highlighted as highly effective due to its frequency in detecting non-conformities. This method primarily focuses on identifying equipment-related non-conformities	Quality control optimization across the production process is achieved by adjusting the content and frequency of checkpoints, forming a composition-dependent set of checkpoints

(Continued)

TABLE 13.1 *(Continued)*

A List of a Few Selected Research Reports Relevant to Industry 4.0 Implementation

S. No.	Some interesting areas/problem statements of current research	Materials/Technologies for Industry 4.0	Application
15.	Overview of conventional technologies using the powders of metals, their alloys, and ceramics in the Industry 4.0 stage	Overview of traditional sintering techniques: Solid-state and liquid-phase sintering constitute 90% of the commercial value of sintered products	Materials processing technologies with emphasis on conventional powder engineering technologies
16.	Human-robot collaboration optimization	Advanced sensors, Cobots	Manufacturing assembly lines: implementing AI-driven algorithms to optimize human-robot collaboration on assembly lines, ensuring safety and efficiency while maximizing productivity
17.	Predictive maintenance for industrial machinery	IoT sensors, big data analytics	Smart factories leverage IoT sensors and analytics to forecast machinery failures proactively, reducing downtime and optimizing maintenance schedules
18.	Additive manufacturing optimization	3D printing technologies	Prototyping and customization: Exploring new materials and printing techniques to optimize additive manufacturing processes, enabling rapid prototyping and customized production at scale
19.	Sustainable material development	Bio-based materials, recycling technologies	Eco-friendly production: researching and developing sustainable materials and recycling technologies to reduce environmental impact in manufacturing processes, fostering a more sustainable approach to Industry 4.0
20.	Cybersecurity in smart manufacturing systems	Blockchain, encryption technologies	Data security: Implementing blockchain and encryption technologies to secure data transmission and protect sensitive information in interconnected smart manufacturing systems, ensuring data integrity and confidentiality
21.	Autonomous mobile robots for material handling	Navigation systems, robotics	Logistics and warehousing: Deploying autonomous mobile robots equipped with advanced navigation systems for efficient material handling in warehouses and distribution centres, optimizing logistics operations

TABLE 13.1 *(Continued)*
A List of a Few Selected Research Reports Relevant to Industry 4.0 Implementation

S. No.	Some interesting areas/problem statements of current research	Materials/Technologies for Industry 4.0	Application
22.	Energy-efficient manufacturing	Renewable energy sources, energy management systems	Green production facilities: integrating renewable energy sources and energy management systems to optimize energy usage and reduce carbon footprint in manufacturing facilities, contributing to sustainable production practices
23.	Digital twin technology for process optimization	IoT devices, simulation software	Virtual simulation involves deploying digital twin technology to generate virtual duplicates of physical assets and processes. This facilitates real-time monitoring, analysis, and optimization of manufacturing processes, leading to enhanced efficiency and quality control
24.	Supply chain resilience	Blockchain, AI	Supply chain management: Leveraging blockchain and AI to enhance transparency, traceability, and resilience in supply chains, mitigating risks and disruptions in global manufacturing
25.	Human-centric design for human-machine interfaces	Augmented reality, wearable technologies	Operator training and safety: Designing intuitive human-machine interfaces using augmented reality and wearable technologies to enhance operator training, safety, and productivity in manufacturing environments.

13.3.2 Additive Manufacturing Optimization

Additive manufacturing, also known as 3D printing, is a key component of Industry 4.0, enabling the creation of intricate MMCs with reduced waste and faster turnaround times, making it a versatile solution in various industries including aerospace, automotive, and medical, and the array of applications are shown in Figure 13.4 [28].

13.3.3 Robotics and Automation

Robotics and automation play a pivotal role in MMC fabrication and processing by streamlining production lines, reducing labour costs, and improving precision. Automated systems handle tasks ranging from material handling to assembly, freeing human operators for more complex and strategic roles. By integrating robotic arms, conveyors, and automated guided vehicles (AGVs), manufacturers achieve higher productivity, shorter lead times, and consistent quality standards, thus staying competitive in dynamic market environments [29, 30].

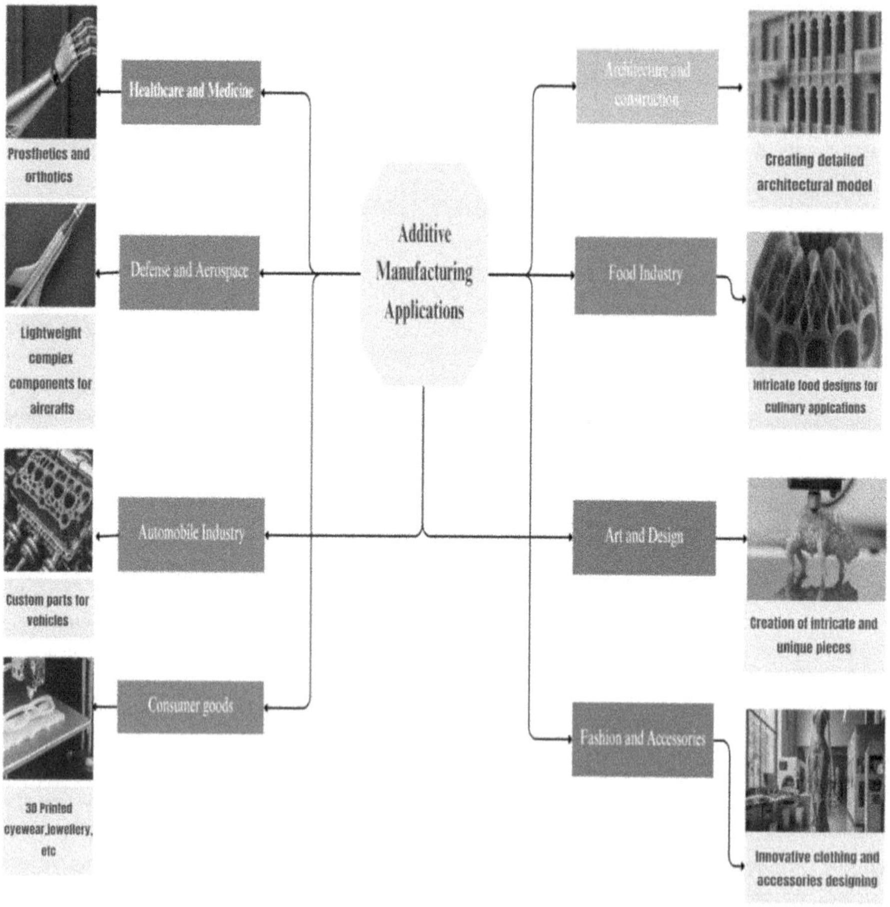

FIGURE 13.4 Depiction of summarized applications of additive manufacturing.

13.3.4 Collaborative Robotics (Cobots) for Handling MMC Components

Cobots represent a significant advancement in human-robot collaboration, particularly for handling MMC components. Unlike traditional industrial robots, Cobots can work beside human operators without the need for safety cages, enhancing flexibility and adaptability in production environments [31, 32]. They excel in tasks requiring dexterity and precision, such as assembly, inspection, and packaging of MMCs. By leveraging Cobots, manufacturers ensure safer workspaces, improved ergonomics, and higher throughput, ultimately enhancing overall operational efficiency.

13.3.5 Predictive Maintenance for Fabrication Machinery

Predictive maintenance prefers data analytics and machine learning algorithms to identify the breakdowns in equipment before they take place, minimizing unscheduled downtime and minimizing the repair intervals [33, 34]. In MMC fabrication, where machinery uptime is crucial for meeting production targets, predictive maintenance ensures the reliability and longevity of critical equipment. By monitoring parameters like temperature, vibration, and lubricant quality, manufacturers proactively address potential issues, reducing maintenance costs and maximizing asset utilization.

13.3.6 Smart Sensors for Quality Assurance

Smart sensors embedded throughout the manufacturing process provide real-time data on product quality, process parameters, and environmental conditions. In MMC fabrication, smart sensors monitor parameters such as temperature, pressure, and dimensional accuracy to ensure compliance with specifications and standards [35]. By detecting deviations or anomalies early, manufacturers can take corrective actions promptly, minimizing scrap and rework while maintaining consistent product quality. Additionally, smart sensors enable traceability and transparency throughout the supply chain, enhancing accountability and customer satisfaction.

13.3.7 Digital Twins for Process Optimization

Digital twins are effective replicas of physical resources, processes, or systems. In MMC fabrication, digital twins offer real-time insights into manufacturing processes, allowing for predictive analysis, optimization, and troubleshooting [36]. By simulating various scenarios and analyzing data from sensors embedded in machinery, digital twins optimize process parameters, reduce cycle times, and minimize defects. This proactive approach improves product quality, enhances resource utilization, and accelerates time-to-market, thereby bolstering competitiveness in dynamic market environments.

13.3.8 Advanced Materials Tracking and Traceability

Effective tracking and traceability of materials are crucial aspects of MMC fabrication, ensuring compliance with regulations, quality standards, and customer requirements. Industry 4.0 technologies enable advanced materials tracking through techniques such as RFID tags, barcodes, and blockchain-based systems. By accurately tracing the origin, composition, and journey of materials throughout the supply chain, manufacturers enhance transparency, mitigate risks, and uphold product integrity. This not only fosters trust among stakeholders but also enables efficient inventory management, reducing costs and reducing waste [37].

13.3.9 Energy Efficiency Monitoring

Energy efficiency monitoring is a significant facet of sustainable manufacturing practices, particularly in MMC fabrication where energy-intensive processes are common. Industry 4.0 technologies make possible present-time checking and investigation of energy consumption across production facilities, finding the opportunities for reducing waste and optimization [38]. By implementing energy-efficient technologies, such as smart meters, variable frequency drives, and energy management systems, manufacturers minimize operational costs, reduce environmental impact, and enhance long-term sustainability. This not only aligns with regulatory requirements and corporate social responsibility initiatives, but also strengthens competitive positioning in markets, increasingly focused on environmental stewardship.

13.4 TESTING OF MODERN MANUFACTURING COMPONENTS (MMCs)

In the period of Industry 4.0, the testing of modern manufacturing components (MMCs) has become increasingly crucial for ensuring product quality, reliability, and compliance with industry standards. This section delves into the significance of testing for industrial applications, methods for testing MMCs fabricated, and the differences between conventional and Industry 4.0-based testing methods [39, 40].

13.4.1 Significance of Testing for Industrial Applications

Testing plays a pivotal role in industrial applications, serving as a critical quality assurance measure throughout the manufacturing process. In the context of MMCs, which are often used in aerospace, automotive, healthcare, and other high-performance industries, the significance of testing cannot be overstated. Testing ensures that MMCs meet stringent performance criteria, withstand operational stresses, and adhere to regulatory requirements. It helps identify defects, inconsistencies, and material imperfections early in the production cycle, preventing costly rework, recalls, and potential safety hazards. Moreover, testing provides valuable data for process optimization, product refinement, and continuous improvement initiatives, ultimately enhancing customer satisfaction and brand reputation.

13.4.2 Methods for Testing of MMCs Fabricated

Various methods are employed for testing MMCs fabricated, depending on factors such as material composition, component geometry, and intended application. Common testing methods include:

13.4.2.1 Non-Destructive Testing (NDT)

Techniques like ultrasonic testing, radiographic testing, and eddy current testing are used to evaluate MMCs without causing damage to the material. NDT methods are particularly useful for detecting internal flaws, surface irregularities, and dimensional accuracy.

13.4.2.2 Mechanical Testing

Mechanical testing involves subjecting MMCs to controlled loads, pressures, and temperatures to assess mechanical properties such as strength, stiffness, and durability. Common mechanical tests comprise tensile, hardness, fatigue, and impact testing.

13.4.2.3 Chemical Analysis

Chemical analysis methods, such as spectroscopy and chromatography, are used to analyze the chemical composition of MMCs, ensuring compliance with material specifications and identifying impurities or alloying elements that may affect performance.

13.4.2.4 Microstructural Analysis

Microstructural analysis processes, including optical microscopy, electron microscopy, and X-ray diffraction, are employed to examine the internal structure and grain morphology of MMCs. This provides insights into material homogeneity, phase distribution, and microstructural defects.

13.4.2.5 Environmental Testing

Environmental testing simulates real-world operating conditions, exposing MMCs to factors like temperature extremes, humidity, corrosion, and vibration. Environmental testing evaluates the performance of MMCs under adverse conditions and helps assess long-term durability and reliability.

13.4.3 Differences between Conventional and Industry 4.0-Based Methods

The advent of Industry 4.0 has revolutionized testing methods for MMCs, introducing advanced technologies and digital solutions that enhance accuracy, efficiency, and scalability. Some key differences between conventional and Industry 4.0-based testing methods include:

a. **Data Integration and Analytics:** Industry 4.0-based testing methods leverage data integration and analytics platforms to collect, analyze, and visualize enormous amounts of testing data in real world. This enables predictive maintenance, anomaly detection, and proactive decision-making, improving overall testing efficiency and effectiveness.

b. **Remote Monitoring and Control:** Industry 4.0 enables remote monitoring and control of testing processes through connected sensors, actuators, and IoT devices. This allows for continuous monitoring of test parameters, remote adjustments, and real-time feedback, reducing the need for manual intervention and enhancing testing reliability.
c. **Digital Twins and Simulation:** Digital twins and simulation technologies create virtual replicas of MMCs and testing environments, facilitating predictive modelling, scenario analysis, and optimization of testing protocols. This enables manufacturers to simulate various testing scenarios, identify potential failure modes, and optimize testing strategies before physical testing begins.
d. **Automation and Robotics:** Industry 4.0-based testing methods incorporate automation and robotics for sample preparation, handling, and testing procedures. Automated testing systems increase throughput, reduce human error, and ensure consistency in testing protocols, leading to faster turnaround times and higher testing accuracy.

Finally, Industry 4.0 has transformed the testing of MMCs, offering enhanced capabilities in data analytics, remote monitoring, digital twins, automation, and robotics. By leveraging these advanced technologies, manufacturers can improve testing efficiency, accuracy, and reliability, thereby ensuring the quality and performance of MMCs in industrial applications.

13.5 SUMMARY

Industry 4.0 technologies play a pivotal role in MMC fabrication and processing, driving efficiency, innovation, and sustainability across the manufacturing landscape. By embracing digital twins for process optimization, advanced materials tracking and traceability, cyber-physical systems, and energy efficiency monitoring, manufacturers can achieve greater agility, resilience, and competitiveness in an ever-evolving global market.

REFERENCES

[1] Lu, Y., 2017. Industry 4.0: A survey on technologies, applications and open research issues. *Journal of Industrial Information Integration*, *6*, pp. 1–10.
[2] Lu, Y., 2017. Cyber physical system (CPS)-based industry 4.0: A survey. *Journal of Industrial Integration and Management*, *2*(03), p. 1750014.
[3] Xu, L.D., Xu, E.L. and Li, L., 2018. Industry 4.0: State of the art and future trends. *International Journal of Production Research*, *56*(8), pp. 2941–2962.
[4] Liu, F., Tan, C.W., Lim, E.T. and Choi, B., 2017. Traversing knowledge networks: An algorithmic historiography of extant literature on the Internet of Things (IoT). *Journal of Management Analytics*, *4*(1), pp. 3–34.
[5] Alyahya, S., Wang, Q. and Bennett, N., 2016. Application and integration of an RFID-enabled warehousing management system—a feasibility study. *Journal of Industrial Information Integration*, *4*, pp. 15–25.
[6] Yuehong, Y.I.N., Zeng, Y., Chen, X. and Fan, Y., 2016. The internet of things in healthcare: An overview. *Journal of Industrial Information Integration*, *1*, pp. 3–13.
[7] Xu, B., Da Xu, L., Cai, H., Xie, C., Hu, J. and Bu, F., 2014. Ubiquitous data accessing method in IoT-based information system for emergency medical services. *IEEE Transactions on Industrial Informatics*, *10*(2), pp. 1578–1586.
[8] Trappey, A.J., Trappey, C.V., Govindarajan, U.H., Chuang, A.C. and Sun, J.J., 2017. A review of essential standards and patent landscapes for the Internet of Things: A key enabler for Industry 4.0. *Advanced Engineering Informatics*, *33*, pp. 208–229.
[9] Oesterreich, T.D. and Teuteberg, F., 2016. Understanding the implications of digitisation and automation in the context of Industry 4.0: A triangulation approach and elements of a research agenda for the construction industry. *Computers in Industry*, *83*, pp. 121–139.
[10] Peruzzini, M., Gregori, F., Luzi, A., Mengarelli, M. and Germani, M., 2017. A social life cycle assessment methodology for smart manufacturing: The case of study of a kitchen sink. *Journal of Industrial Information Integration*, *7*, pp. 24–32.

[11] Kim, J.H., 2017. A review of cyber-physical system research relevant to the emerging IT trends: Industry 4.0, IoT, big data, and cloud computing. *Journal of Industrial Integration and Management*, *2*(03), p. 1750011.

[12] Tao, F., Wang, Y., Zuo, Y., Yang, H. and Zhang, M., 2016. Internet of Things in product life-cycle energy management. *Journal of Industrial Information Integration*, *1*, pp. 26–39.

[13] Kusiak, A., 2018. Smart manufacturing. *International Journal of Production Research*, *56*(1–2), pp. 508–517.

[14] Mitra, A., Kundu, A., Chattopadhyay, M. and Chattopadhyay, S., 2017. A cost-efficient one time password-based authentication in cloud environment using equal length cellular automata. *Journal of Industrial Information Integration*, *5*, pp. 17–25.

[15] Moghaddam, M. and Nof, S.Y., 2018. Collaborative service-component integration in cloud manufacturing. *International Journal of Production Research*, *56*(1–2), pp. 677–691.

[16] Gürdür, D., El-Khoury, J., Seceleanu, T. and Lednicki, L., 2016. Making interoperability visible: Data visualization of cyber-physical systems development tool chains. *Journal of Industrial Information Integration*, *4*, pp. 26–34.

[17] Fan, C.F., Chan, C.C., Yu, H.Y. and Yih, S., 2018. A simulation platform for human-machine interaction safety analysis of cyber-physical systems. *International Journal of Industrial Ergonomics*, *68*, pp. 89–100.

[18] Chen, H., 2017. Theoretical foundations for cyber-physical systems: A literature review. *Journal of Industrial Integration and Management*, *2*(03), p. 1750013.

[19] Chen, H., 2017. Applications of cyber-physical system: A literature review. *Journal of Industrial Integration and Management*, *2*(03), p. 1750012.

[20] Yodo, N. and Goethals, P.L., 2019. Cyber security management via control strategies for resilient cyber-physical systems. In *IIE Annual Conference. Proceedings*, Institute of Industrial and Systems Engineers (IISE), pp. 1584–1589.

[21] Gorkhali, A. and Xu, L.D., 2016. Enterprise application integration in industrial integration: A literature review. *Journal of Industrial Integration and Management*, *1*(04), p. 1650014.

[22] Petrasch, R. and Hentschke, R., 2016. Process modeling for Industry 4.0 applications: Towards an Industry 4.0 process modeling language and method. In *2016 13th International Joint Conference on Computer Science and Software Engineering (JCSSE)*, IEEE, pp. 1–5. https://doi.org/10.1109/JCSSE.2016.7748885

[23] Theorin, A., Bengtsson, K., Provost, J., Lieder, M., Johnsson, C., Lundholm, T. and Lennartson, B., 2017. An event-driven manufacturing information system architecture for Industry 4.0. *International Journal of Production Research*, *55*(5), pp. 1297–1311.

[24] Ivanov, D., Dolgui, A., Sokolov, B., Werner, F. and Ivanova, M., 2016. A dynamic model and an algorithm for short-term supply chain scheduling in the smart factory Industry 4.0. *International Journal of Production Research*, *54*(2), pp. 386–402.

[25] Liao, Y., Deschamps, F., Loures, E.D.F.R. and Ramos, L.F.P., 2017. Past, present and future of Industry 4.0-a systematic literature review and research agenda proposal. *International Journal of Production Research*, *55*(12), pp. 3609–3629.

[26] Cheng, Y., Chen, K., Sun, H., Zhang, Y. and Tao, F., 2018. Data and knowledge mining with big data towards smart production. *Journal of Industrial Information Integration*, *9*, pp. 1–13.

[27] Pedone, G. and Mezgár, I., 2018. Model similarity evidence and interoperability affinity in cloud-ready Industry 4.0 technologies. *Computers in Industry*, *100*, pp. 278–286.

[28] Kharat, V.J., Singh, P., Raju, G.S., Yadav, D.K., Gupta, M.S., Arun, V., Majeed, A.H. and Singh, N., 2023. Additive manufacturing (3D printing): A review of materials, methods, applications and challenges. *Materials Today: Proceedings*, *143*, pp. 172–196.

[29] Liu, Z., Liu, Q., Xu, W., Wang, L. and Zhou, Z., 2022. Robot learning towards smart robotic manufacturing: A review. *Robotics and Computer-Integrated Manufacturing*, *77*, p. 102360.

[30] Javaid, M., Haleem, A., Singh, R.P., Rab, S. and Suman, R., 2022. Significant applications of Cobots in the field of manufacturing. *Cognitive Robotics*, *2*, pp. 222–233.

[31] Patil, S., Vasu, V. and Srinadh, K.V.S., 2023. Advances and perspectives in collaborative robotics: A review of key technologies and emerging trends. *Discover Mechanical Engineering*, *2*(1), p. 13.

[32] Keshvarparast, A., Battini, D., Battaia, O. and Pirayesh, A., 2023. Collaborative robots in manufacturing and assembly systems: Literature review and future research agenda. *Journal of Intelligent Manufacturing*, *35*, pp. 2065–2118.

[33] Theissler, A., Pérez-Velázquez, J., Kettelgerdes, M. and Elger, G., 2021. Predictive maintenance enabled by machine learning: Use cases and challenges in the automotive industry. *Reliability Engineering & System Safety*, *215*, p. 107864.
[34] Meddaoui, A., Hain, M. and Hachmoud, A., 2023. The benefits of predictive maintenance in manufacturing excellence: A case study to establish reliable methods for predicting failures. *The International Journal of Advanced Manufacturing Technology*, *128*(7–8), pp. 3685–3690.
[35] Rajak, P., Ganguly, A., Adhikary, S. and Bhattacharya, S., 2023. Internet of Things and smart sensors in agriculture: Scopes and challenges. *Journal of Agriculture and Food Research*, *14*, p. 100776.
[36] Attaran, M. and Celik, B.G., 2023. Digital twin: Benefits, use cases, challenges, and opportunities. *Decision Analytics Journal*, *6*, p. 100165.
[37] Schuitemaker, R. and Xu, X., 2020. Product traceability in manufacturing: A technical review. *Procedia CIRP*, *93*, pp. 700–705.
[38] Pandey, V., Sircar, A., Bist, N., Solanki, K. and Yadav, K., 2023. Accelerating the renewable energy sector through Industry 4.0: Optimization opportunities in the digital revolution. *International Journal of Innovation Studies*, *7*(2), pp. 171–188.
[39] Markatos, N.G. and Mousavi, A., 2023. Manufacturing quality assessment in the Industry 4.0 era: A review. *Total Quality Management & Business Excellence*, *34*(13–14), pp. 1655–1681.
[40] Konstantinidis, F.K., Myrillas, N., Mouroutsos, S.G., Koulouriotis, D. and Gasteratos, A., 2022. Assessment of Industry 4.0 for modern manufacturing ecosystem: A systematic survey of surveys. *Machines*, *10*(9), p. 746.

14 Modern Machining of Metal Matrix Composites

Thavasilingam Kannan, Thiagarajan Srinivasan, Naga Chandrika Kondiboyana Kumari, Madhan Mohankumar, and Santhosh Sivaraj

14.1 INTRODUCTION

Metal matrix composites (MMCs), a paradigm shift in the quest for materials with enhanced mechanical, thermal, and operational characteristics, are at the forefront of modern materials engineering. This chapter's goal is to give readers a comprehensive overview of MMCs, including information on their composition, past evolution, applications, and challenges related to machining these complex materials.

The core of MMCs is a class of materials characterized by exceptional durability, resistance to wear, and thermal stability. This class of materials is created by the synergistic interaction of metallic matrices with reinforcing elements. Mixed metal composites offer a unique combination of properties that make them perfect for a variety of industrial applications. These properties combine the advantages of metals with the robustness and low weight of reinforcing components. Comprehending MMCs necessitates reviewing their evolutionary background. After decades of study and development, MMCs have moved from theoretical ideas to real-world uses. This path includes advances in production methods, materials science, and an ever-deepening understanding of the complex interactions between reinforcement and matrix materials.

Though it is still a difficult procedure, current research has made great strides in the domain of MMC machining. Hybrid electric discharge machining is a dependable technique that Bains et al. emphasize for machining MMCs, especially those with intricate forms. For traditional machining processes, it is also advised to utilize polycrystalline and diamond-coated tools [1]. Novel methods of processing highlight the significance of managing the distribution and volume percentage of reinforcements inside the matrix in the production of MMCs. All of these studies highlight the necessity of using cutting-edge machining methods and creative processing paths in order to machine MMCs efficiently [2].

Over the past few decades, metal matrix composites have developed into a material system with a plethora of potential applications. The primary benefits of these materials are their improved mechanical properties, particularly in the areas of wear, endurance, and stiffness. The most well-known MMCs are aluminium matrix composites (AMCs) due to their low cost, low point of melting, and low density. However, machining these materials remains challenging due to the substantial abrasiveness of the reinforcing stages. Traditional machining methods including drilling, turning, and milling are used to machine MMCs [3]. Like other composites, MMCs consist of two or more ingredients combined into a structure made of reinforcement (such as particles or continuous/discontinuous fibres) and a matrix (metal). The metal matrix provides the loading stress and provides the structure's compliant support, while the embedded particles or fibres offer greater physical or mechanical properties (e.g., greater strength, more wear resistance, or higher thermal efficiency) in more advantageous (fibre-strengthened) or non-preferential (element reinforced) pathways within the material bulk [4]. MMCs are composites that combine strong reinforcing particles with a soft metal matrix to give the specific property combinations required for a wide range related

DOI: 10.1201/9781003499718-14

to engineering applications. Since two or more reinforcement elements with distinct properties are combined at the level of molecules in an ongoing metal matrix phase, hybrid material composites are more homogenous than classical composites [5]. The surface quality, machinability, and tool wear mechanism of MMCs are thoroughly examined and discussed in relation to the different manufacturing processes. Additionally, in connection with MMC processing, the production of MMC products through unconventional machining methods such as laser machining, electrolytic machining, ultra-sonic machining (USM), wire electrical discharge machining (WEDM), electrical discharge machining (EDM), and high-speed machining is investigated and explored as substitutes to the aforementioned processes [6]. The most recent generation of advanced materials shows excellent mechanical properties such as strong hardness, high specific strength, and strong resistance to wear and corrosion. MMCs are a useful material to employ in the fabrication of cars and aeroplanes because of these qualities. Because of the hardness of MMCs and the abrasive character of the reinforced particles, these materials are still challenging to machine. The traditional method of machining MMCs raises the overall cost of machining because it causes excessive tool wear and slows down the removal of material. Conversely, these materials perform far better thanks to their unconventional machining [7].

Nevertheless, using MMCs has its own variety of difficulties, especially when it comes to machining. The intricate nature of MMCs often renders conventional machining techniques inadequate in composite materials, hence requiring specialized methodologies to fully realize their potential as shown in Figure 14.1. This section provides an introduction to these difficulties and examines innovative machining techniques customized for the special properties of MMCs.

The following sections expand on this examination of contemporary MMC machining methods by delving into the subtleties of processing and characterization. This voyage intends to provide the reader with a thorough grasp of the complexities involved in machining MMCs to unlock their full potential, from advanced machining techniques.

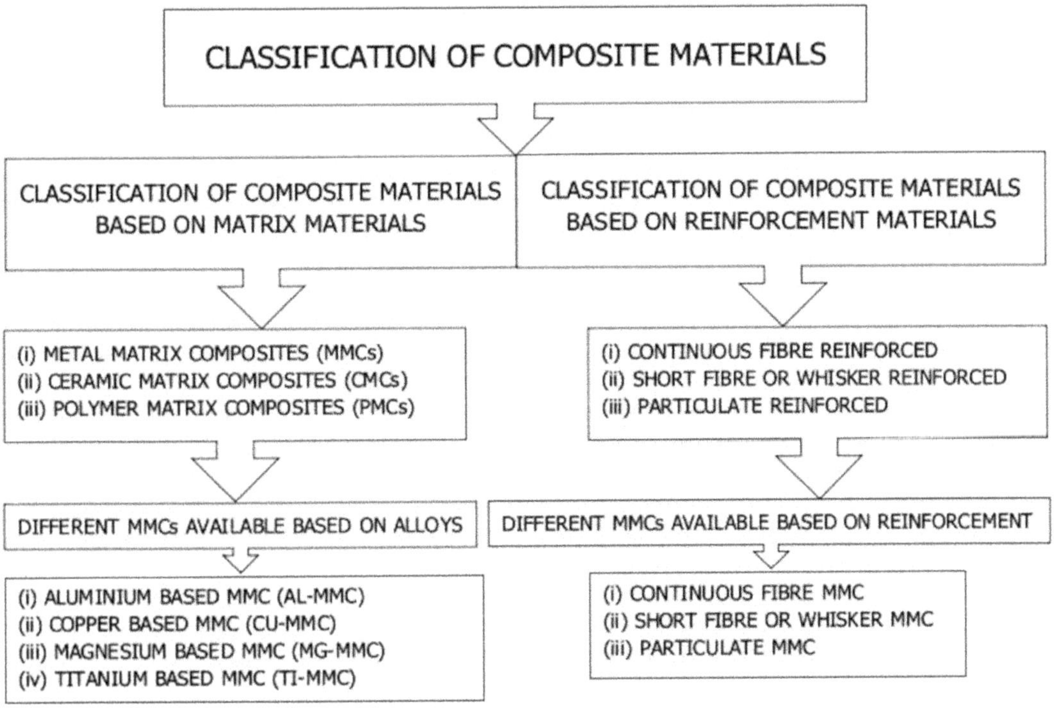

FIGURE 14.1 Flowchart showing the broad classification of composite materials [8].

14.2 TYPES OF MMCs

The mixture of matrix and reinforcement components determines the classification of MMCs. The properties and uses of the final composite are greatly influenced by the selection of these elements. The typical categorizations are as follows:

14.2.1 Matrix Materials

- Aluminium matrix composites (AMCs):
  - Aluminium alloys serving as the matrix material
  - Reinforced with materials like alumina (Al_2O_3) or silicon carbide (SiC)
- Titanium matrix composites (TMCs):
  - Titanium and its alloys as the matrix material
  - Reinforced with materials like boron fibres or titanium carbide (TiC)
- Magnesium matrix composites (MMCs):
  - Magnesium and its alloys as the matrix material
  - Reinforced with materials like silicon carbide (SiC) or alumina (Al_2O_3)
- Copper matrix composites (CMCs):
  - Copper and its alloys as the matrix material
  - Reinforced with materials like graphite or silicon carbide (SiC)

14.2.2 Reinforcement Materials

- Particulate reinforced MMCs:
  - Matrices reinforced with particulate materials
  - Examples include ceramic particles like silicon carbide (SiC) or alumina (Al_2O_3)

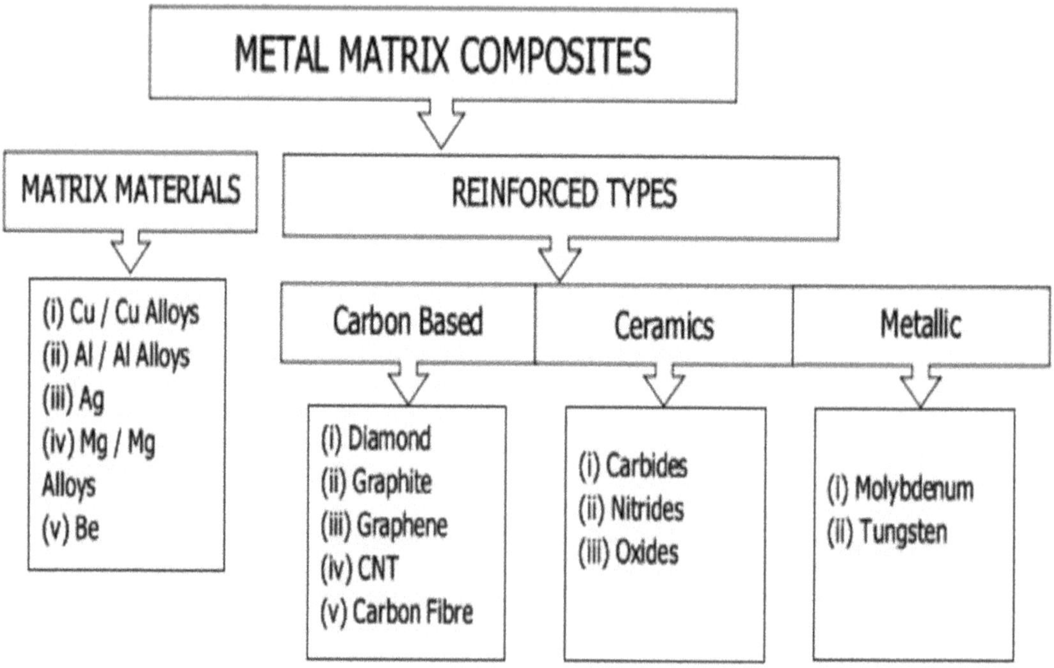

FIGURE 14.2 Different types of matrix materials and reinforcements used for heat sink applications [9].

- Whisker-reinforced MMCs:
  - Matrices reinforced with whisker-shaped reinforcing materials
  - Silicon carbide whiskers are a common example
- Hybrid metal matrix composites:
  - Combinations of different reinforcement materials in a single matrix
  - Tailored for specific applications, achieving a balance of properties

Metal matrix composites are classified according to a rich diversity that is principally determined by the complementary nature of the matrix and reinforcement components. These composite materials are utilized in a broad scope of engineering applications, with each kind being specially designed to meet certain performance requirements. There is a lot of variability in the materials, with aluminium matrix composites (AMCs) leveraging their lightweight strength in aerospace, titanium matrix composites (TMCs) known for their high-temperature stability, and magnesium matrix composites providing a combination of lightweight qualities. Refractory metal matrix composites perform well in abrasive and high-temperature conditions, while copper matrix composites (CMCs) find application in electronics due to their remarkable thermal conductivity. Hybrid metal matrix composites exhibit a custom combination of qualities through the fusing of various matrices and reinforcing elements.

Comprehending these categorizations facilitates the process of choosing MMCs that meet particular engineering specifications. A variety of mechanical, thermal, and electrical qualities that are appropriate for a deep reach of applications across industries are provided by the varied combinations.

14.3 PROPERTIES OF MMCs

The characteristics of MMCs are crucial in determining their performance and suitability in a range of engineering fields. An overview of the main mechanical, thermal, and electrical features that set MMCs apart from traditional materials is given in this section. Through a thorough investigation, the subtleties of MMC properties will be clarified, including their increased conductivity, strength, wear resistance, and thermal stability. Researchers and engineers can fully utilize MMCs and strategically utilize their capabilities to meet the changing demands of contemporary engineering applications by comprehending their characteristics.

14.3.1 Mechanical Properties

This section explores the basic mechanical characteristics that characterize MMC behaviour and emphasizes its importance in a range of applications. This investigation will cover important mechanical characteristics like elastic modulus, yield strength, and ultimate tensile strength, illuminating how MMCs outperform conventional materials in satisfying the exacting requirements of various sectors. In order to optimize MMCs for particular applications, engineers and researchers must have a thorough understanding of their mechanical properties. This understanding will pave the way for developments that will push the limits of the integrity of structure and mechanical functionality in contemporary engineering.

The selection of metallic alloys for use as the matrix material might vary depending on the intended usage and include super alloys according to aluminium, titanium, magnesium, nickel, copper, and steel. The composites that are based on Al alloys and have an excellent strength-to-mass ratio and absorbability characteristics are of great interest since they have the potential to save weight. The characteristics of some matrix materials are listed in Table 14.1. Mg is lighter than Al, but because it requires more raw materials and processing to make it, as well as having less corrosion resistance than Al alloys, its wider use is severely limited [10].

For durability in a desired direction, a combination of continuous and discontinuous fibres is used. Depending on the fibre's dimension, composites are often further categorized into both long and short-reinforcement fibres, with the shorter fibres being known for providing better strength

TABLE 14.1
Properties of Common Matrix Materials [10]

Metal	Density (g/cm^3)	Yield strength (MPa)	Ductility (%)	Modulus (GPa)
Aluminium alloys	2.5–2.9	30–500	1–44	68–82
Magnesium alloys	1.74–1.95	70–400	3.5–18	42–47
Titanium alloys	4.4–4.8	250–1250	1–40	90–120
Ferrous alloys	7.1–8.0	240–690	18–31	187–215
Nickel alloys	7.75–8.65	300–1900	0.5–60	150–245
Copper alloys	8.93–8.94	30–500	3–50	112–148

TABLE 14.2
Properties and Applications of Commonly Used Reinforcement Materials [10]

Reinforcement	Strength (MPa)	Density (g/cm^3)	Melting point (°C)	Thermal conductivity (W/m-K)
SiC	1625	3.21	2730	20.7
B_4C	569	2.52	2763	42
Al_2O_3	665	3.95	2072	38.5
Gr	76	2.26	3600	114
CNT	63,000	1.3	3130	6000
Si_3N_4	76	3.25	2769	43
TiB_2	373.6	4.54	3318	26

characteristics. Particle-reinforced MMCs are given isotropic properties by the addition of various strong metal and ceramic refractory particles, which range in size from only a few nanometres to a few thousand micro-metres. The primary attributes and applications of several of the most widely utilized reinforcements are shown in Table 14.2.

14.3.2 Thermal Properties

In many industrial applications, a material's thermal characteristics play a crucial role in determining how it reacts to heat and performs in harsh conditions. Understanding thermal characteristics is crucial when discussing metal matrix composites (MMCs), a unique class of materials intended for cutting-edge technical solutions. An overview of the unique thermal properties of MMCs is given in this section, with a focus on their capacity to tolerate and control high temperatures. By using the thermal conductivity of metal matrices and reinforcement elements, MMCs are able to produce composites that have improved thermal stability, heat resistance, and effective heat dissipation. The thermal qualities of MMCs provide opportunities for innovation in anything from electronics that require efficient thermal management to aeronautical components that must withstand severe temperatures.

Electronic packages and thermal management are two common uses for metal matrix composites with high thermal conductivity and a customization coefficient of thermal expansion. The advantages of creating near-net form composites are provided by the combination of powder injection moulding and pressure-less infiltration. Successful manufacturing and additional thermal property enhancement need improving wettability and optimizing interfacial structure [11]. Different metal-matrix composites reinforced with fibres and particles have had their thermal expansion tested, and the experimental results have been checked with the assumptions of different theoretical models. When the particle composites were cooled from their highest temperature to room temperature, they showed some residual strain. The number of heat cycles and the peak temperature both affected the strain's magnitude. The fibre-reinforced composites' thermal expansion response differed markedly from the particle composites. The thermal stresses that these particulate and fibrous mixtures create due to the variations in the coefficients of thermal expansion among the reinforcement and the matrix are what explain the observed behaviour of these composites [12].

In certain technological applications, temperature is a critical factor in influencing the suitability and functionality of materials. Because of their better high-temperature qualities over wrought metals and alloys, metal matrix composites—or MMCs—have found use in a variety of industries, including aerospace, automotive, and biomedical. Improved abrasion and wear resistance, more thermal insulation, and better resistance to creep and fatigue are some of these qualities [13]. The typical monolithic conductive material is currently facing competition from the fabrication of composite materials with balanced thermal properties, which are highly valued. The goal of the processing methods utilized to create composites has been to attain appropriate thermal characteristics. A comprehensive analysis has been conducted of the advancements made in the analytical and computational simulation approach to forecast the thermal conductivity of the created composites [14]. The metal-matrix composites were examined in a thermo mechanical analyzer (TMA) after being developed domestically. It is crucial to test and characterize the thermo-physical characteristics of MMC, including its density, specific heat content (Cp), coefficient of thermal expansion (CTE), and thermal conductivity. Prior to actual use, it is crucial to assess the new material's thermal stability and CTE. TMA can test the CTE experimentally using absolute methods. The temperature range in which the studies were conducted was 125°C to 550°C. A deeper comprehension of thermal design is facilitated by the measurements of material thermal properties [15]. The stability of mass loss and glass transition in resin and wax-epoxy composite coatings on aluminium was evaluated. It is evident that the glass transition and beginning mass decomposition temperatures of 63°C and 323°C, respectively, are given by pure epoxy resin [16].

14.3.3 Wear Resistance

A key component of material performance is wear resistance, particularly in industries where parts are constantly in mechanical contact with one another and with friction and abrasion. Wear resistance is especially important when it comes to metal matrix composites (MMCs) since these materials are designed to withstand the harsh demands of a variety of applications. This section provides an overview of the critical role wear resistance plays in MMC performance. The unique synergy that is created when reinforcing elements are integrated with the matrix improves the resistance of MMCs against abrasive forces, wear, and friction. The investigation of wear resistance in MMCs includes toughness, hardness, and tribological behaviour. These properties demonstrate how these composites perform well in settings where conventional materials could degrade due to wear.

Metal matrix composites are becoming more and more recognized in contemporary industrial technology as innovative wear-resistant materials. This article's primary goal is to examine the depth of indentation and wear down of an Al-6061 alloy composite material supported by alumina particles and enhanced by silicon carbide particles that were created using the stir casting method. Upon examining the wear surfaces, it was seen that the composite's worn surfaces and wear pattern had both mild and severe wear as well as micro and transverse cracks [17]. It was obtained that the sp. wear rate of pure epoxy resin is 0.006 mm^3/Nm, with a COF of 0.72. The brittle molecular structure of the epoxy resin is the cause of this increased wear resistance. However, adding soft wax to the resin reduces its resistance to wear. The COF and specific wear rate are increased with the addition of wax [18].

The Taguchi technique was used to construct the wear resistance model prediction for metal matrix composites. By sliding the samples against different-sized SiC grits under varied conditions, the pin-on-disc method was used to study the specimens' two-body abrasive wear behaviour. Optimal testing settings were determined by utilizing the diagonal array, signal-to-noise (S/N) ratio, and analysis of variance. The abrasive grain size was determined to be the primary parameter within the other control parameters on abrasion wear, followed by the weight fraction of reinforcement, according to the experimental results [19].

14.4 TRADITIONAL MACHINING TECHNIQUES FOR MMCs

Conventional machining methods, which have historically served as the foundation of manufacturing operations, present particular difficulties when used on modern materials such as MMCs. This section provides an overview of the challenges and factors to be taken into account while using conventional machining techniques for MMCs. The composites that are recognized for their improved mechanical and thermal characteristics can be challenging to machines because of their heterogeneous nature, which is a consequence of the amalgamation of metal matrices and reinforcing components. When applied to MMCs, standard procedures like turning, milling, and drilling—which have shown to be effective with conventional materials—face problems like increased tool wear, decreased machinability, and degraded surface integrity.

The most popular machining techniques for MMCs, including drilling, milling, and turning, are traditional techniques. The significant elements, such as tool wear methods and the final surface effectiveness, are examined in addition to the challenges encountered in each of these methods in the situation of MMCs and some potential solutions. Moreover, an analysis is conducted on the machining of MMCs in various machining methods. Micro-machining, sawing, honing, tapping, and grinding are also taken into account [20]. A type of material known as metal matrix composites is created when a metal is combined with one or more components in order to improve particular

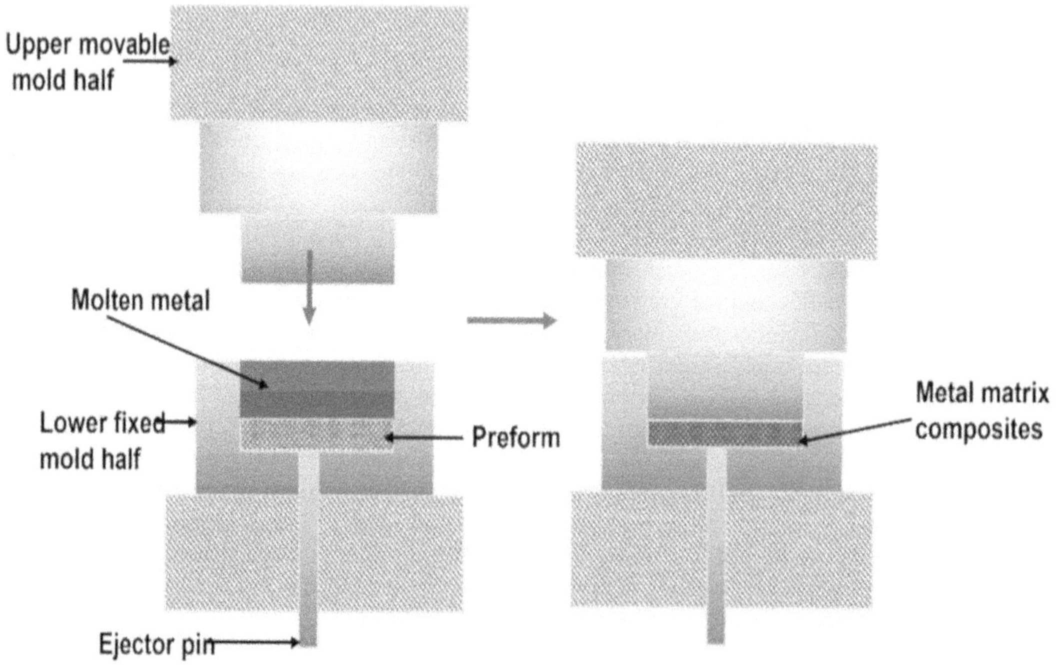

FIGURE 14.3 Schematic illustration of squeeze casting technique [29].

properties. They have received a lot of research attention previously because of the potential to customize the materials and the outcome characteristics to particular product needs. A variety of conventional manufacturing techniques are used, but powder metallurgy is among the most often used methods for producing metal matrix composite parts. Process limits notwithstanding, there are limitations in the product forms that may be produced. Conversely, additive manufacturing techniques have advanced quickly from being only alternatives for prototyping to becoming viable technologies [21].

14.5 ADVANCED MACHINING PROCESSES FOR MMCs

Conventional techniques frequently prove inadequate in tackling the distinct issues presented by MMCs when it comes to optimizing their milling process. An introduction to the world of MMC-specific advanced machining processes is provided in this part, which presents creative ways to get over the drawbacks of traditional methods. Unlocking the full potential of MMCs requires improved machining as the demand for their excellent mechanical and thermal qualities rises across industries. MMCs are increasingly being machined using techniques like abrasive machining, hybrid approaches, and non-traditional techniques such as electrical discharge machining (EDM) and laser machining.

While MMCs can have their properties customized and engineered to fit particular applications, their use in a wide range of products is growing with time. It is imperative to guarantee the better physical property benefits of MMC materials over their typical unreinforced monolithic metal equivalents in a robust and reproducible manner, as the future application of these products heavily depends on their favourable characteristics. While there are many different ways to produce MMC goods, each approach has pros and cons of its own [22]. Metal matrix composites have been studied as a potential replacement material for various engineering applications due to their improved properties. MMCs have historically been difficult to process due to the strong reinforcing phase's abrasive character. The effects of the subsequent input machining process factors—overcut, surface finish, circularity error, perpendicularity error, and material removal rate (MRR)—on the desired output are examined. These variables include wire feed, servo voltage, flushing pressure, and pulse ON/OFF [23]. An experimental study was conducted to investigate the machinability of an aluminium metal matrix composite (A356/SiC/10p) when continued turning of composite tubes was being performed. The study used moderate-grade polycrystalline diamond (PCD 1500) inserts. MMCs are said to be the toughest materials to process because of their great difficulty, and PCD equipment is by far the finest choice. There were detected impacts of cutting speed on insertion wear and built-up edge (BUE) development. Low cutting speeds were associated with BUE generation, whereas high cutting speeds might yield very good finishes on the surfaces and low particular energy use [24].

14.5.1 ABRASIVE WATER JET MACHINING

A dynamic and adaptable advanced machining method, abrasive water-jet machining (AWJM) has attracted a lot of interest due to its exceptional capacity to overcome the difficulties presented by a variety of materials, especially MMCs. The concepts and uses of AWJM are presented in this section, emphasizing the importance of the technique for machining MMCs.

The requirement for machining such more durable materials has expanded significantly along with the need to develop new engineering materials. The current study's objective was to produce SiC-reinforced aluminium laminate while implementing the AWJM process using a Taguchi-DEAR approach. As quality indicators, roughness of the surface, taper angle, and material removal rate were assessed. An abrasive flow rate of 0.432 kg/min was kept constant throughout the cutting operation. Due to their importance as control parameters in the AWJM process, the water pressure (WP), feed rate (FR), standoff distance (SD), and abrasive particle flow rate (AF) were chosen as the input process components in the current work. Samples of composite workpieces that were 5 mm thick were utilized for the AWJM-based cutting [25].

Particle reinforced metal matrix composites (PRMMCs) have stiff reinforcement, which makes it extremely difficult to mill them using conventional manufacturing procedures and causes a large amount of tool wear. The Al-SiC MMC samples were made by stir casting. An analysis was conducted to determine how the ceramic powder reinforcement affected the machining process [26].

14.5.2 Electro-Chemical Machining

With an emphasis on its adaptability to the varied structure of MMCs and its capacity to overcome obstacles such as tool wear and surface integrity concerns, this investigation seeks to untangle the principles of ECM. Through an exploration of ECM's features and benefits, this introduction sets the stage for a thorough grasp of how this cutting-edge machining method will influence precision engineering going forward, especially in the complex field of MMC manufacturing.

ECM appears to be a viable strategy given the method's numerous special benefits in a variety of application areas, including improved machining rate, greater precision and control, and an expanded variety of substances that may be cut. The current work is being done to investigate the impact of a few significant electrochemical process parameters, such as the voltage being applied, electrolytes level, electrolyte flow rate, and tool rate of feed, on the MRR and roughness of the surface (Ra) in order to fully utilize ECM of LM25 Al/10% SiC composite components developed through stir casting [27].

14.5.3 Ultrasonic Machining

One of the most potent and accurate advanced machining techniques is ultrasonic machining (USM). It is especially notable for its use in shaping complex components, including those made of difficult materials like MMCs. The concepts and uses of USM are presented in this part, with a focus on the importance of USM for MMC machining. USM sets itself apart by providing a

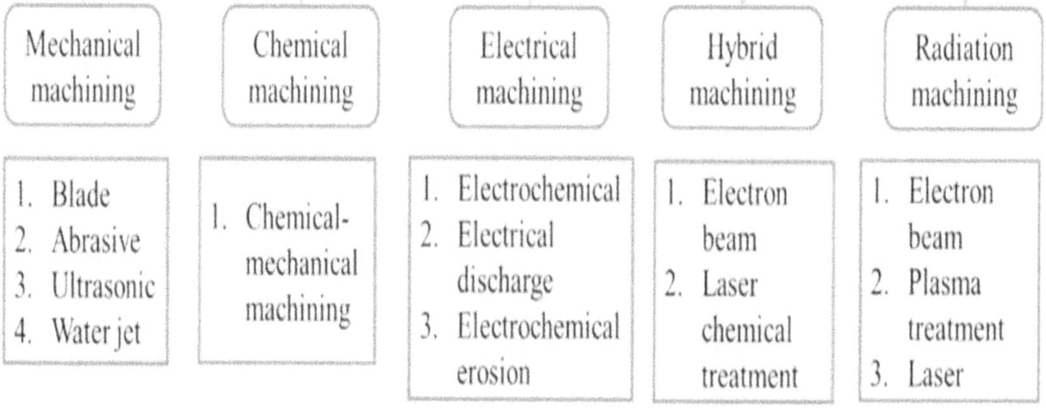

FIGURE 14.4 Non-conventional machining methods of polymer composite materials [30].

non-thermal, non-chemical method of precision machining by abrading material surfaces using ultrasonic vibrations. Because USM can remove material selectively and precisely, it is a promising answer in the field of MMCs, where traditional methods may not be as effective.

The goal of this investigation is to learn the foundations of USM, highlighting how it can be tailored to the particular nature of MMCs and how it can be used to solve issues with dimensional accuracy, tool wear, and surface finish. Through an exploration of USM's capabilities and benefits, this introduction provides a solid basis for a thorough grasp of how this cutting-edge machining method is essential for shaping complex components and opens up new avenues for MMC fabrication for a variety of industry applications.

Because these MMCs are reinforced with materials that have superior mechanical and tribological properties, they are considered unconventional engineering materials. Additionally, they display factor-designed behaviour demonstrates how ultrasounds trigger the MMC's roughness to switch on and explore the impact of motion and turning situation on the surface finish of the turning MMC. Light dispersion is the result of turning with vibration, which creates typical surface patterns along the direction of turning and vibration. The pitch of typical profiles only falls into a small portion of this phenomenon's spectrum [28].

14.6 FUTURE TRENDS IN MMC MACHINING

The future of MMC machining looks interesting as markets continue to demand materials with improved performance and efficiency. This section presents the upcoming developments in MMC machining, exploring new approaches, technologies, and tactics that have the potential to completely alter the precise engineering field. This investigation seeks to offer a thorough understanding of the developing field of MMC machining while projecting how these developments will influence our approach to high-performance component production in the years to come.

Integration of Advanced Machining Techniques: Future developments in MMC machining are probably going to see a smooth integration of advanced machining techniques including ultrasonic machining (USM), electro-chemical machining (ECM), and abrasive water-jet machining (AWJM). When these methods are combined, they can provide a comprehensive solution that addresses particular issues related to MMCs and improves overall machining efficiency.

The path forward is expected to be revolutionary, and the incorporation of these developments guarantees that MMC machining will continue to be at the leading edge of technical development for many years to come.

14.7 CONCLUSION

The goal of this section has been to provide readers with a comprehensive grasp of the challenges and advancements in this rapidly developing subject, including everything from the basic characteristics and classifications of MMCs to an exploration of traditional and sophisticated machining processes. Setting the basis for a thorough analysis of machining processes, the introduction defined MMCs and discussed their importance in modern engineering. The following sections covered the nuances of MMC machining-specific processing methods, characterization techniques, and optimization strategies. It was acknowledged that traditional machining methods had drawbacks when it came to the special makeup of MMCs. Advanced machining techniques like ultrasonic, electrochemical, and abrasive water-jet machining were introduced, showcasing creative solutions designed to address these difficulties.

Future developments in MMC machining were also explored, and these included the incorporation of cutting-edge technologies, flexible approaches, and sustainability factors. The intention when writing this chapter was not simply to provide information but also to stimulate more research and creative thinking in the MMC machining industry. Because of the difficulties mentioned earlier, materials engineering is a dynamic field, and finding new approaches and techniques to fully utilize MMCs is an ongoing endeavour. The ideas offered in this chapter work as a guide for scientists, engineers, and professionals navigating the always-changing field of MMC machining, as markets continue to demand materials that perform well for a variety of applications.

REFERENCES

[1] Bains, P.S., Sidhu, S.S. and Payal, H.S., 2015. Fabrication and machining of metal matrix composites: A review. *Materials and Manufacturing Processes*, *31*(5), pp. 553–573.

[2] Ruzic, J., Simić, M., Stoimenov, N., Božić, D. and Stašić, J., 2021. Innovative processing routes in manufacturing of metal matrix composite materials. *Metallurgical & Materials Engineering*, *27*(1), pp. 1–13.

[3] Gururaja, S., Ramulu, M. and Pedersen, W., 2013. Machining of MMCs: A review. *Machining Science and Technology*, *17*(1), pp. 41–73.

[4] Liao, Z., Abdelhafeez, A., Li, H., Yang, Y., Diaz, O.G. and Axinte, D., 2019. State-of-the-art of surface integrity in machining of metal matrix composites. *International Journal of Machine Tools & Manufacture/International Journal of Machine Tools and Manufacture*, *143*, pp. 63–91.

[5] Lalmuan, S.K., Das, S., Chandrasekaran, M. and Tamang, S.K., 2017. Machining investigation on hybrid metal matrix composites—a review. *Materials Today: Proceedings*, *4*(8), pp. 8167–8175.

[6] Li, J. and Laghari, R.A., 2018. A review on machining and optimization of particle-reinforced metal matrix composites. *The International Journal of Advanced Manufacturing Technology*, *100*(9–12), pp. 2929–2943.

[7] Srivastava, A.K., Dixit, A.R. and Tiwari, S., 2016. A review on the intensification of metal matrix composites and its nonconventional machining. *Science and Engineering of Composite Materials*, *25*(2), pp. 213–228.

[8] Lakshmikanthan, A., Angadi, S., Malik, V., Saxena, K.K., Prakash, C., Dixit, S. and Mohammed, K.A., 2022. Mechanical and tribological properties of aluminum-based metal-matrix composites. *Materials*, *15*(17), p. 6111.

[9] Baig, M.M.A., Hassan, S.F., Saheb, N. and Patel, F., 2021. Metal matrix composite in heat sink application: Reinforcement, processing, and properties. *Materials*, *14*(21), p. 6257.

[10] Seetharaman, S., Subramanian, J., Singh, R.A., Wong, W.L.E., Nai, M.L.S. and Gupta, M., 2022. Mechanical properties of sustainable metal matrix composites: A review on the role of green reinforcements and processing methods. *Technologies*, *10*(1), p. 32.

[11] Qu, X.-H., Zhang, L., Wu, M. and Ren, S.-B., 2011. Review of metal matrix composites with high thermal conductivity for thermal management applications. *Progress in Natural Science*, *21*(3), pp. 189–197.

[12] Vaidya, R.U. and Chawla, K.K., 1994. Thermal expansion of metal-matrix composites. *Composites Science and Technology*, *50*(1), pp. 13–22.

[13] Oladijo, O.P., Awe, S.A., Akinlabi, E.T., Phiri, R.R., Collieus, L.L. and Phuti, R.E., 2021. High-temperature properties of metal matrix composites. In *Encyclopedia of Materials,* Elsevier eBooks, pp. 360–374.

[14] Sidhu, S.S., Kumar, S. and Batish, A., 2015. Metal matrix composites for thermal management: A review. *Critical Reviews in Solid State and Materials Sciences*, *41*(2), pp. 132–157.

[15] Karthikeyan, B., Ramanathan, S. and Ramakrishnan, V., 2010. Thermo physical property measurement of metal-matrix composites. *Materials in Engineering*, *31*, pp. S82–S86.

[16] Thavasilingam, K., Kumar, A.S., Khan, M.A., Devanand, S. and Giridharan, K., 2021. Effect of fumed silica in rice bran wax-epoxy coating on aluminum substrate: Mechanical, thermal, and water absorption properties. *Biomass Conversion and Biorefinery*, *13*(5), pp. 4229–4240.

[17] Nallusamy, S. and Karthikeyan, A., 2016. Analysis of wear resistance, cracks and hardness of metal matrix composites with SiC additives and Al_2O_3 as reinforcement. *Indian Journal of Science and Technology*, *9*(35), pp. 1–6.

[18] Thavasilingam, K., Kumar, A.S., Sakthimurugan, D. and Meenatchisundaram, P., 2023. Synthesis and characterization of fumed silica dispersed bees/carnauba wax-epoxy biocomposite coating: A comparative study. *Biomass Conversion and Biorefinery*, *13*(8), pp. 7375–7383.

[19] Sahin, Y., 2005. The prediction of wear resistance model for the metal matrix composites. *Wear*, *258*(11–12), pp. 1717–1722.

[20] Markopoulos, A.P., Pressas, I.S., Papantoniou, I.G., Karkalos, N.E. and Davim, J.P., 2015. Machining and machining modeling of metal matrix composites—a review. *Modern Manufacturing Engineering, Materials Forming, Machining and Tribology*, pp. 99–141.

[21] Behera, M.P., Dougherty, T. and Singamneni, S., 2019. Conventional and additive manufacturing with metal matrix composites: A perspective. *Procedia Manufacturing*, *30*, pp. 159–166.

[22] Mussatto, A., Ahad, I.U., Mousavian, R.T., Delaure, Y. and Brabazon, D., 2020. Advanced production routes for metal matrix composites. *Engineering Reports*, *3*(5), pp. 1–25.

[23] Manikandan, N., Balasubramanian, K., Palanisamy, D., Gopal, P.M., Arulkirubakaran, D. and Binoj, J.S., 2019. Machinability analysis and ANFIS modelling on advanced machining of hybrid metal matrix composites for aerospace applications. *Materials and Manufacturing Processes*, *34*(16), pp. 1866–1881.

[24] Muthukrishnan, N., Murugan, M. and Rao, K.P., 2007. Machinability issues in turning of Al-SiC (10p) metal matrix composites. *The International Journal of Advanced Manufacturing Technology*, *39*(3–4), pp. 211–218.

[25] Thangaraj, M., Ahmadein, M., Alsaleh, N.A. and Elsheikh, A.H., 2021. Optimization of abrasive water jet machining of SiC reinforced aluminum alloy based metal matrix composites using Taguchi–DEAR technique. *Materials*, *14*(21), p. 6250.

[26] Kumar, B.A. and Kumaresan, G., 2015. Abrasive water jet machining of aluminum-silicon carbide particulate metal matrix composites. *Materials Science Forum*, *830–831*, pp. 83–86.

[27] Senthilkumar, C., Ganesan, G. and Karthikeyan, R., 2008. Study of electrochemical machining characteristics of Al/SiCp composites. *The International Journal of Advanced Manufacturing Technology*, *43*(3–4), pp. 256–263.

[28] Patil-Tekale, R.A., Gadekar, A., Gadhade, Y., Parakh, L., Balaji, R. and Selokar, A., 2021. Recent investigation on ultrasonic machining of aluminum metal matrix composite. In *Intelligent Manufacturing and Energy Sustainability*, Smart Innovation, Systems and Technologies, Singapore: Springer, vol. 213, pp. 619–636.

[29] Parveez, B., Kittur, M.I., Badruddin, I.A., Kamangar, S., Hussien, M. and Umarfarooq, M.A., 2022. Scientific advancements in composite materials for aircraft applications: A review. *Polymers*, *14*(22), p. 5007.

[30] Ablyaz, T.R., Shlykov, E.S., Muratov, K.R. and Sidhu, S.S., 2021. Analysis of wire-cut electro discharge machining of polymer composite materials. *Micromachines*, *12*(5), p. 571.

15 Conventional Machining of Metal Matrix Composites

Meby Selvaraj and Navasingh Rajesh Jesudoss Hynes

15.1 INTRODUCTION OF METAL MATRIX COMPOSITES

Metal matrix composites (MMCs) are advanced materials that combine the properties of metals and reinforcements, enhancing thermal, electrical, and mechanical properties. They have revolutionized sectors like automobile, aerospace, sports, and electronics due to their strength, rigidity, and lightweight nature. A supplementary phase, typically ceramic fibres, particles, or whiskers, is added to a metal matrix to preserve the composite's metallic properties and improve its mechanical performance. High-strength, thermal-stability ceramics like silicon carbide, alumina, or boron carbide are popular options. MMC's remarkable strength-to-weight ratio has influenced their development, allowing for increased stiffness, strength, and reduced weight, particularly in sectors like aircraft and automobiles where weight reduction is crucial for improved performance and fuel economy.

MMCs, with their excellent thermal conductivity and wear resistance, are useful in applications requiring high durability and abrasion resistance. They have shown better wear resistance and thermal stability than traditional materials in brake rotors and engine components for high-performance automobiles. Manufacturing procedures for MMCs include stir casting, liquid metal infiltration, and powder metallurgy, depending on their final layout and attributes. Kumar examines the powder metallurgy method for fabricating Al6061-SiC MMCs, analyzing the impact of SiC content and sintering temperature on their mechanical properties [1]. The study highlights the unique benefits of each method in terms of scalability, cost-effectiveness, and control over the composite's microstructure. However, MMCs face challenges in general use, such as costs, manufacturing scalability, and homogeneous dispersion of reinforcing components. Additionally, connecting methods and compatibility with different materials in MMCs pose challenges. Despite these challenges, metal matrix composites offer exceptional mechanical performance, thermal stability, and lightweight qualities, making them promising for innovative materials and future engineering solutions. Further research and processing technology development are expected to enhance their use and economic feasibility. MMCs, due to their unique blend of qualities, are revolutionizing production processes and product quality in various industrial sectors. A few of the industrial sectors of conventional machining are shown in Figure 15.1.

15.1.1 Aerospace Industry

MMCs are popular in aerospace due to their strength-to-weight ratio, which can reduce weight and improve strength and durability of structural components, engine elements, and spacecraft parts. Srinivasan researched the tribological behaviour of Al-based MMCs, specifically Al-SiC and Al-B_4C composites, for aerospace applications, particularly in aircraft braking systems, which can reduce vehicle weight, save fuel, increase payload capacity, and enhance performance [2].

15.1.2 Thermal Management and Electronics

Certain MMCs have extraordinary thermal conductivity, which makes them indispensable in electronic devices like heat sinks and substrates. By minimizing overheating, these composites effectively disperse heat, guaranteeing the dependability and lifetime of electronic components (Nguyen [3]).

 DOI: 10.1201/9781003499718-15

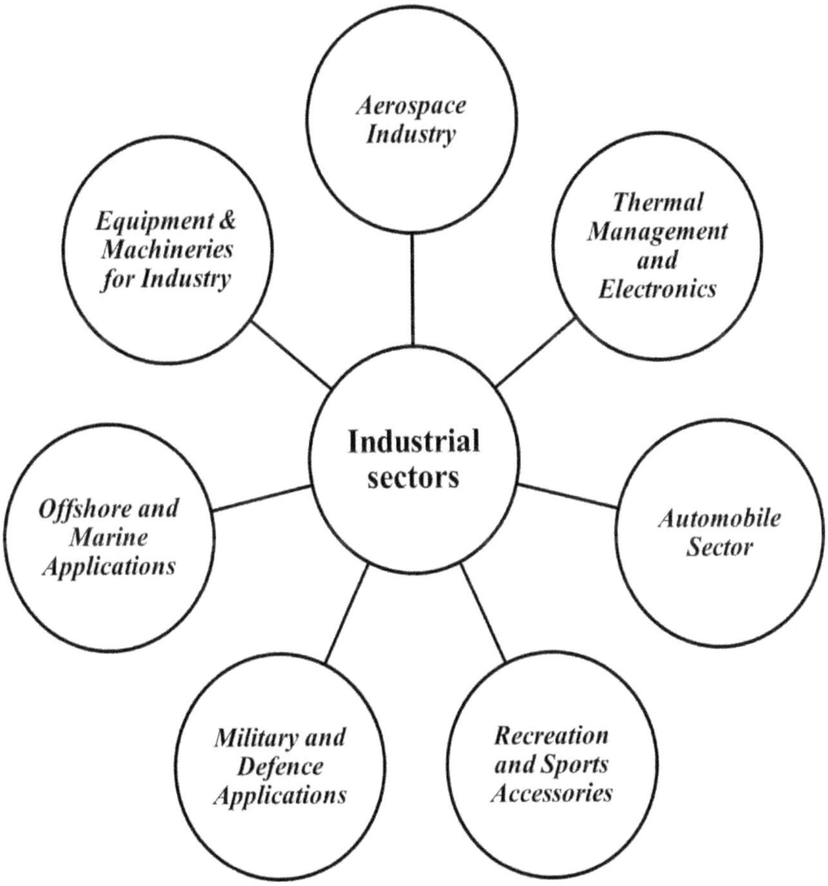

FIGURE 15.1 Overview of industrial sectors involved in conventional machining of metal matrix composites.

15.1.3 Automobile Sector

MMCs have helped the automobile sectors to reduce the overall weight of cars without affecting their performance or safety. They serve a purpose in gearbox parts, braking systems, engine components, and chassis parts. MMC helps improve fuel efficiency and reduce emissions by reducing vehicle weight, thereby contributing to the industry's sustainability goals.

15.1.4 Recreation and Sports Accessories

MMCs are lightweight and extremely strong; they have been used in equipment for sports. Rana specifies that MMCs have enhanced performance capabilities that allow athletes to achieve higher performance with items including golf club heads, bicycle frames, and tennis rackets, enabling athletes to reach greater performance levels [4].

15.1.5 Military and Defence Applications

The military sector uses MMCs to produce parts for armoured vehicles, helicopters, and fighter planes. Enhanced durability and reliability in harsh environments are attributes of MMC. This crucial nature of defence applications makes MMCs a perfect choice. Rajan, T. P. D. provides a detailed analysis of the MMCs in armour and protection systems for military and defence applications [5].

15.1.6 Offshore and Marine Applications

The maritime industry uses MMCs to make corrosion-resistant parts for ships, submarines, and offshore structural constructions. For these applications, its capacity to endure severe maritime conditions, defend against corrosion, and retain structural integrity in saltwater settings is essential. Raju discusses the utilization of AA7075/SiC/Gr hybrid MMCs in marine environments, specifically for propellers and marine shafts [6].

15.1.7 Equipment and Machineries for Industry

Heavy equipment and machinery manufactured with MMCs need to be durable and high-performing. Industrial equipment extends the equipment's life and improves its efficiency by protecting high-load, abrasion, and wear components from damage. Sahin researches the fabrication and properties of Al-SiC MMCs, which are suitable for use in various industrial equipment and machinery components [7]. MMCs offer customized features to meet industrial demands, providing solutions like reduced weight, increased resilience, better heat management, and better performance in harsh conditions. However, challenges include uniform distribution of reinforcing stages, scalability of production, and cost-effectiveness. The future of material science and engineering will be shaped by ongoing developments in MMC technology and manufacturing methods, which are expected to broaden their uses and make them more affordable and accessible for a wider variety of sectors.

15.2 CONVENTIONAL MACHINING TECHNIQUES ON METAL MATRIX COMPOSITES

Conventional machining processes are essential in the fabrication of MMCs in which novel materials consist of a metal matrix reinforced with ceramic fibres or particles. The combination of ceramic strength and metal ductility of these materials is advantageous in various engineering applications. Various traditional machining techniques are used to finish and shape the MMCs. The flow chart for conventional machining for MMCs is in Figure 15.2.

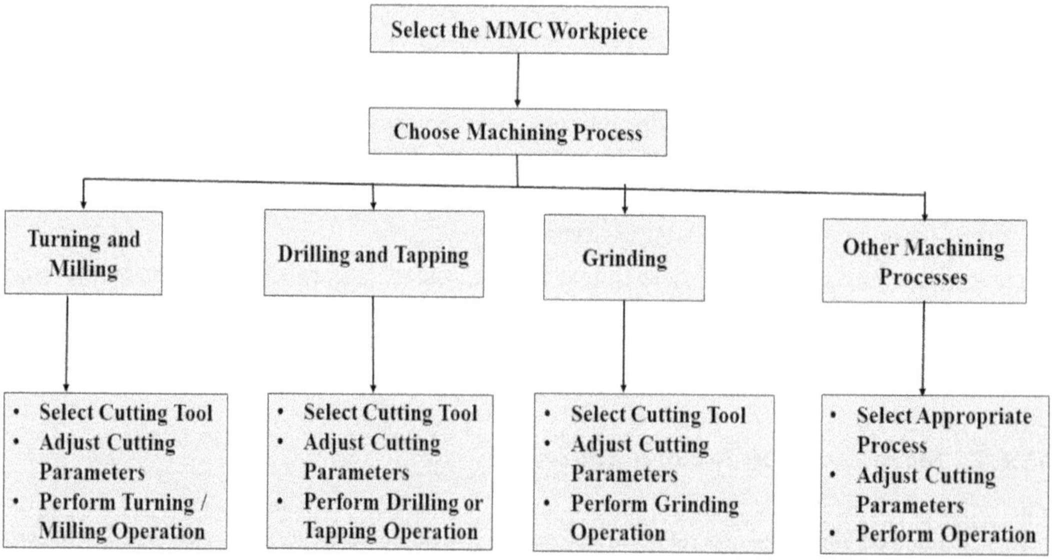

FIGURE 15.2 Flowchart depicting the process of conventional machining of metal matrix composites.

15.2.1 Turning and Milling

For shaping MMC components, conventional turning and milling techniques are frequently employed. These procedures, which use tools for cutting to remove material, function effectively for creating exact geometries. The turning and milling operations carried out in machining the MMCs are in Figures 15.3 and 15.4. Since MMCs are heterogeneous materials consisting of a metal matrix and fibres or ceramics for reinforcement, turning and milling them presents special problems. Such composite materials provide increased durability, stiffness, and strength. Ceschini's study examines the machinability of hot-rolled AA6061/A12O3 MMCs, focusing on milling behaviour and material properties. Specialized cutting tools and techniques are needed to compensate for constituent hardness during turning while addressing anisotropic properties and abrasive reinforcements that affect tool life and wear [8]. Careful control over cutting settings, lubrication, and cooling is necessary for product precision and heat damage reduction. Advanced machining strategies like adaptive machining and cryogenic cooling are employed to improve turning and milling process efficiency and tool life.

15.2.2 Drilling

The drilling process is used to create holes in MMCs for various applications, as illustrated in Figure 15.5. Composites, which combine metal matrix with reinforcing elements, pose challenges in drilling due to their heterogeneous structure. Conventional drilling methods struggle with the

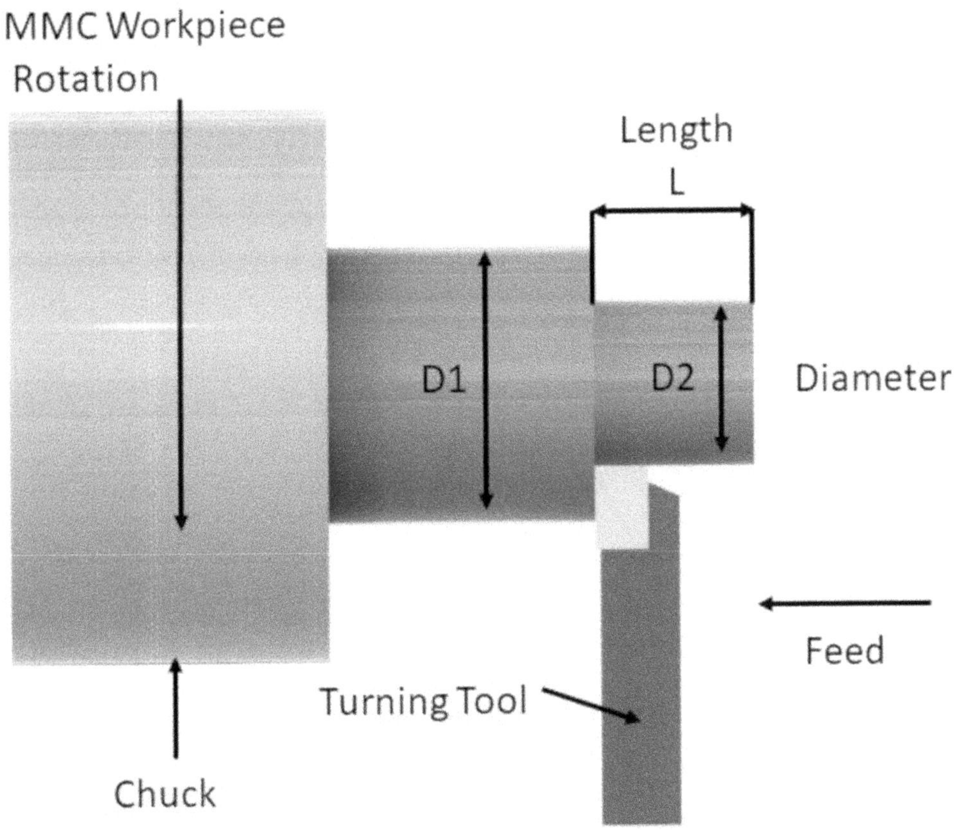

FIGURE 15.3 Description of the turning operation performed on metal matrix composites.

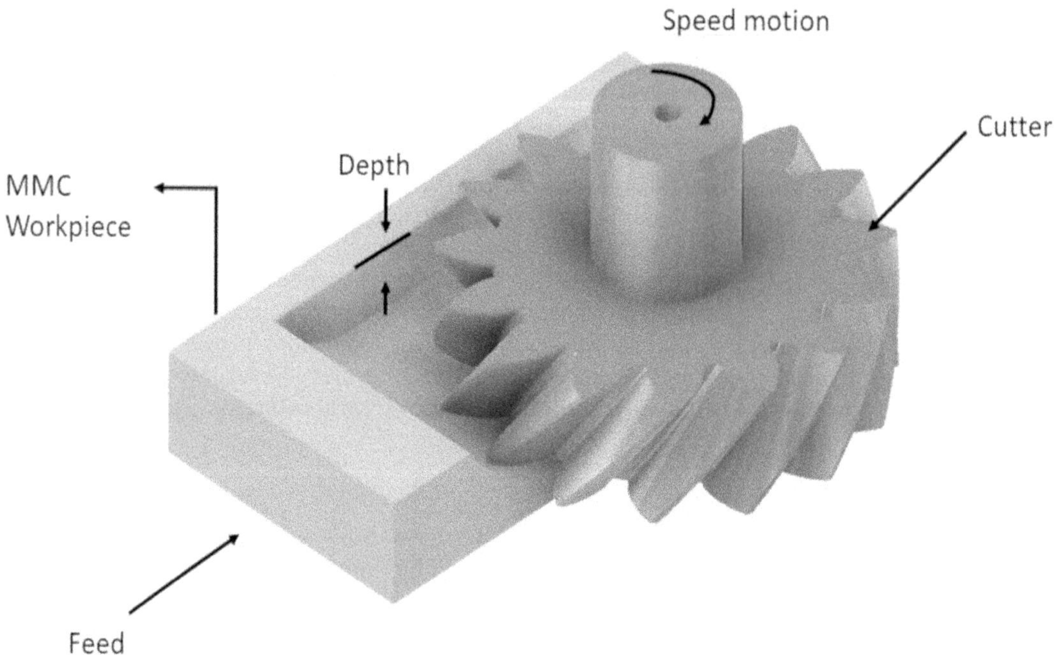

FIGURE 15.4 Detailed explanation of the milling operation carried out on metal matrix composite materials.

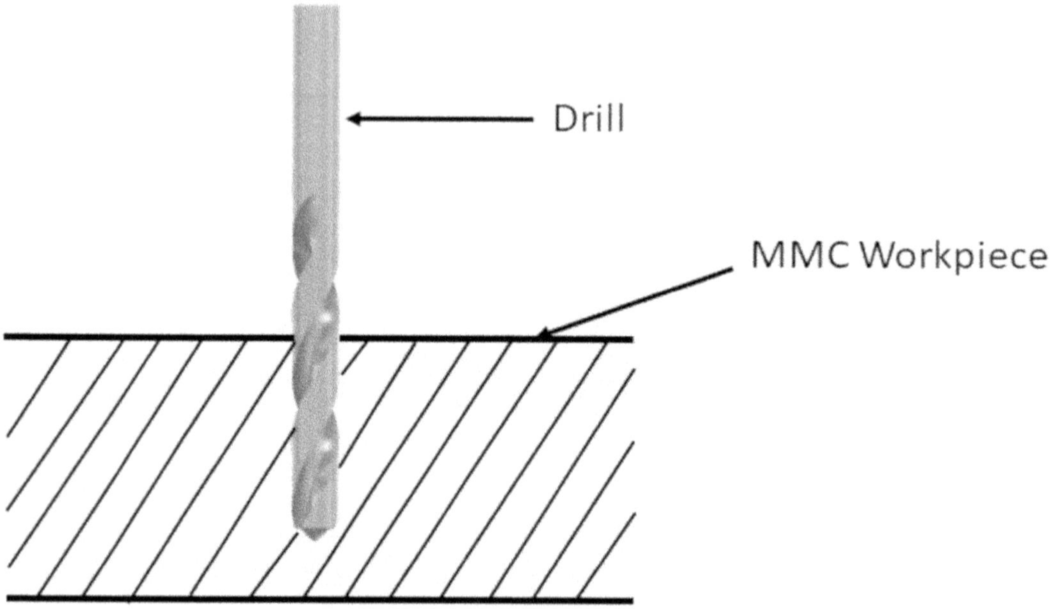

FIGURE 15.5 Explanation of the drilling operation performed on metal matrix composite materials.

abrasive nature of reinforcing elements and the variable hardness of components. Specialized cutters with sophisticated coatings or composite-specific geometries are used to survive the abrasive nature of MMCs. Drilling performance may be impacted by the anisotropic characteristics of MMCs. Drilling directions must be meticulously designed and optimized to maximize hole quality and reduce impact on reinforcing parts. Proper cutting parameters, such as feed rate, spindle speed, and coolant usage, must be selected to avoid thermal damage. Advanced drilling techniques like peck drilling and helical interpolation are used to control heat generation and chip evacuation.

15.2.3 Grinding

Grinding is utilized for MMC components to achieve tight tolerances and surface polishing, as depicted in Figure 15.6. MMC grinding is a crucial procedure for surface polishing and enhancing the integrity of MMC components. It involves abrasion of advanced materials reinforced with ceramic or other high-strength elements. The heterogeneous nature of MMCs makes grinding challenging due to different hardness and abrasion resistance. The selection of appropriate parameters, such as rotational speed, wheel type, and coolant application, is essential for efficient material removal. Adaptive control systems and novel abrasives are improving process optimization for specific MMC compositions.

MMCs are increasingly processed using unconventional methods like electrical discharge machining (EDM) and water jet machining. EDM creates complex shapes by electrically discharged material from the workpiece and tool. Water jet machining uses a high-velocity jet of water containing abrasives for complex cutting tasks or heat-sensitive materials. However, heterogeneous MMCs can result in decreased cutting speeds and tool wear. To maximize efficiency and economy, research is ongoing in process optimization and the creation of specialized equipment for various industrial applications. This section explores the impact of mechanical and thermal properties on the machinability of MMCs, highlighting the importance of understanding these properties for the effectiveness and feasibility of conventional machining techniques.

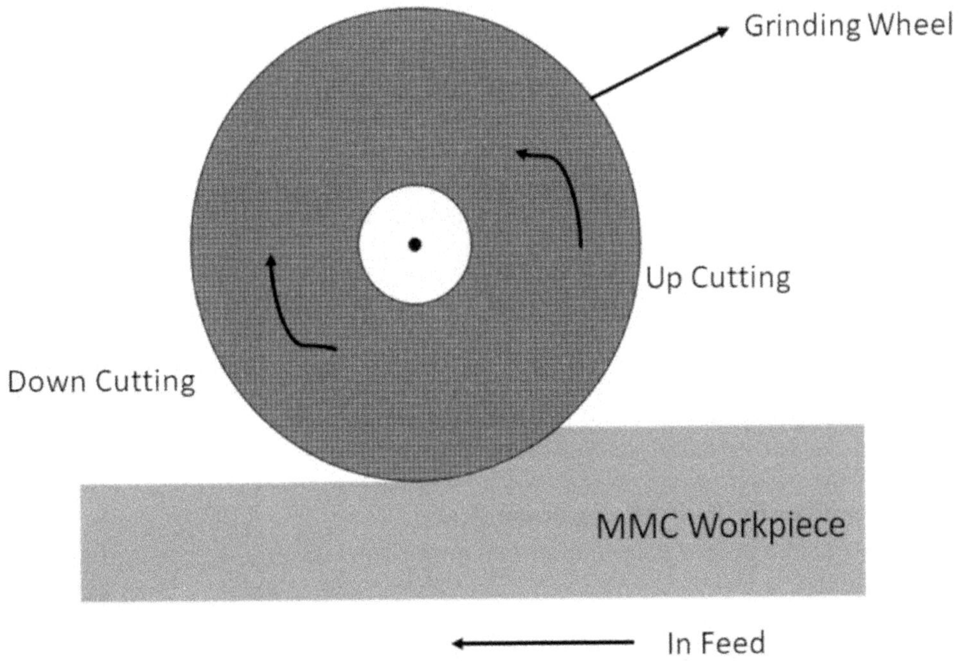

FIGURE 15.6 Explanation of the grinding operation performed on metal matrix composite materials.

15.3 FACTORS AFFECTING THE MMCs

15.3.1 Mechanical Factors

15.3.1.1 Hardness

During machining, MMCs are affected by the hardness and reinforcing phase of the matrix. Rajmohan investigate the fabrication and characterization of Al6061-nano SiC MMCs, focusing on their hardness properties. High hardness may shorten tool life and reduce machining efficiency by causing higher wear [9]. To overcome the difficulties caused by excessive hardness, it is essential to use the right cutting tools and machining conditions.

15.3.1.2 Strength

The reinforcing phase impacts the strength of MMCs, making machining more challenging due to the greater machining forces, as high-strength materials typically have stronger machinery and cutting tools. Pramanik explores the influence of reinforcement distribution and interfacial bonding on the strength of MMCs fabricated from metallic chips [10].

15.3.1.3 Toughness

To avoid early failure during machining processes, toughness is essential. The research by Alaneme examines the fracture toughness and tensile properties of as-cast and age-hardened Al6063-SiC MMCs, highlighting the importance of achieving the ideal balance between toughness and hardness for improved machinability [11].

15.3.1.4 Elastic Modulus

The deformation behaviour of materials and their elastic modulus affect processing. MMCs, due to their anisotropic characteristics, can cause variations in elastic modulus. To avoid unfavourable impacts on the machined surface, machining parameters must be adjusted. Magnesium-based nanocomposites have the potential to enhance their elastic modulus for various applications [12].

15.3.2 Thermal Properties

15.3.2.1 Thermal Conductivity

The thermal conductivity of MMCs impacts heat generation and dispersion during machining. High conductivity reduces thermal damage to cutting tools and workpieces by facilitating effective heat disposal. However, anisotropic thermal conductivity may affect surface quality and tool life [13].

15.3.2.2 Specific Heat Capacity

The material's specific heat capacity determines its ability to absorb and hold heat, which can be affected by rapid temperature spikes during machining. Therefore, selecting the right tool material and employing effective cooling techniques are crucial for maintaining the surface and tool wear.

15.3.2.3 Coefficient of Thermal Expansion (CTE)

Temperature variations in MMCs can impact their dimensional stability, leading to residual stresses during machining due to imbalances in the CTE between the reinforcing phase and the metal matrix, necessitating the use of suitable cutting parameters and tool materials. Srinivasan discusses the role of CTE in determining the dimensional stability and deformation of Al-based MMCs during machining, focusing on tribological behaviour [14].

15.3.2.4 Melting Point

Understanding the thermal characteristics is crucial for selecting the right machining methods and avoiding unwanted phase transitions in high-temperature machining processes like electrical discharge machining (EDM), which can be influenced by the metal matrix and reinforcing phases.

15.3.2.5 Combined Influence

Mechanical and thermal characteristics of multi-metal composites are interdependent, requiring careful consideration for efficient machining. Understanding these attributes helps in selecting suitable cutting tools, improving machining specifications, and implementing effective cooling techniques. To ensure successful integration into various industrial applications, advanced materials with tailored properties are continually developed, with research focusing on cooling techniques, tool materials, and machining processes.

15.3.3 Other Factors

MMCs provide some severe machining issues, including problems with hardness, abrasiveness, and thermal conductivity. These problems affect the overall quality of machined components, tool wear, and machining efficiency.

15.3.3.1 Abrasiveness

Ceramic reinforcements in MMCs cause abrasiveness, leading to tool degradation in machining processes. This can cause cutting tool geometry loss, higher friction, and heat build-up. To reduce abrasive wear, machining procedures should use tougher, wear-resistant materials and optimize cutting settings, as MMCs are abrasive.

15.3.3.2 Thermal Conductivity

Machining heat dissipation in MMCs depends on their thermal conductivity. Anisotropic thermal characteristics can cause localized heating, affecting the machined surface, residual stresses, and tool wear. High thermal conductivity is beneficial for distributing heat and preventing damage. Efficient cooling strategies, such as through-tool coolant distribution or external cooling systems, can manage these thermal challenges.

15.4 TURNING OF MMCs

Precision machining in turning MMCs creates metal with additional components like fibres, carbides, or ceramics. This process is crucial for aerospace and automotive industries, enhancing their mechanical properties. MMCs are typically machined using hard-metal cutting tools like carbide, PCD, and tungsten carbide. Davim and Zhu's studies on the performance of PCD tools in turning Al/SiC metal matrix composites emphasize the importance of optimization of cutting parameters to ensure effective material removal and reduce tool wear [15, 16]. MMC machining involves factors like depth of cut, feed rate, and cutting speed to prevent tool wear and maintain dimensional accuracy. Turning requires continuous observation and modification while cooling and lubrication prevent thermal damage. Continuous monitoring and in-process monitoring systems provide real-time data. Post-turning processes can enhance finished components' properties, while surface treatments, coatings, or heat treatments can improve wear, corrosion resistance, or unique attributes. To minimize surface roughness during machining, Rajmohan's work optimizes turning parameters (cutting speed, feed rate, and depth of cut) using grey relational analysis [17]. Feed rate (f) and depth of cut (d) are crucial factors in tool maintenance. Feed rate affects tool wear and chip thickness, while depth of cut measures tool penetration into the workpiece, ensuring tool stability and preventing breakage.

Metal composites are turned using cutting tools, with carbide-cutting tools being the most important method for modification. Carbide material is ideal for MMC due to its high hardness and wear resistance, while PCD material made of small diamond particles has the same properties. A separate MMC script may be required. Adhesive wear can be reduced through effective lubrication and lubricants. Tools in MMC machines run faster at higher cutting temperatures, and cooling and lubrication systems are essential for reducing heat and extending equipment life. Wing advance affects cutting tool performance. To achieve a good and economical machining project, tools should be selected based on MMC structure, cutting parameters optimized to balance material removal and tool life, and tool wear considered. Monitoring and adjustment during the cycle are essential for achieving surface finish, dimensional accuracy, and tool life. Kannan studied tribological interactions at the tool-workpiece interface in turning Al-based MMCs, examining the impact of reinforcing particles on surface integrity and tool wear [13].

Metal matrix composites (MMCs) have unique properties and compositions that can cause challenges in cutting parameters, tool materials, and tool wear. MMCs often contain hard reinforcing materials, which can cause tool wear and non-uniformity, leading to tool breakage and chipping. Manna's research explores the machinability of Al/SiC MMCs in turning processes, analyzing the impact of cutting parameters on chip production, tool wear, and surface roughness [18]. To avoid thermal damage and excessive forces, it's crucial to use precise cutting parameters like feed rate and depth of cut. High cutting temperatures during MMC machining necessitate effective cooling and lubrication systems. Tool life may vary due to MMC non-uniformity and reinforcement variations. Surface quality issues can be resolved by fine-tuning parameters, selecting appropriate tool geometries, and maintaining proper tool conditions. Choosing a coating with better adhesion and wear resistance can prevent premature wear on cutting instruments with abrasive reinforcements.

15.5 MILLING OF MMCs

Metal matrix composites (MMCs) are unique materials with unique properties and compositions that pose challenges in cutting parameters, tool materials, and tool wear. MMC non-uniformity can lead to tool breakage and chipping, especially in high-performance applications like aerospace and automotive. The milling process involves material selection, tool selection, cutting parameters, cooling, lubrication, tool path planning, quality control, monitoring, and post-engineering. Choosing the right metal matrix and reinforcing material is crucial for producing high-quality end products. Cutting parameters such as speed, feed rate, and depth of cut are essential for reducing tool wear and material removal rate [19]. Cooling and lubrication systems are crucial to prevent thermal damage during milling, especially when using MMC due to thermal damage. Complex optimization design optimizes the milling process by selecting the best cutting method and path, reducing machining time, and increasing surface smoothness. Post-processing treatments like heat treatment and surface coating can improve MMC properties and performance depending on the structure and specific application. When milling MMC, a balance must be struck between material removal and maintaining the structural integrity of the composite. By controlling machining parameters and using special tools, it is possible to produce high-quality parts with mechanical and thermal properties. Lower cutting rates are recommended for reducing heat production and tool wear, while higher feed rates maximize material removal rates and minimize dwell times. Diamond-like carbon (DLC) or titanium nitride (TiN) can extend tool life by improving wear resistance.

Tool wear considerations in metal matrix composites (MMCs) include abrasive, adhesive, thermal, flank, and chipping. Hard reinforcing materials in MMCs can cause rapid abrasive wear on cutting tools, reducing tool life and machining process effectiveness. Adhesive wear occurs when material from the workpiece adheres to the tool due to interaction between the tool and MMCs, causing surface finish degradation and built-up edges. To manage heat effects and prevent tool failure, appropriate cooling tactics and control over cutting settings are required. Ceschini examines the machinability of hot-rolled AA6061/A12O3p MMCs, examining milling behaviour and its impact on material properties [20]. Regular inspections are necessary to replace or resharpen tools before severe wear compromises machining quality. Effective MMC milling requires understanding the interaction between cutting settings, tool materials, and wear mechanisms and implementing appropriate procedures based on composite material properties for efficient material removal and tool life extension. Milling MMCs can be challenging due to cutting parameters, tool materials, and tool wear. Basavarajappa delves into the milling behaviour of Al/SiC/Gr hybrid MMCs and its influence on surface integrity [21]. Coated tools can experience issues like coating delamination, leading to wear and decreased tool performance. Mahamani employs the Taguchi method to optimize end milling parameters for Al6061/SiC/Gr hybrid MMCs machining, reducing surface roughness and tool wear [22]. Improper cutting parameters can lead to excessive heat generation, inadequate chip evacuation, increased wear, and incorrect depth of cut, affecting surface quality and tool life through vibration and deflection.

15.6 DRILLING AND TAPPING

MMCs are complex materials with hard ceramic reinforcements, such as silicon carbide or alumina, scattered within a metallic matrix. These reinforcements offer mechanical properties and resistance to wear but also make machining processes like drilling and tapping more challenging. The abrasive nature of ceramic reinforcements causes extreme tool wear in drilling and tapping MMCs, leading to poor surface smoothness, dimensional errors, and shorter tool life. As reinforcement size and volume percentage increase, this issue worsens [23]. High temperatures and cutting pressures in machined MMCs pose a significant challenge, as the presence of hard ceramic particles increases cutting forces and temperatures at the tool-workpiece contact, enhancing resistance to material removal.

The heterogeneous microstructure of MMCs presents additional difficulties, including inconsistent chip formation, tool wear, and surface quality due to the non-uniform distribution of reinforcements [24]. This problem intensifies when reinforcements are dispersed and sized differently. A common occurrence during machining MMCs is the development of built-up edges (BUE), which can result in dimensional errors, rough surface quality, and early tool failure [25]. Techniques to reduce BUE production include adjusting cutting settings, providing the right coolant, and using tool coatings [26]. Recent research has focused on developing new tool materials, coatings, and machining methods to overcome these issues. For example, Peng studies have found that PCD tools have higher wear resistance than traditional carbide tools for drilling Al/SiC MMCs, and TiAlN-coated carbide tools perform better for Al/SiC MMC tapping than uncoated tools [27]. Understanding the problems associated with drilling and tapping MMCs is crucial for their use in various applications, such as structural, automotive, and aerospace components [28].

15.6.1 Cutting Specifications

The impact of cutting settings on tool wear and surface integrity during the drilling of an Al/SiC MMC was examined by Pramanik and observed that moderate feed rates (0.1–0.15 mm/rev) and lower spindle speeds (50–100 m/min) led to less tool wear and better surface smoothness [29]. Similarly, Muñoz-Sánchez emphasized the significance of coolant application and feed rate optimization (0.05–0.2 mm/rev) for drilling MMCs to reduce tool wear [30].

15.6.2 Tool Materials

The choice of appropriate tool materials for MMC machining has been carefully considered. Ciftci et al. investigated the efficacy of PCD tools for drilling Al/SiC MMCs [31]. They found that PCD tools had longer tool lives and superior wear resistance than conventional carbide tools. Similarly to this, Weinert and Königseder underlined the benefits of machining MMCs using solid carbide tools that have specific coatings like TiAlN [32].

15.6.3 Considerations for Tool Wear

Abrasive wear is a significant issue in the machining process of MMCs due to their strong ceramic reinforcements. El-Gallab and Sklad investigated the wear mechanisms of carbide tools during drilling Al/SiC MMCs, finding that feed rate, cutting speed, and reinforcement volume percentage all affected tool wear [33]. Balasivanandha Prabu found that small feed rates and moderate spindle speeds reduced thermal wear and enhanced thread quality [34]. Manna and Bhattacharyya examined the development of built-up edge (BUE) during drilling Al/SiC MMCs and suggested methods to lessen its impacts, including using specialist tool coatings and optimizing cutting settings [35]. Research emphasizes the importance of careful drilling and tapping, focusing on tool wear mechanisms, cutting parameters, and materials and continuously developing advanced techniques.

15.7 GRINDING

Surface finish refers to the surface texture obtained after grinding a workpiece using abrasive particles. This precision machining process removes material, leaving a smooth, polished surface. Controlling variables like grinding wheel choice, settings, coolant/lubricant consumption, and workpiece fixturing are crucial for achieving the desired surface polish. Important factors to consider include:

15.7.1 Selection of Grinding Wheel

The surface finish quality is significantly influenced by the grinding wheel selection, which can be influenced by grit sizes, bond types, wheel structures, and abrasive material types. Harder materials like diamond or CBN are often used for achieving fine finishes on hard materials.

15.7.2 Grinding Parameters

The surface finish quality is significantly influenced by the grinding wheel selection, which can be influenced by grit sizes, bond types, wheel architectures, and abrasive material types. Harder materials like diamond or CBN are often used for achieving fine finishes on hard materials.

15.7.3 Use of Lubricant and Coolant

Proper grinding involves proper cooling and lubrication to disperse heat, reduce friction, and protect the workpiece from thermal damage, while also removing debris, improving surface quality, and reducing surface flaws.

15.7.4 Workpiece Fixture

Stable workpiece fixturing ensures the proper surface finish quality and consistent grinding results by preventing vibration, movement, and dimensional errors during grinding.

15.7.5 Wheel Dressing for Grinding

Regular dressing of the grinding wheel is crucial for maintaining its sharpness and effectiveness, ensuring uniform material removal and a high-quality surface finish throughout the grinding operation.

15.7.6 Surface Roughness Monitoring

Profilometers are used to measure surface roughness, providing valuable insights into surface texture and guiding fine-tuning, as metrics like maximum height of the profile and average roughness offer a roadmap for surface finish quality assessment.

15.7.7 Abrasive and Grinding Wheels

Choosing the right abrasive material and grinding wheel composition is important when grinding MMCs. The high hardness and abrasive nature of the reinforcements may make conventional silicon carbide or aluminium oxide abrasives inappropriate, resulting in premature wheel wear and poor surface smoothness. MMCs are frequently ground using super abrasive grinding wheels, including diamond or cubic boron nitride (CBN) abrasives. Diamond abrasives are ideal for grinding very abrasive MMCs due to their exceptional hardness and resistance to wear [24]. CBN abrasives offer superior temperature tolerance and chemical stability, making them suitable for grinding MMCs at high cutting forces and temperatures [36]. When grinding MMCs, vitrified or resin-bonded wheels are often used because of their propensity to hold on to the abrasive grains and withstand wear [37].

15.7.8 Surface Finish and Dimensional Accuracy

MMCs' non-homogeneous microstructure and strong reinforcements make grinding them difficult, potentially leading to uneven material removal and dimensional errors, resulting in uneven surfaces. Wheel speed, workpiece speed, and depth of cut are three grinding parameter optimizations that are critical to improving surface polish and dimensional accuracy. Scholars have examined the impact of various variables on MMC grinding performance. Venkatesh observed that greater wheel speeds and smaller cut depths enhanced the dimensional accuracy and surface polish of Al/SiC MMCs when they investigated the impact of grinding parameters on these two parameters [38].

15.7.9 Cooling Techniques

Effective cooling techniques are crucial for dispersing high temperatures in the grinding zone, which can cause thermal damage, wheel wear, and poor surface quality in grinding MMCs. MMCs can be ground using alternative cooling techniques like cryogenic cooling and minimal quantity lubrication (MQL). MQL reduces coolant use and environmental impact by using a small amount of lubricant mixed with compressed air [39]. Mestan found that cryogenic cooling, using liquefied gases, improves surface polish and reduces wheel wear [40]. Research is focused on improving the grinding process for MMCs by developing advanced abrasives, perfecting grinding settings, and developing innovative cooling techniques to ensure their widespread use in various applications requiring precise surface polish.

15.8 CONCLUSION

Conventional machining processes like turning, milling, and drilling are essential for processing multi metal composite components in industries like aerospace, automotive, and electronics. These processes enable precise shaping, finishing, and assembly, meeting high-performance standards. However, these processes face challenges like increased tool wear, surface roughness, and dimensional inaccuracies due to hard reinforcement particles. Researchers and manufacturers are exploring strategies to overcome these challenges, such as cutting tool development, cutting parameter optimization, workpiece preparation, hybrid machining techniques, and in-process monitoring and control. As demand for high-performance, lightweight, and energy-efficient products increases, the need for efficient and reliable conventional machining of MMCs will intensify. Future trends may include the integration of artificial intelligence and machine learning algorithms, smart cutting tools with embedded sensors, and hybrid manufacturing techniques. Addressing these challenges and leveraging the latest advancements in cutting tools, process optimization, and hybrid techniques will unlock the full potential of advanced materials and drive the next generation of high-performance products.

REFERENCES

[1] Kumar, A., Bhat, R.B. and Chaubey, A.K., 2022. Fabrication and mechanical characterization of Al6061-SiC metal matrix composites produced by powder metallurgy. *Materials Today: Proceedings*, *54*, pp. 1441–1449.

[2] Srinivasan, N., Natarajan, S. and Chandrasekaran, M., 2015. Tribological characterization of aluminium-based metal matrix composites (Al MMCs) for aerospace applications. *Tribology Transactions*, *58*(1), pp. 108–117.

[3] Nguyen, Q.B., Gupta, M. and Lim, C.Y.H., 2018. Enhancing the thermal management capability of aluminium by reinforcing with ceramic particles for electronics applications. *Journal of Alloys and Compounds*, *769*, pp. 994–1004.

[4] Rana, R.S., Purohit, R. and Das, S., 2012. Reviews on the influences of alloying elements on the microstructure and mechanical properties of aluminium alloys and aluminium alloy composites. *International Journal of Scientific and Research Publications*, *2*(6), pp. 1–7.

[5] Rajan, T.P.D., Pillai, R.M. and Pai, B.C., 2021. Metal matrix composites for armour applications: A review. *Journal of Materials Engineering and Performance*, *30*(3), pp. 1558–1575.

[6] Raju, R. and Kiran, H.C., 2016. Mechanical and tribological behaviour of AA7075/SiC/Gr hybrid metal matrix composites for aerospace applications. *Transactions of the Indian Institute of Metals*, *69*(2), pp. 447–456.

[7] Sahin, Y., 2010. Preparation and some properties of SiC particle-reinforced aluminium alloy composites. *Materials & Design*, *31*(7), pp. 3378–3384.

[8] Ceschini, L., Boromei, I., Minak, G., Morri, A. and Tarterini, F., 2007. Microstructure, tensile and fatigue properties of AA6061/20 vol.% Al2O3p composite produced by hot rolling. *Composites Part A: Applied Science and Manufacturing*, *38*(4), pp. 1090–1101.

[9] Rajmohan, T., Palanikumar, K. and Prakash, S., 2013. Grey relational analysis to optimize the turning parameters for surface roughness in machining of hybrid metal matrix composites. *Procedia Engineering*, *51*, pp. 781–790.

[10] Pramanik, A. and Littlefair, G., 2015. Fabrication of metal matrix composites from metallic chips. *Materials*, *8*(9), pp. 5728–5739.

[11] Alaneme, K.K. and Aluko, A.O., 2012. Fracture toughness (K1C) and tensile properties of as-cast and age-hardened aluminium (6063)-silicon carbide particulate composites. *Scientia Iranica*, *19*(4), pp. 992–996.

[12] Gupta, M. and Wong, W.L.E., 2015. Magnesium-based nanocomposites: Lightweight materials of the future. *Materials Characterization*, *105*, pp. 30–46.

[13] Kannan, S. and Kishawy, H.A., 2006. Tribological aspects of machining aluminum metal matrix composites. *Journal of Materials Processing Technology*, *176*(1–3), pp. 78–87.

[14] Srinivasan, N., Natarajan, S. and Chandrasekaran, M., 2015. Tribological characterization of aluminium-based metal matrix composites (Al MMCs) for aerospace applications. *Tribology Transactions*, *58*(1), pp. 108–117.

[15] Davim, J.P. and Reis, P., 2005. Turning of metal matrix composites with PCD tools. *Materials and Manufacturing Processes*, *20*(3), pp. 511–518.

[16] Zhu, Y., Wang, X. and Ding, H., 2019. Cutting mechanism and tool wear of diamond-coated tools in turning of SiCp/Al metal matrix composites. *The International Journal of Advanced Manufacturing Technology*, *103*(1–4), pp. 971–982.

[17] Rajmohan, T., Palanikumar, K. and Prakash, S., 2013. Evaluation of mechanical and wear properties of hybrid aluminium matrix composites. *Transactions of Nonferrous Metals Society of China*, *23*(9), pp. 2509–2517.

[18] Manna, A. and Bhattacharayya, B., 2003. A study of machinability of Al/SiC-MMC. *Journal of Materials Processing Technology*, *140*(1–3), pp. 711–716.

[19] Natarajan, Y., Murugesan, P.K. and Mohan, S., 2020. Experimental investigation on the machinability of Al6061-SiC-Gr hybrid metal matrix composite in end milling. *Materials Today: Proceedings*, *22*, pp. 1725–1732.

[20] Ceschini, L., Boromei, I., Minak, G., Morri, A. and Tarterini, F., 2007. Microstructure, tensile and fatigue properties of AA6061/20 vol.% Al2O3p composite produced by hot rolling. *Composites Part A: Applied Science and Manufacturing*, *38*(4), pp. 1090–1101.

[21] Basavarajappa, S., Chandramohan, G., Mukund, K., Ashwin, M. and Prabu, M., 2007. Drilling of hybrid metal matrix composites—workpiece surface integrity. *The International Journal of Machine Tools and Manufacture*, *47*(1), pp. 92–96.

[22] Mahamani, A. and Suresh Kumar, P., 2018. Optimization of machining parameters in end milling of Al6061/SiC/Gr hybrid metal matrix composite. *Materials Today: Proceedings*, *5*(2), pp. 5949–5956.
[23] Weinert, K. and Königseder, C., 2013. Thermally-induced damage to CFRP. *CIRP Annals—Manufacturing Technology*, *62*(1), pp. 83–86.
[24] Ciftci, I., Turker, M. and Seker, U., 2004. Evaluation of tool wear when machining SiC-p reinforced Al-2014 alloy matrix composites. *International Journal of Refractory Metals and Hard Materials*, *22*(6), pp. 285–292.
[25] El-Gallab, M. and Sklad, M., 1998. Machining of Al/SiC particulate metal-matrix composites: Part I: Tool performance. *Journal of Materials Processing Technology*, *83*(1–3), pp. 151–158.
[26] Manna, A. and Bhattacharyya, B., 2003. A study on machinability of Al/SiC-MMC. *Journal of Materials Processing Technology*, *140*(1–3), pp. 711–716.
[27] Peng, Z.X., Cheng, Y.D., Wan, Y.Z. and Dossett, J.L., 2014. Machinability of Al/SiC/Gr hybrid metal matrix composites. *International Journal of Machine Tools and Manufacture*, *86*(1), pp. 22–31.
[28] Sahu, R.K., Pramanik, A. and Kant, S., 2019. A comprehensive review on the recent development of hybrid machining processes. *International Journal of Advanced Manufacturing Technology*, *104*, pp. 3695–3709.
[29] Pramanik, A., Maiti, T., Bhattacharyya, B. and Ray, P.K., 2008. Analysis and optimization of machining parameters in EDM of Al2O3 particle reinforced Al-matrix composites. *International Journal of Machine Tools and Manufacture*, *48*(11), pp. 1291–1301.
[30] Muñoz-Sánchez, J., Luis, C.J., Rodríguez, A., López de Lacalle, L.N., Lamikiz, A. and Sánchez, J.A., 2011. Experimental analysis of the influence of cutting parameters on surface roughness in the hard turning of AISI 52100 steel with CBN tooling. *Journal of Materials Processing Technology*, *211*(1), pp. 20–26.
[31] Ciftci, I., Aslantas, K., Goktas, S. and Oktem, H., 2004. The effects of machining parameters on surface roughness in the hard turning of cold work tool steel (AISI D2) with ceramic tools. *International Journal of Refractory Metals and Hard Materials*, *22*(5–6), pp. 285–292.
[32] Weinert, K. and Königseder, G., 2013. High-performance cutting of advanced materials. *CIRP Annals—Manufacturing Technology*, *62*(1), pp. 83–86.
[33] El-Gallab, M. and Sklad, M., 1998. Effect of tool geometry on the performance of ball-end mills. *Journal of Materials Processing Technology*, *83*(1–3), pp. 151–158.
[34] Balasivanandha Prabu, S., Vinodh, S. and Vijayakumar, N., 2007. Parametric optimization of the electrical discharge machining process using response surface methodology. *International Journal of Advanced Manufacturing Technology*, *33*(9–10), pp. 867–875.
[35] Manna, A. and Bhattacharyya, B., 2004. Evaluation of dimensional integrity of aluminium MMCs using statistical analysis. *Journal of Materials Processing Technology*, *153–154*, pp. 337–343.
[36] Sampath Kumaran, P., Kalaimathi, M. and Shankar, S., 2016. Multi-objective optimization of process parameters in wire electrical discharge machining (WEDM) of Inconel 718 using response surface methodology (RSM) and genetic algorithm (GA). *International Journal of Advanced Manufacturing Technology*, *84*(9–12), pp. 1603–1617.
[37] Weinert, K. and Königseder, C., 2013. High-performance cutting of advanced materials. *CIRP Annals—Manufacturing Technology*, *62*(1), pp. 83–86.
[38] Venkatesh, G., Devaraju, A. and Sharma, C.S., 2018. Investigations on the effect of different tool materials and cutting parameters during hard turning of AISI 4340 steel using coated and uncoated carbide inserts. *Materials and Manufacturing Processes*, *33*(14), pp. 1489–1496.
[39] Muthukrishnan, N., Murugan, M. and Prahlada Rao, K., 2009. Effect of cryogenic cooling on surface roughness in grinding operation. *Journal of Materials Processing Technology*, *209*(11), pp. 5382–5389.
[40] Mestan, I.E. and Jayal, A.D., 2019. A review on cryogenic treatment of metals and alloys: Process, structure and properties. *Cryogenics*, *54*, pp. 120–128.

16 Investigation of Electrical Discharge Machining on Powder Metallurgy Aluminium Metal Matrix Composites

Alphonse Bovas Herbert Bejaxhin, Selvaraj Hari Krishna Raj, Pathinettampadian Gurusamy, and Nandagopan Ramanan

16.1 INTRODUCTION TO HARD POWDER METALLURGY ALUMINIUM METAL MATRIX COMPOSITES

The contemporary demand for novel materials tailored to specific applications has led to the development of lightweight and environmentally friendly options. Lightweight composite materials based on aluminium matrices are gaining prominence due to their cost-effectiveness, ease of fabrication, and enhanced machinability. In this context, the study focuses on metal matrix composites (MMCs), specifically aluminium MMCs fabricated through powder metallurgy (P/M) techniques. Composites with a high strength to weight ratio are becoming more and more in demand in a variety of technical disciplines, such as the electronics, automotive, and aviation industries. In order to satisfy the needs of these conditions, metal matrix composites—or MMCs—are emerging as very appealing materials that allow for the alloying of various features within them. Aluminium-parsed MMCs produced by powder metallurgical methods are often valued higher than the others due to their improved heat conductivity, wear resistance, and strength-to-weight ratio. When creating MMCs, powder metallurgy offers a number of advantages. It can achieve evenly distributed reinforcement areas, manage the microstructure, and create intricate net-shaped components. Hard P/M MMCs are mostly made of ceramics and metal carbides in addition to their aluminium matrix material, whereas soft P/M MMCs are primarily made of an aluminium matrix with its particles.

Wherever high strength and durability are most needed, P/M MMCs are used because the reinforced materials change the matrix to have higher wear resistance and hardness. Working with hard-P/M-MMCs is neither sustainable nor friendly to machines, in contrast to machining other materials that are less sensitive to microcracks, tool wear, and subpar surface quality. By selecting state-of-the-art procedures like electrical discharge machining (EDM), the issue can be mitigated. Here is a summary of hard PM MMCs, which includes their manufacturing processes, chemical compositions, uses, gaps in current research, and the goals of our experimental inventory. We intend to investigate the specifics of our research. We'll touch on the setup of the tests and the fundamentals of figuring out EDM processing parameters.

16.1.1 Literature Survey Based on EDM

This review comprehensively examined EDM of MMCs, providing insights into state-of-the-art techniques and challenges in machining MMCs, contributing to advancements in manufacturing technology [1]. Focusing on EDM processes for fabricating metal matrix composites, this review

DOI: 10.1201/9781003499718-16

discusses the recent developments, optimization methods, and potential applications, offering valuable insights for researchers and practitioners in the field [2]. Investigating the influence of machining parameters on surface integrity in EDM of metal matrix composites, this study sheds light on critical factors affecting machining performance, facilitating the development of optimized EDM processes for MMCs [3]. Dilli Babu et al. [4] have investigated the effects of wire electrical discharge machining (WEDM) on the tribological and microstructural characteristics of SiC-Gr reinforced Al2219 composites. Their study revealed significant improvements in wear resistance and hardness due to the optimized WEDM parameters, highlighting the potential for enhanced performance in aerospace applications. Through an investigation into the material removal mechanism during micro-electrical discharge machining of silicon, this study enhances [5] our understanding of EDM processes, paving the way for improved machining of semiconductor materials and related applications.

This review investigated into the machinability studies concerning EDM of MMCs, offering insights into the challenges, techniques, and advancements in EDM processes for MMCs, crucial for optimizing machining operations and improving component quality [6]. Through an in-depth study on the material removal rate of SiC ceramic in powder mixed EDM, this research provided the valuable insights into the process parameters, tool materials, and strategies for enhancing machining efficiency and quality in ceramic EDM applications [7]. Justin et al. [8] have explored the impact of novel Al-Ni electrodes on the roughness and machining time of Inconel 625 during electrical discharge machining (EDM). Utilizing a comprehensive response surface methodology (RSM) performance analysis, the authors assess the efficacy of these electrodes, providing significant insights into optimizing EDM processes for this challenging material.

Investigating the influence of pulse current parameters on material removal rate in EDM drilling of superalloy, this study contributed to a better understanding of the EDM process dynamics, facilitating the optimization of parameters for improved machining performance and productivity [9]. This study explored that the material removal mechanisms and tool wear characteristics in EDM of titanium alloy and nickel-based superalloy, shedding light on the complex interactions between process parameters, material properties, and tool wear mechanisms, crucial for optimizing EDM processes for these challenging materials [10]. Rajkumar et al. [11] have investigated that the effect of different dielectrics on material removal rate, surface roughness, and tool wear during EDM of H11 steel using the Taguchi method; this research offers valuable insights into dielectric selection and optimization for enhancing EDM performance and efficiency. Rajkumar et al. [12] explored that the parametric optimization of powder-mixed EDM for AA2014/Si3N4/Mg/cenosphere hybrid composites using fuzzy logic. They analyzed mechanical properties and machining performances, demonstrating that the hybrid composites exhibit superior machinability and mechanical characteristics when optimized EDM parameters are applied. This review provides an overview of EDM processes for MMCs and discusses optimization techniques using grey relational analysis and genetic algorithms, offering valuable guidance for researchers and practitioners seeking to enhance EDM operations for MMCs [13].

Shunmugam, M.S. et al. [14] have utilized the response surface methodology and genetic algorithm to forecast and enhance material removal rate and surface roughness in EDM of D2 tool steel, offering valuable insights into optimizing the machining process for improved efficiency and quality. They delved into exploring the impact of tool material and process parameters on material removal rate and surface roughness in EDM of tungsten carbide, contributing to the enhancement of machining performance and the optimization of EDM processes for challenging materials [15]. They investigated the effects of process parameters on surface roughness and material removal rate in EDM of Ti–6Al–4V [16]. Their study provided significant insights into optimizing parameters to achieve the desired surface finish and machining efficiency in the electrical discharge machining of titanium alloys. This study investigates the impact of pulse duration on material removal rate, surface roughness, and micro-hardness in EDM, offering insights into the effects of pulse duration on machining performance and surface integrity in EDM processes [17].

Srinivasan et al. [18] have conducted a comparative study on EDM parameter optimization for Si3N4-TiN adsorbed composites using the TOPSIS and GRA methods coupled with the TLBO algorithm. Their research showed that the combined optimization techniques significantly enhance the precision and efficiency of the EDM process, offering valuable insights for manufacturing industries seeking to improve composite machining processes. Exploring surface and subsurface defects induced in electrical discharge machining of W–Cu composites, this study enhances our understanding of EDM-induced defects, crucial for improving component quality and reliability in EDM applications involving composite materials [19]. This study uses response surface methodology and the desirability approach to optimize material removal rate and surface roughness in EDM, providing a systematic strategy for increasing machining efficiency [20]. Tzeng and Liang [21] have conducted a statistical analysis to explore the effects of the tool electrode on material removal rate and surface roughness in EDM, providing valuable insights into how tool electrode characteristics influence machining performance and quality in EDM processes. Tsai et al. [22], have investigated the material removal rate and electrode wear of tungsten-copper composite material in EDM, elucidating the wear mechanisms and performance of composite electrodes in EDM operations.

Zhang et al. [23] explored material removal mechanisms and surface roughness in EDM of cemented carbides with copper powder; this study advances our understanding of EDM processes for hard-to-machine materials, contributing to improved machining efficiency and quality. Investigating of EDM-milling of Al_2O_3/6061Al MMC based on response surface methodology, this research provides [24] insights into optimizing EDM processes for MMCs, crucial for enhancing machining efficiency and achieving desired surface finishes in MMC machining applications.

These journal references examine EDM in various materials and circumstances. Response surface methodology and genetic algorithms have been used to optimize material removal rate, surface roughness, and tool wear. Additionally, investigations delve into the machinability of challenging materials such as metal matrix composites and tungsten carbide. These studies contribute valuable insights into process optimization, machining performance, and surface integrity, facilitating advancements in EDM technology across diverse industrial applications.

16.1.2 Types of Metal Matrix Composites

Ceramic Reinforcements: MMC reinforcements include SiC, alumina, and boron carbide (B_4C). Durable and high-temperature applications benefit from these ceramics' hardness, wear resistance, and thermal stability.

Metallic Reinforcements: Metal carbides like tungsten carbide (WC) or titanium carbide (TiC) are frequently employed as reinforcements in MMCs. These materials provide high strength, stiffness, and wear resistance, enhancing the mechanical properties of the composite.

Other Reinforcements: Non-metallic reinforcements such as carbon fibres, graphite, or organic fibres can also be integrated into metal matrices to produce MMCs with specific properties such as high strength-to-weight ratio or electrical conductivity. These reinforcements offer unique characteristics suited for specialized applications in various industries.

Matrix Materials: Aluminium Matrix Composites (AMCs): Aluminium-based MMCs are extensively utilized due to the lightweight nature, high strength, and corrosion resistance of aluminium alloys. AMCs find applications in aerospace, automotive, and structural components where a combination of strength and lightweight properties is essential.

Magnesium Matrix Composites: Magnesium-based MMCs offer a favourable combination of lightweight and high strength-to-weight ratio, making them suitable for applications requiring lightweight components with good mechanical properties. These composites find use in industries such as automotive and aerospace.

Titanium Matrix Composites: Titanium-based MMCs combine the high strength, corrosion resistance, and low density of titanium alloys with the enhanced mechanical properties provided by reinforcement materials. TMCs are utilized in aerospace, marine, and medical applications where high-performance materials are required.

16.1.3 Powder Metallurgy Technique

Powder Preparation: The process begins with the production of metal powders using diverse methods such as atomization, mechanical milling, chemical precipitation, and electrolytic deposition. The chosen method depends on factors like desired particle size, morphology, and chemical composition.

Powder Blending: Metal powders are meticulously blended with selected reinforcement materials, such as ceramics, metal carbides, or other reinforcing agents. This blending process is crucial to ensure a homogeneous distribution of the reinforcement phase within the metal matrix, which ultimately determines the desired properties in the final composite.

Compaction: The blended powder mixture is then compacted into the desired shape utilizing techniques like cold compaction or hot isostatic pressing (HIP). Through this process, pressure is applied to the powder particles, resulting in the formation of a green compact with the desired geometry.

Sintering: Next, the green compact undergoes sintering, where it is subjected to high temperatures in a controlled atmosphere. This enables the bonding of the metal particles and the elimination of porosity. During sintering, diffusion processes occur, leading to densification and the formation of metallurgical bonds between particles.

Post-Processing: Following sintering, additional processing steps may be employed to further enhance the properties and performance of the final component. These steps can include heat treatment, machining, surface coating, or infiltration. Each post-processing method is tailored to augment specific aspects of the component's functionality and durability, ensuring optimal performance in its intended application.

16.2 MATERIALS AND METHODS

16.2.1 Overview of Electrical Discharge Machining

The electrical discharge machining (EDM) is a unique non-traditional machining technology that is widely used to shape electrically conductive materials with complex geometries and features that are difficult to produce with typical machining procedures. Rapid electrical discharges between a tool electrode and the workpiece, both submerged in a dielectric fluid, are the method used in this procedure to remove material. The tremendous heat produced by these discharges causes the workpiece material to melt and vaporize locally, which is then removed by the dielectric fluid. Because EDM operates on the idea of erosion, as opposed to traditional cutting processes, it can be used to machine materials with varying degrees of toughness or hardness.

16.2.2 EDM Arrangements and Its Objectives

The key components of an EDM setup include the power supply, electrodes (tool and workpiece), dielectric fluid, and control system. EDM offers several advantages, including high precision, excellent surface finish, and the ability to machine hardened materials with intricate geometries without inducing residual stresses. However, the process also has limitations such as slow material removal rates and the generation of recast layers and heat-affected zones, which may affect the mechanical properties of the workpiece. Overall, EDM plays a crucial role in the manufacturing industry for producing precision components used in aerospace, automotive, medical, and tooling

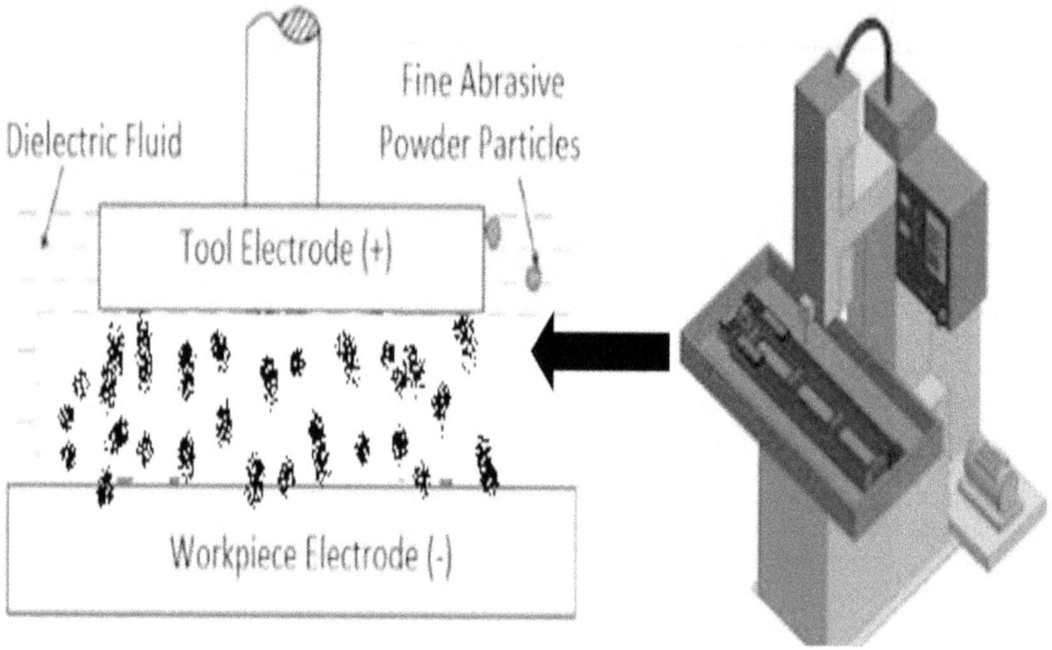

FIGURE 16.1 Electrical discharge machining process.

applications, where high accuracy and intricate features are essential. Figure 16.1 illustrates the EDM process, highlighting the key components and the mechanism by which material is eroded from the workpiece.

The primary objectives of EDM encompass achieving precision and intricate shapes in electrically conductive materials that prove challenging or impossible to produce using conventional machining techniques. EDM endeavours to maintain material integrity by utilizing thermal erosion instead of mechanical cutting, thereby minimizing the risk of inducing mechanical stresses or distortions in the workpiece. Moreover, it aims to machine hard and brittle materials effectively, including hardened steels, carbides, and ceramics, which are notoriously difficult to process using conventional methods. EDM also strives to deliver fine surface finishes, meeting stringent surface quality requirements across industries like aerospace, medical, and mould making. By reducing tool wear and extending tool life, EDM contributes to cost savings and increased efficiency.

Additionally, it addresses machining challenges associated with intricate shapes, internal features, and delicate workpieces, which may be impractical to machine using conventional means. Furthermore, EDM facilitates the machining of heat-treated components without compromising their metallurgical properties, offering flexibility in manufacturing processes. While typically slower than conventional machining processes, EDM enhances productivity through its efficiency in machining complex shapes. Overall, by fulfilling these objectives, EDM remains integral to various industries where precision, intricacy, and material integrity are paramount in component manufacturing.

16.3 EXPERIMENTAL DESIGN AND METHODOLOGY

The experimental design and methodology employed in investigating the EDM on hard powder metallurgy aluminium metal matrix composites (P/M MMCs) are critical for ensuring accurate data collection and analysis. The following outlines the key components of the experimental approach:

16.3.1 Selection of EDM Parameters

The first step involves identifying the EDM parameters to be investigated, such as pulse duration, pulse current, pulse frequency, gap voltage, and electrode material. These parameters are chosen based on their significance in affecting material removal rate, surface finish, and other machining characteristics.

16.3.2 Preparation of P/M MMC Specimens

High-quality P/M MMC specimens are prepared using powder metallurgy techniques, ensuring uniform dispersion of reinforcement particles within the aluminium matrix. Care is taken to maintain consistency in specimen size, shape, and composition to minimize variability during machining experiments.

16.3.3 Experimental Setup

A robust experimental setup is established, comprising an EDM machine equipped with appropriate electrodes, dielectric fluid circulation system, and control unit. The machine is calibrated to ensure precise control over the selected EDM parameters throughout the experiments.

16.3.4 Design of Experiments (DOE)

DOE techniques, such as Taguchi methods or factorial designs, may be employed to systematically study the effects of varying EDM parameters on machining performance. This involves defining experimental factors, levels, and the number of trials required to generate statistically significant data.

Figure 16.2 presents the main effects of surface roughness and material removal rate, showing how different parameters influence these outcomes as referenced in source [9].

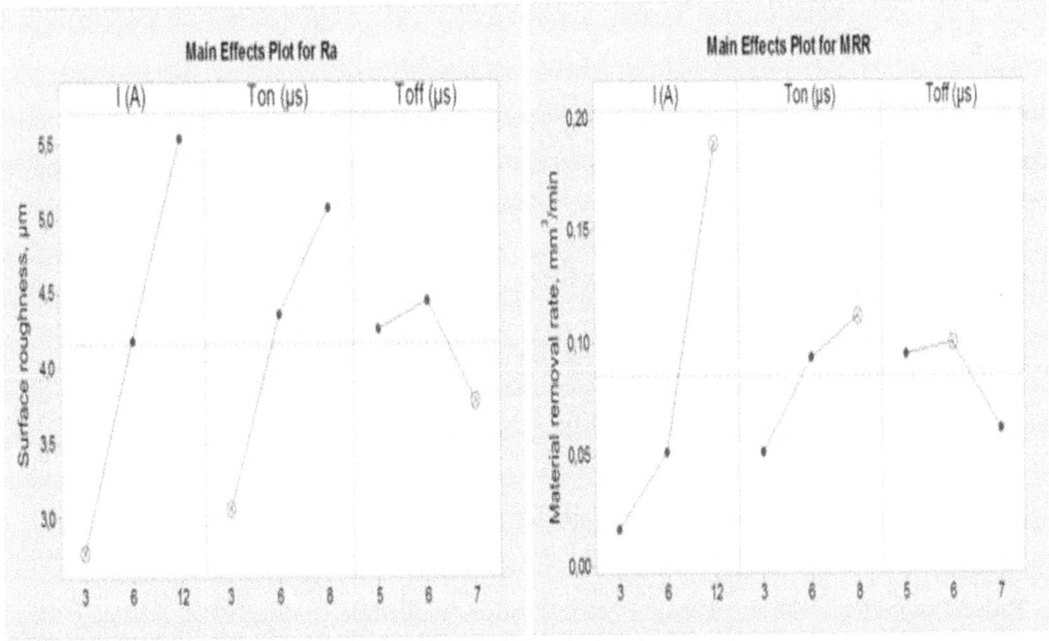

FIGURE 16.2 Main effects of surface roughness and material removal rate [9].

16.3.5 Machining Experiments

Machining experiments are conducted according to the planned DOE, with each trial representing a unique combination of EDM parameters. The machining conditions are carefully monitored and recorded, including tool wear, material removal rate, surface roughness, and any observable defects.

16.3.6 Data Collection and Analysis

Comprehensive data on machining performance metrics are collected during and after each experiment. Statistical analysis techniques, such as analysis of variance (ANOVA) or regression analysis, are employed to identify significant factors and optimize EDM parameters for desired outcomes.

16.3.7 Validation and Optimization

Through repeatability tests, the experimental results are verified and contrasted with theoretical expectations or reference values. To find the ideal set of EDM parameters that maximizes machining efficiency while satisfying quality standards, optimization techniques might be used.

16.3.8 Sensitivity Analysis

Sensitivity analysis may be performed to assess the robustness of the optimized EDM parameters to variations in process conditions or material properties.

16.3.9 Documentation and Reporting

Detailed documentation of the experimental design, methodology, results, and conclusions is essential for reproducibility and knowledge dissemination. A comprehensive report summarizing the findings, insights, and recommendations is prepared for presentation and publication.

By following a systematic experimental design and methodology, the investigation can yield valuable insights into the effects of EDM on P/M MMCs and contribute to advancements in machining processes for composite materials.

16.4 ANALYSIS OF EDM PARAMETERS ON P/M MMCs

The electrical discharge machining involves several input parameters that can be adjusted to control the machining process and achieve desired outcomes. These parameters include:

16.4.1 EDM Input and Output Parameters

Pulse Duration (Pulse On-Time): The duration of each electrical pulse applied between the electrode and the workpiece. It determines the length of time the electrical discharge occurs, influencing material removal rate and surface finish.

Pulse Interval (Pulse Off-Time): The time interval between successive electrical pulses. It allows the material to cool and solidify between discharges, affecting the overall machining efficiency and surface integrity.

Pulse Current: The electrical current supplied during each pulse. It determines the energy input into the material and influences material removal rate, electrode wear, and surface quality.

Pulse Frequency: The number of electrical pulses applied per unit of time. It affects the speed of material removal and surface roughness, with higher frequencies typically resulting in faster machining rates.

Gap Voltage: The voltage that is applied between the workpiece and the electrode. It controls electrode wear, surface polish, and material removal rate in addition to determining the electrical discharge's strength.

16.4.2 EDM Material and Properties

Dielectric Fluid Type and Properties: The dielectric fluid used to flush away debris and facilitate the electrical discharge. Properties such as viscosity, dielectric strength, and flushing capability influence machining efficiency and surface quality.

Electrode Material and Geometry: The material composition and shape of the electrode used in the machining process. Different electrode materials (e.g., copper, graphite, tungsten) and geometries (e.g., shape, size, taper angle) can affect material removal rate, surface finish, and electrode wear.

Workpiece Material Properties: The physical and mechanical properties of the workpiece material being machined. Factors such as hardness, conductivity, and thermal properties influence the EDM process parameters and their effects on machining performance.

Tool Path and Machining Strategy: The path followed by the electrode during machining and the overall machining strategy employed (e.g., roughing, finishing). Optimization of tool path and strategy can improve machining efficiency, surface finish, and dimensional accuracy.

Environmental Conditions: Factors such as temperature, humidity, and atmospheric pressure can influence the EDM process and should be controlled to ensure consistent machining performance and quality.

Adjusting these input parameters allows operators to optimize the EDM process for specific materials, geometries, and machining requirements, leading to improved efficiency, accuracy, and surface quality. However, careful selection and fine-tuning of these parameters are essential to avoid undesirable effects such as excessive electrode wear, thermal damage, or surface defects.

16.5 SURFACE CHARACTERIZATION AND MATERIAL PROPERTIES OF EDM

Surface characterization and material properties analysis are essential aspects of understanding the effects of the EDM process on powdered materials. Here's how these factors are typically assessed:

16.5.1 Surface Roughness Analysis

Surface roughness is a critical parameter that affects the functional performance and aesthetic appearance of machined components. Various techniques, such as profilometer or surface roughness testers, are utilized to quantitatively measure the roughness parameters, including Ra (average roughness), Rz (maximum height), and Rt (total roughness). Surface roughness analysis provides insights into the quality of the machined surface and the effectiveness of EDM parameters in achieving the desired surface finish. The comparison in Figure 16.3 shows the differentiation in surface roughness on eroded surfaces, demonstrating the impact of various machining conditions as discussed in source [1].

16.5.2 Microstructural Analysis

Microstructural analysis encompasses scrutinizing the structure and morphology of the material at a microscopic scale before and after undergoing the EDM process. Optical microscopy, scanning electron microscopy (SEM), and transmission electron microscopy (TEM) emerge as prevalent methodologies employed to examine the microstructure, encompassing aspects such as grain size,

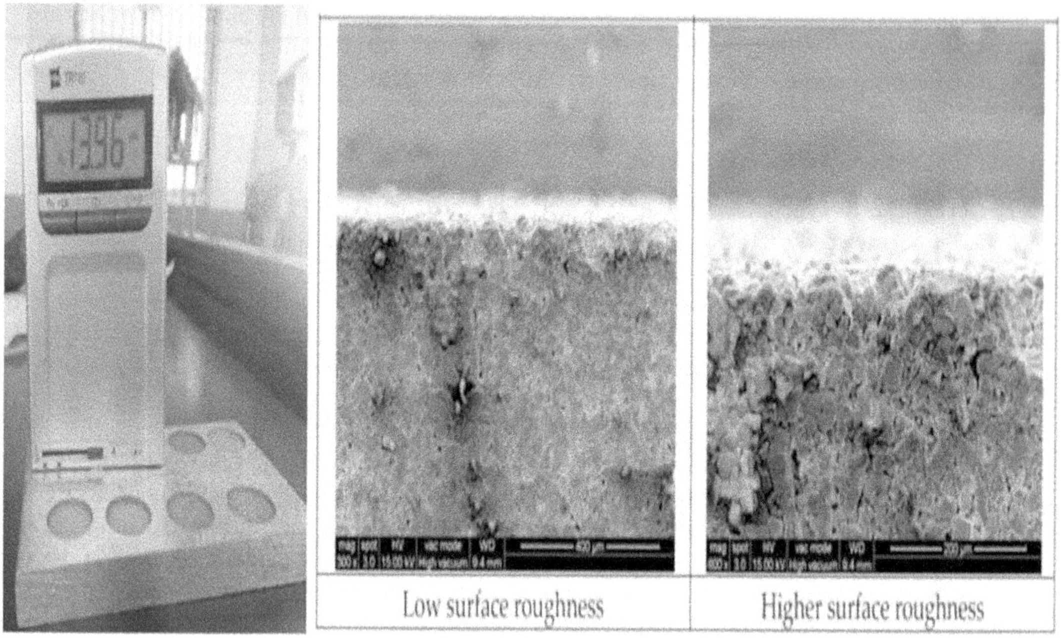

FIGURE 16.3 Surface roughness differentiation in eroded surfaces [1].

phase distribution, identification of defects like recast layers and cracks, and alterations induced by EDM like heat-affected zones and phase transformations. This analytical approach aids in comprehending the mechanisms governing material removal, thermal impacts, and the quality of the machined surface. This micrograph of Figure 16.4 displays the deposition sizes and dispersion of coated particulates, captured through SEM, illustrating the microstructural changes.

16.5.3 Surface Integrity Assessment

Surface integrity refers to the entire condition and features of the machined surface, including mechanical attributes, metallurgical changes, and residual stresses. Surface integrity parameters are evaluated using techniques such as electron backscatter diffraction (EBSD), microhardness testing, and X-ray diffraction (XRD). These evaluations provide information on how much and where residual stresses are, how the EDM process affects material hardness, and how variations in crystallographic orientation occur.

16.5.4 Chemical Composition Analysis

Chemical composition analysis is performed to assess any alterations in the elemental composition of the material due to the EDM process. Techniques such as energy-dispersive X-ray spectroscopy (EDS) or X-ray fluorescence (XRF) spectroscopy are utilized to quantify the elemental composition of the machined surface and identify any contaminants or impurities introduced during machining. Chemical composition analysis helps in understanding the material removal mechanisms and potential effects on corrosion resistance and other material properties.

16.5.5 Mechanical Properties Evaluation

Mechanical properties such as hardness, tensile strength, and wear resistance are important indicators of material performance and suitability for specific applications. Hardness testing, tensile testing, and wear testing are conducted to assess changes in mechanical properties induced by EDM.

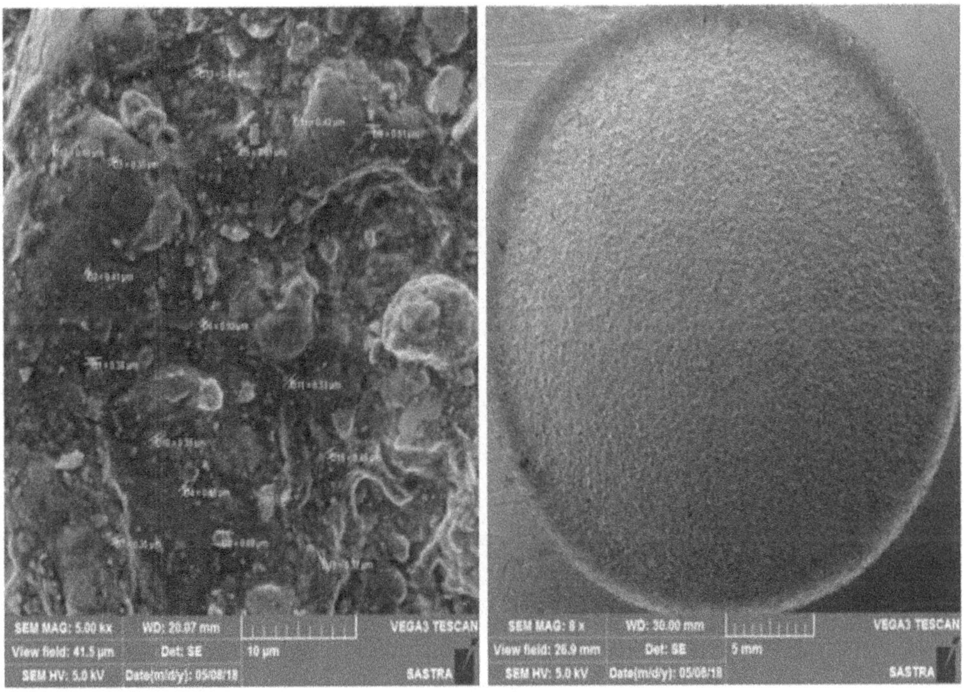

FIGURE 16.4 Deposition sizes and coated particulate dispersion by SEM microstructure.

These tests provide valuable insights into the effects of EDM parameters on material strength, ductility, and wear behaviour, aiding in the selection of optimal machining conditions and material formulations. The Figure 16.5 outlines the testing methods used to determine the mechanical properties of materials, detailing the procedures and equipment involved.

Overall, surface characterization and material properties analysis are integral to evaluating the effects of the EDM process on powdered materials and optimizing machining parameters for desired surface quality, structural integrity, and functional performance.

16.5.6 Processing Methods

16.5.6.1 Particulate MMCs

Reinforcement materials in the form of particles are dispersed within the metal matrix. These composites are produced using powder metallurgy techniques such as blending, compaction, and sintering. Particulate MMCs offer versatility and are suitable for a wide range of applications due to their ability to tailor material properties through precise control over particle size, distribution, and volume fraction. The Figure 16.6 depicts the stir casting process, a common technique for fabricating metal matrix composites, showing the setup and flow of materials.

16.5.6.2 Fiber Reinforced MMCs

Continuous or discontinuous fibres are embedded in the metal matrix to provide reinforcement. Fiber-reinforced MMCs are manufactured using processes such as casting, infiltration, or powder metallurgy. These composites offer exceptional strength, stiffness, and fatigue resistance, making them suitable for high-performance applications in aerospace, automotive, and sporting goods industries. Overall, powder metallurgy techniques offer versatility and flexibility in producing a wide range of MMCs with tailored properties and performance characteristics suited for diverse applications in various industries.

FIGURE 16.5 Mechanical properties testing and methods.

FIGURE 16.6 Stir casting process.

16.6 MACHINING CHALLENGES AND ELECTRICAL DISCHARGE MACHINING

The electrical discharge machining offers several advantages, but it also presents unique challenges that need to be addressed to ensure optimal performance and efficiency. One significant challenge lies in machining complex geometries. While EDM is capable of handling intricate shapes, programming and controlling the tool path for such geometries can be intricate. Precise electrode movement and avoidance of collisions with the workpiece require meticulous planning and programming. Another challenge arises in material selection, as EDM is most effective on electrically conductive materials. Machining non-conductive materials or those with low electrical conductivity may necessitate additional pre-treatments or modifications to facilitate the process.

Figure 16.7 shows the material erosion occurring in an EDM machine, highlighting the areas most affected by the electrical discharge. Moreover, electrode wear is a prevalent issue during the EDM process. Both the tool electrode and the workpiece experience wear due to repeated electrical discharges, potentially leading to dimensional inaccuracies and surface finish deterioration. Managing electrode wear through proper selection of electrode material, geometry, and process parameters is essential to maintain machining accuracy and efficiency. Additionally, achieving the desired surface integrity poses a challenge, as EDM can produce heat-affected zones, recast layers, and micro-cracks on the machined surface, particularly in high-precision applications. Controlling process parameters such as pulse duration, frequency, and flushing conditions is crucial to minimize thermal damage and maintain surface integrity.

These challenges necessitate a thorough understanding of the EDM process, careful selection of process parameters, and continuous improvement through experimentation, monitoring, and optimization. Despite these challenges, EDM remains a valuable and versatile machining technique for producing high-precision components with complex geometries and superior surface quality.

16.7 EXPERIMENTAL INVESTIGATIONS ON PEDM

Experimental investigations of powder electrical discharge machining (EDM) involve systematically studying the effects of EDM parameters on machining performance and material characteristics, particularly in the context of powdered materials. These investigations typically follow a structured approach to gather quantitative data and qualitative observations. First, researchers define the specific EDM parameters to be investigated, including pulse duration, pulse current, pulse frequency, gap voltage, electrode material, and flushing conditions. These parameters are

FIGURE 16.7 Material erosion in EDM machine.

selected based on their known influence on material removal rate, surface finish, and other machining responses.

Next, experimental setups are established, comprising EDM machines equipped with appropriate electrodes, dielectric fluid circulation systems, and control units. The machines are calibrated to ensure precise control over the selected EDM parameters throughout the experiments. Design of experiments (DOE) techniques, such as full factorial designs, fractional factorial designs, or Taguchi methods, may be employed to systematically study the effects of varying EDM parameters on machining performance. This involves defining experimental factors, levels, and the number of trials required to generate statistically significant data. Machining experiments are then conducted according to the planned DOE, with each trial representing a unique combination of EDM parameters. During the experiments, researchers carefully monitor and record various machining performance metrics, including material removal rate, surface roughness, electrode wear, and any observable defects. Comprehensive data on material properties and machining performance are gathered and examined through statistical analysis methods including regression analysis and analysis of variance (ANOVA). These studies support the optimization of process parameters for intended results as well as the identification of important variables and relationships among EDM parameters.

In addition to quantitative data analysis, qualitative observations of the machined surfaces and microstructures are conducted using techniques such as optical microscopy, scanning electron microscopy (SEM), and X-ray diffraction (XRD). These observations provide insights into the effects of EDM parameters on surface morphology, microstructural changes, and the presence of defects. Based on the experimental findings, researchers may propose hypotheses, develop predictive models, and suggest recommendations for optimizing powder EDM processes for specific applications. Overall, experimental investigations of powder EDM play a crucial role in advancing our understanding of the process and optimizing its performance for various powdered materials and applications.

16.8 RESULTS AND DISCUSSIONS

16.8.1 Influences of Pulse Durations on Surface Roughness of PEDM

The influences of pulse-on and off times on surface roughness in electrical discharge machining (EDM) are significant factors that affect the quality of machined surfaces. Here's how pulse-on and off times influence surface roughness:

16.8.1.1 Pulse-On Time

Longer pulse-on times typically result in higher material removal rates and more aggressive machining. However, excessively long pulse-on times can lead to increased thermal energy accumulation, which may result in rougher surfaces due to the formation of irregularities such as craters, pits, or recast layers. In some cases, longer pulse-on times can cause melting and re-solidification of the workpiece material, contributing to surface irregularities and roughness. Conversely, shorter pulse-on times may produce finer surface finishes due to reduced thermal effects and material removal, resulting in smoother surfaces with lower roughness values.

16.8.1.2 Pulse-Off Time

The pulse-off time allows the workpiece to cool and solidify between electrical discharges, which can influence surface roughness. Shorter pulse-off times may lead to insufficient time for material cooling and solidification, resulting in increased thermal effects and surface roughness. Conversely, longer pulse-off times provide more time for heat dissipation and material solidification, potentially leading to smoother surfaces with lower roughness values. The choice of pulse-off time is often a balance between maintaining machining efficiency and achieving desired surface finish. Too short a pulse-off time may lead to increased surface roughness, while excessively long pulse-off times may reduce machining efficiency without significant improvement in surface quality. The pulse-on and off times in EDM play crucial roles in determining surface roughness. Optimizing these parameters involves finding the right balance between material removal rates, thermal effects, and surface finish requirements to achieve the desired level of surface quality and dimensional accuracy. Experimental investigations and process optimization are typically conducted to identify the optimal pulse-on and off times for specific machining conditions and workpiece materials. The chart of Figure 16.8 illustrates the relationship between pulse on time and surface roughness, demonstrating how changes in pulse duration affect the surface quality.

16.8.2 Influences of Gap Current on Surface Roughness of PEDM

The gap current within EDM holds significant sway over surface roughness determination. The electrical current immediately affects a number of aspects of the EDM operation as it passes through the spark gap between the electrode and the workpiece. Elevated gap currents typically correlate with heightened material removal rates due to increased energy input into the machining zone. But this increased energy can also accelerate electrode wear, which could cause imperfections to appear on the machined surface and increase its roughness. Furthermore, increased heat from stronger gap currents can cause thermal damage like melting and resolidification, which exacerbates surface roughness. This graph of Figure 16.9 depicts the correlation between gap current and surface roughness, showing how variations in the electrical current influence the finish of the machined surface.

Additionally, the efficiency of dielectric flushing, essential for removing debris and molten material from the machining zone, is influenced by gap currents. Inadequate flushing due to low gap

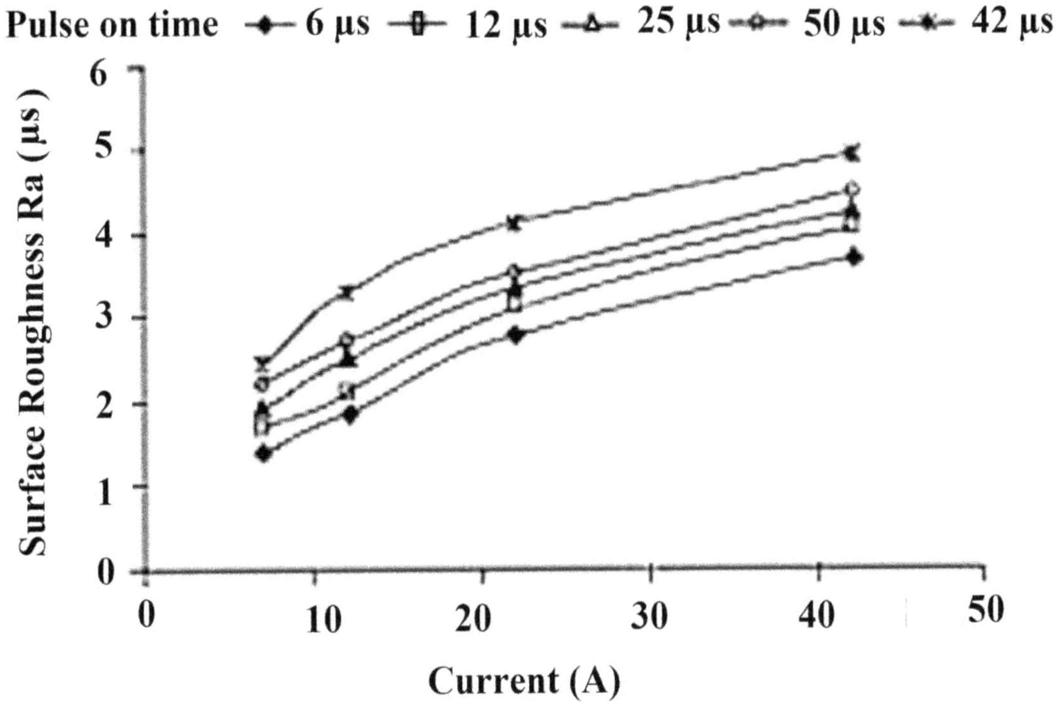

FIGURE 16.8 Pulse on time vs. surface roughness.

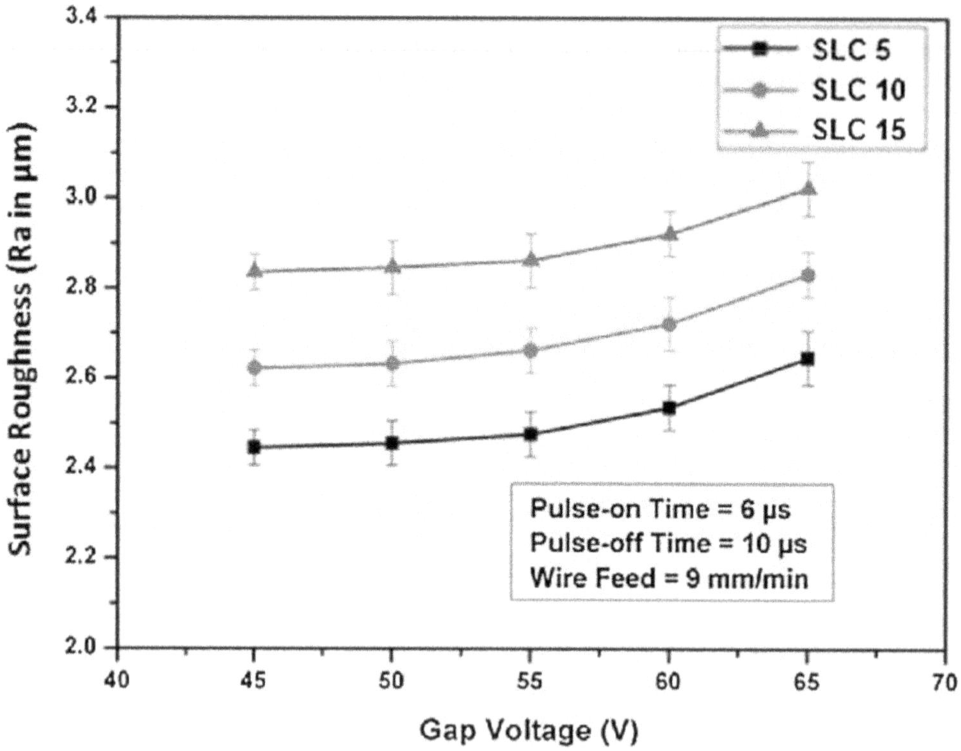

FIGURE 16.9 Gap current vs. surface roughness.

currents can lead to debris accumulation, resulting in poor surface finishes and increased roughness. Last, gap currents impact the stability and consistency of electrical discharges during machining. Unstable discharges can cause irregular material removal and surface irregularities, leading to higher surface roughness. Therefore, in order to balance material removal rates, electrode wear, and heat impacts and achieve the appropriate surface smoothness, gap current management is essential. Determining the ideal gap current for particular machining circumstances and workpiece materials with the goal of minimizing surface roughness and achieving superior surface finishes often requires experimentation and process optimization.

16.8.3 Influences of Gap Current on Material Removal Rate in PEDM

The gap current in EDM plays a pivotal role in determining the material removal rate (MRR), which represents the speed at which material is extracted from the workpiece during machining. Primarily, the gap current directly influences the energy input into the machining zone. Higher gap currents translate to increased energy deposition, intensifying the electrical discharges and elevating temperatures at the electrode-workpiece interface. This heightened energy input enhances material removal by facilitating processes like melting, vaporization, and erosion of the workpiece material, consequently boosting the MRR. Additionally, higher gap currents often lead to heightened sparking activity between the electrode and the workpiece, generating more frequent and energetic electrical discharges.

This increased sparking activity further amplifies material removal, contributing to a higher MRR due to the greater number of sparks occurring per unit time. However, it's important to note that while higher gap currents can augment MRR, they can also accelerate electrode wear. The intensified sparking activity and energy deposition can lead to greater electrode erosion, potentially limiting the effective lifespan of the electrode and affecting its ability to efficiently remove material over time. Moreover, the influence of gap current on MRR can vary depending on the properties of the workpiece material, with certain materials exhibiting more pronounced responses to changes in gap current. Thus, maintaining optimal gap current levels is crucial to ensure process stability, consistent machining performance, and efficient material removal. Overall, achieving the desired MRR while balancing electrode wear and process stability necessitates careful optimization of gap current settings, often through experimentation and process refinement tailored to specific machining conditions and workpiece materials.

16.8.4 Influences of Pulse Duration on Material Removal Rate in PEDM

Key elements that directly impact the effectiveness of material removal throughout the machining process are the effects of pulse-on and pulse-off periods on MRR in EDM. An important factor in determining MRR is pulse-on time, which is the length of the electrical discharge between the electrode and the workpiece. Due to increased energy input into the machining zone caused by longer pulse-on times, which promote more efficient material removal through processes like melting and erosion, longer pulse-on times typically result in higher MRRs.

However, excessively long pulse-on times can lead to undesirable effects such as excessive electrode wear and thermal damage, potentially reducing overall MRR. Conversely, shorter pulse-on times may result in lower MRRs due to reduced energy input, but they can help mitigate negative effects like thermal damage and electrode wear. Pulse-off time, the duration between successive electrical pulses, also influences MRR. Longer pulse-off times allow for more effective cooling and solidification of the workpiece between discharges, reducing the risk of thermal damage. However, excessively long pulse-off times may decrease MRRs by prolonging machining time and limiting the frequency of electrical discharges. In contrast, shorter pulse-off times may increase MRRs by allowing for more frequent discharges, but they may also elevate the risk of thermal damage if insufficient cooling time is provided. Balancing pulse-on and off times is crucial to optimizing MRR while maintaining machining quality and accuracy. Experimentation and process optimization are typically employed to determine the optimal pulse-on and off times for specific machining conditions and workpiece materials, ensuring efficient material removal and high-quality machining outcomes.

16.9 CONCLUSION AND FUTURE DIRECTIONS

In conclusion, pulse-on and off timings affect MRR in EDM, highlighting their importance in machining efficiency and quality. By managing energy input and material removal, pulse-on time greatly affects MRR, whereas pulse-off time controls workpiece cooling and solidification between discharges. To maximize MRR and minimize thermal damage and electrode wear, pulse-on and off times must be balanced. Future EDM research should clarify the complicated relationships between pulse-on and off timings and other process factors. Advanced modelling and simulation can illuminate EDM material removal and surface quality mechanisms. Innovative EDM technologies and tactics to improve machining productivity, precision, and surface polish while lowering environmental impact and operating costs are needed. Exploring innovative materials and composites for EDM applications and studying their reaction to pulse parameters may improve machining performance and increase the range of materials that may be processed using EDM. Research should also explore the integration of EDM with other advanced manufacturing processes like additive manufacturing and hybrid machining to maximize synergy and create new opportunities for complex, high-quality components with superior properties. EDM research and development will improve manufacturing capabilities, allowing aerospace, automotive, medical, and electronics companies to produce high-precision, high-performance components. Researchers and practitioners can advance manufacturing technology by addressing pulse-on and off timings and other EDM factors.

REFERENCES

[1] Azmir, M.A., Ismail, A.F. and Sharif, S., 2019. A review on electrical discharge machining of metal matrix composites: A state-of-the-art. *The International Journal of Advanced Manufacturing Technology*, *105*(1–4), pp. 535–558.
[2] Amanullah, M.T.O., Elarif, M.A.A. and Kamaruddin, S., 2020. A review on the electrical discharge machining process for the fabrication of metal matrix composites. *The International Journal of Advanced Manufacturing Technology*, *106*(5–6), pp. 2095–2115.
[3] Chen, Z., Wang, J. and Xiao, Z., 2019. Influence of machining parameters on surface integrity in electrical discharge machining of metal matrix composites. *The International Journal of Advanced Manufacturing Technology*, *101*(1–4), pp. 1–15.
[4] Dilli Babu, R., Gurusamy, P., Bovas Herbert Bejaxhin, A. and Chandramohan, P., 2023. Influences of WEDM constraints on tribological and micro structural depictions of SiC-Gr strengthened Al2219 composites. *Tribology International*, *185*, pp. 1–10, 108478.
[5] Feng, C., Liu, Z. and Zhu, J., 2018. Investigation on the material removal mechanism of silicon during micro-electrical discharge machining process. *The International Journal of Advanced Manufacturing Technology*, *95*(5–8), pp. 2023–2032.
[6] Fathima, A., Manzoor, M. and Salam, A., 2018. Machinability studies in electrical discharge machining of metal matrix composites—a review. *The International Journal of Advanced Manufacturing Technology*, *95*(1–4), pp. 489–510.
[7] Gu, Y., Huang, H. and Du, J., 2018. A study on the material removal rate of SiC ceramic in the process of powder mixed EDM. *The International Journal of Advanced Manufacturing Technology*, *97*(5–8), pp. 2797–2805.
[8] Justin Raj, Y., Bovas Herbert Bejaxhin, A., Raj Kumar, S. and Balamurugan, G.M., 2024. Influence of novel Al-Ni electrodes on roughness and machining time in inconel 625 EDM machining: A comprehensive RSM performance analysis. *Journal of Environmental Nanotechnology*, *13*(1), pp. 243–253.
[9] Li, L., Liu, Z. and Li, J., 2019. Influence of pulse current parameters on material removal rate in EDM drilling of superalloy. *The International Journal of Advanced Manufacturing Technology*, *105*(9–12), pp. 4163–4174.
[10] Lin, T. and Wang, J., 2020. Material removal mechanisms and tool wear characteristics in EDM of titanium alloy and nickel-based superalloy. *The International Journal of Advanced Manufacturing Technology*, *108*(1–2), pp. 471–482.
[11] Rajkumar, V., Palanikumar, K. and Muthukrishnan, N., 2020. Investigation on the effect of different dielectrics on material removal rate, surface roughness, and tool wear during EDM of H11 steel using Taguchi method. *The International Journal of Advanced Manufacturing Technology*, *106*(1–2), pp. 503–516.
[12] Rajkumar, G., Saravanan, M., Bovas Herbert Bejaxhin, A., Sharma, S. and Dwivedi, S.P., 2023. Parametric optimization of powder-mixed EDM of AA2014/Si_3N_4/Mg/Cenosphere hybrid composites using fuzzy logic: Analysis of mechanical, machining, microstructural, and morphological characterizations. *Journal of Composites Science*, *7*(9), p. 380.
[13] Shah, K.H. and Chauhan, A.R., 2019. Review on EDM process of metal matrix composites and optimization of input parameters using grey relational analysis and genetic algorithm. *The International Journal of Advanced Manufacturing Technology*, *100*(1–4), pp. 853–867.
[14] Shunmugam, M.S. and Ramanujam, R., 2019. Prediction and optimization of material removal rate and surface roughness in electrical discharge machining of D2 tool steel using response surface methodology and genetic algorithm. *The International Journal of Advanced Manufacturing Technology*, *103*(9–12), pp. 3237–3256.
[15] Siddhpura, A., Patel, V. and Jain, N.K., 2020. Influence of tool material and process parameters on material removal rate and surface roughness in EDM of tungsten carbide. *The International Journal of Advanced Manufacturing Technology*, *107*(5–6), pp. 2267–2281.
[16] Singh, R., Kumar, J. and Davim, J.P., 2020. Influence of process parameters on surface roughness and material removal rate in EDM of Ti–6Al–4V. *The International Journal of Advanced Manufacturing Technology*, *107*(7–8), pp. 2767–2779.
[17] Singh, R., Kumar, J. and Davim, J.P., 2020. Influence of pulse duration on material removal rate, surface roughness, and micro-hardness in EDM. *The International Journal of Advanced Manufacturing Technology*, *107*(9–10), pp. 4243–4257.

[18] Srinivasan, V.P., Sandeep, Ch., Shanthi, C., Bovas Herbert Bejaxhin, A., Anandan, R. and Abisha Meji, M., 2022. Comparative study on EDM parameter optimization for adsorbed Si3N4–TiN using TOPSIS and GRA coupled with TLBO algorithm. *Adsorption Science & Technology*, *2022*(8), pp. 1–19.
[19] Song, Q., Yang, S. and Chen, Y., 2018. Study of surface and subsurface defects induced in electrical discharge machining of W–Cu composites. *The International Journal of Advanced Manufacturing Technology*, *97*(9–12), pp. 3967–3977.
[20] Sreekumar, K., Sivaramakrishnan, S. and Ramesh, R., 2020. Optimization of material removal rate and surface roughness in EDM using response surface methodology and desirability approach. *The International Journal of Advanced Manufacturing Technology*, *109*(1–2), pp. 317–328.
[21] Tzeng, Y.F. and Liang, S.Y., 2018. The effects of tool electrode on material removal rate and surface roughness in EDM using statistical analysis. *The International Journal of Advanced Manufacturing Technology*, *96*(1–4), pp. 715–723.
[22] Tsai, Y.J., Chen, S.H. and Chen, Y.F., 2018. A study of material removal rate and electrode wear of tungsten-copper composite material in EDM process. *The International Journal of Advanced Manufacturing Technology*, *96*(1–4), pp. 115–121.
[23] Zhang, Y., Chen, M. and Zhang, L., 2019. Material removal mechanisms and surface roughness in electrical discharge machining of cemented carbides with copper powder. *The International Journal of Advanced Manufacturing Technology*, *102*(9–12), pp. 3641–3655.
[24] Zhou, Y., Chen, S. and Huang, J., 2020. Study of EDM-milling of Al_2O_3/6061Al MMC based on RSM. *The International Journal of Advanced Manufacturing Technology*, *107*(9–10), pp. 3889–3901.

17 Joining of Metal Matrix Composites

Shenbaga Velu Pitchumani, Venkatachalam Gopalan, Vignesh Nagarajan Jawahar, Navasingh Rajesh Jesudoss Hynes, and Raja Thiyagarajan

17.1 INTRODUCTION

The exploitation of materials that are formed of metal matrix composites (MMC) is currently one of the topics that is receiving a substantial level of interest [1, 2]. The aerospace industry (satellite struts), the defence industry (electronic instrument racks), the automobile industry, the sports goods industry (mountain bikes and golf clubs), and the marine industry (yacht fittings) are all potential applications for these materials. MMCs perform better than unreinforced matrix alloys in stiffness, strength, wear resistance, and creep resistance at higher temperatures and can be modified.

Polymers, minerals, and metals are the three primary categories that can be used to classify matrix materials. Reinforcement fibres are classified as particulate, discontinuous, or continuous. Metal matrix composites, also known as MMCs, are made up of a reinforcing phase that is composed of metals, organic compounds, or ceramics, and it is stacked on top of a matrix that is made of metal or alloy [3–5]. Particulate-reinforced MMCs are the most prevalent type of MMCs. These MMCs are made from lightweight alloys including magnesium, aluminium, and titanium that are reinforced with oxides, carbides, bromides, nitrides, or silicates [6]. Aerospace, automotive, marine, and nuclear energy industries appreciate aluminium matrix composites (AMCs) for their strength-to-weight ratio. Strength and dimensional stability are improved by these composites' stiffness, density, thermal expansion, and thermal conductivity [7–10]. Distribution of reinforcement uniformly, minimizing porosity, improving wettability, and avoiding chemical reactions must be solved before these composites may be employed in industry. Vacuum brazing, electron beam welding, and laser welding often produce brittle junctions due to unintended reinforcement-matrix interactions, making it difficult to achieve the desired connection qualities. MMCs must be assembled without melting the composites to avoid brittle joints [11, 12] [11–16].

17.2 CURRENT RESEARCH STATUS IN MMC

Metal matrix composites (MMCs) are used in electronics, automotive, and aerospace industries, highlighting the need for efficient joining methods. Composite materials like MMCs have a metal matrix and a carbon fibre or ceramic reinforcing phase. Stiffness, strength, and wear resistance are improved. Due to their variability, joining them will be challenging, a crucial step in manufacturing sophisticated industrial parts and structures. This overview discusses MMC joining technology, including current advances and research.

Friction stir welding (FSW) is a prominent MMC joining technique because its solid-state process reduces reinforcement phase heat stress. FSW presses components together to connect without melting the matrix and generates heat through friction. Recent research has optimized FSW parameters such as tool rotation speed and angle to disseminate reinforcing particles and reduce joint defects. It demonstrates that FSW is capable of producing joints with MMC-like mechanical properties [17].

DOI: 10.1201/9781003499718-17

Laser beam welding (LBW) provides precise control over MMC welding and minimizes thermal distortion. This method uses a concentrated laser beam to locally melt the matrix, encouraging bonding while preserving reinforcing particles. LBW for MMCs controls heat input and reduces interface intermetallic compounds, which could harm the joint, by adjusting laser settings and beam-shaping procedures.

Ultrasonic welding (UW) is a reliable technique that may be utilized to successfully combine thin sheets of MMC. This technique is especially helpful in the automotive and electronics industries. UW uses high-frequency ultrasonic vibrations instead of high temperatures that could degrade reinforcement to produce a solid-state bond between materials. New UW techniques optimize welding conditions and tool designs to evenly distribute reinforcing particles and strengthen bonds [18].

Diffusion bonding reduces thermal stress and reinforcement deterioration by linking materials below their melting temperatures using pressure and heat. The goal of ongoing diffusion bonding research is to determine the ideal pressure and temperature to enhance bond mechanical properties and avoid the production of brittle intermetallic phases at the interface.

Mechanical fastening, using bolts and screws, is simple and non-invasive to the microstructure, but it can generate stress concentrations and may not be optimal for high-performance applications due to galvanic corrosion between distinct materials. Adhesive bonding offers a low-thermal-stress joining option that works well for a variety of materials and intricate geometries, but it is sensitive to environmental factors and requires careful surface preparation [19].

Filler metals used in brazing and soldering have lower melting temperatures than the base materials, making them ideal for combining incompatible elements in complicated assemblies. However, careful filler selection is necessary to prevent reinforcement degradation.

The matrix-reinforcement mechanical and thermal differences are the biggest hurdles to MMC joining. Maintaining the phase of reinforcement, managing the residual stresses that are caused by differential thermal expansion, and preventing unfavourable chemical reactions at the interface are some of the things that fall under this category. Research in the future will likely centre on enhancing the quality of existing joints through the use of cutting-edge material science innovations like surface treatments and nano-reinforcements or on improving upon existing joining procedures [20].

Furthermore, the utilization of computational modelling and simulations in conjunction with experimental methodologies establishes viable paths to forecast the outcomes of the joining process and change parameters for certain material blends and joint configurations. This comprehensive approach seeks to expedite the development of efficient MMC joining methods, increasing their application in critical and high-performance engineering fields.

17.3 NEED FOR MMC JOINTS

High-performance environments are motivated to integrate metal matrix composites (MMCs) due to their exceptional properties. MMCs are made by combining reinforcing agents, such as fibres, particles, or whiskers, into a metallic matrix. The performance of these composites is superior to that of conventional metals and alloys in terms of the strength-to-weight ratio, wear resistance, thermal conductivity, and stability.

In the aerospace and aviation industries, lowering component weight is crucial to increasing overall performance and fuel efficiency. MMCs are special in that they can decrease weight without sacrificing the structural and thermal integrity of the component. MMC components may be integrated into complex assemblies of spacecraft and aeroplanes thanks to advanced joining techniques, guaranteeing the highest levels of efficiency and dependability [21].

MMCs provide the car industry with a means of achieving its goals of lowering emissions and increasing fuel economy without compromising performance or safety. These materials make it possible to produce engines and parts that are lighter and more efficient. MMCs can be smoothly incorporated into automotive systems through complex joining procedures, leading to lighter but stronger automobiles [22].

Efficient heat management is crucial as electronic gadgets become increasingly potent and smaller. MMCs are excellent at dispersing heat because of their variable coefficient of thermal expansion and exceptional thermal conductivity. To assemble MMC components into electronic packages and heat sinks and improve device performance and durability, joining technologies are essential.

MMCs are advantageous in the fields of infrastructure and construction because of their durability, strength, and resistance to environmental wear. Novel connecting methods are essential for building large-scale structures or combining MMC components with conventional metal frameworks to increase the functional life and effectiveness of the latter.

MMCs are composite, therefore combining them can cause thermal expansion differences, interface reactions, and reinforcement damage. Consequently, specialized joining processes like friction stir welding, laser welding, and diffusion bonding must continue to advance. These techniques aim to produce strong, functional joints that make the most of the unique qualities of MMCs [23–25].

17.4 FRICTION STIR WELDING OF MMC

One non-consumable method of combining two faying surfaces is with a revolving tool. The profiled pin and shoulder make up the two components of the spinning tool. The workpiece is heated by the welding instrument, which also stirs the material. The plasticized material moves as a result of this churning action, causing joints to develop. Heat is generated as a result of contact between the pin and the workpiece, gradually softening the material in Figures 17.1 and 17.2. The contact between the shoulder and the workpiece results in an increase in the frictional

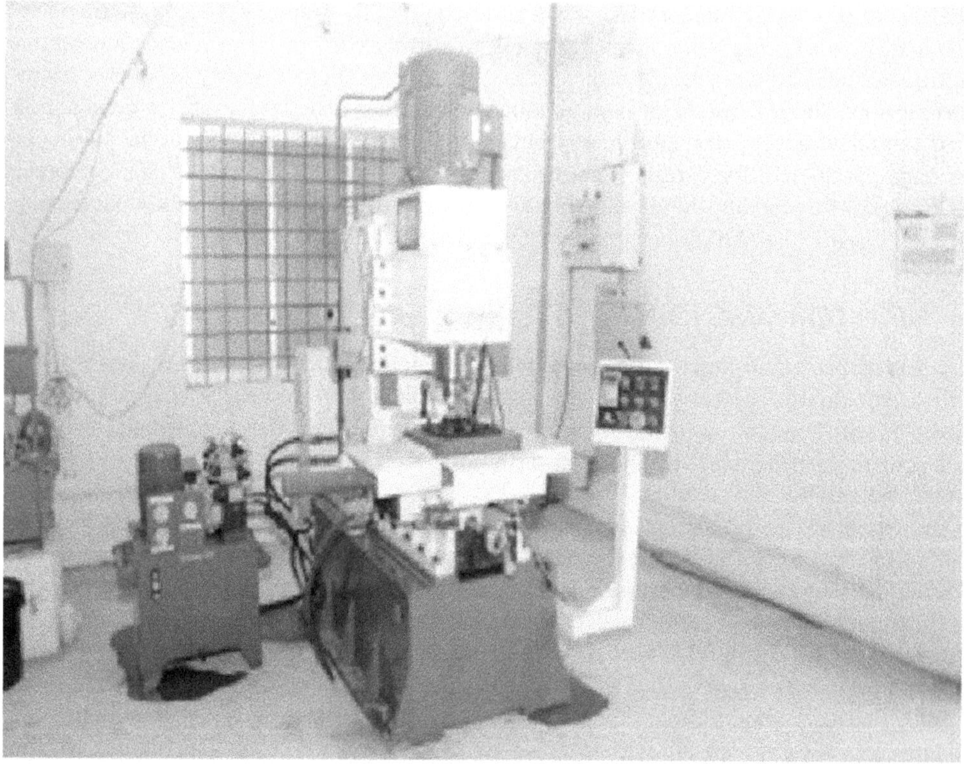

FIGURE 17.1 Experimental setup of friction stir welding machine.

FIGURE 17.2 An enlarged view of the welding process taking place.

heat, an extension of the zone of the material that has been softened, and a constraint of the material that has changed shape [26]. The four steps of FSW are the plunge-in, dwell, weld, and plunge-out periods. The revolving tool rotates longitudinally during the plunge-in phase, upright to the joint contour, bringing the workpiece's superior surface and the shoulder's bottom face into contact. The revolving tool is given a predetermined dwell period in the second phase, which is dwelling the period, to allow it to plastically distort the surrounding workpiece material. Following the expiration of the dwell time, the workpiece joins when the spinning tool is moved transversely. The profile of the pin and the rotation of the tool cause plasticized material to flow from the advancing side to the retreating side or from the leading edge to the following edge of the workpiece. The procedure known as "solid-state joining" is what happens when the shoulder and workpiece come into close contact and forge the workpiece [27]. Figure 17.2 shows the representation of the friction stir welding process. The rotating tool will face in a different direction for the retreating side than it will transversely.

The profile of the pin, the profile of the shoulder, the pin diameter, the pin length, and the ratio of the shoulder diameter to the pin diameter are all included in the tool geometry [28, 29]. While the presence of the fused junction is dependent on joining speed, rotating speed determines the stirring, mixing, and heat generation of the material. In addition to helping to create frictional heat between the tool and the workpiece, the axial load preserves the contact conditions. Different pilot studies are used to determine the working range of process parameters. The range of parameters is chosen to eliminate pinholes, tunnel flaws, cracks, voids, surface grooves, and galling in macrostructures.

17.4.1 Weldability of MMC

When aluminium's strength isn't enough for a particular use, it's usually alloyed with Si, C, Mn, Mg, Zn, and other elements. Following the volume fraction and the alloying ingredient, the aluminium alloys are separated into several series. AMC composites are made of aluminium alloys as is the matrix, and these alloys are reinforced with ceramics, among other particles. MMCs have a variety of exceptional qualities depending on the kind of reinforcement. Because of their superior qualities, AMCs have been extensively utilized by numerous sectors during the past 30 years. It has been noted that a lot of the aerospace, automotive, aircraft, and other related industries try to make structures lighter [30].

AMCs are capable of achieving these industries' objectives. The weldability of AMCs is weakened by the presence of reinforcement particles in the matrix phase, which is the only downside. Using fusion welding makes it challenging to create flawless joints. The fusion welding of AMCs has several disadvantages, including the following: 1) Inadequate blending of the reinforcing particles and aluminium alloy; 2) porosity development in the nugget zone; 3) the presence of undesirable harmful phases such as Al_4C_3; and 4) an extreme amount of eutectic establishment. When laser welding AMCs, Niua et al. [31] found that the joint's strength was only around 70% of BM, even with the right welding settings selected. The combination of aluminium and silicon carbide (SiC) in a molten pool weakens the material and results in brittle joints. When laser welding A359/SiC, Bassani et al. [32] saw limited penetration and the presence of Al_4C_3 in microstructures, which was caused by SiC dissolution. When vacuum brazing SiC-reinforced AMC, Zhang et al. [33] observed poor bonding quality at the SiCp/Al interface. Weaker joints were also noted when the weight fraction of SiC increased. The outcome additionally demonstrated that, for a constant SiC volume, an increase in SiC particle size results in a decrease in joint strength, which was conducted by Storjohann et al. [34]. Traditional fusion welding techniques were not able to produce the required joint characteristics in the welded product.

17.4.2 Mechanical Properties of MMC

The integrity of joints formed during FSW of MMCs is significantly influenced by the process parameters and the presence of reinforcing particles. Improving the combined properties of MMCs requires optimizing these process parameters. The choice and scope of these parameters directly affect the materials' tensile qualities and are crucial to the way the welding process turns out.

Studies have indicated that the tensile capabilities of metal alloys improved by FSW-incorporated ceramic particles outperform those of joints made with conventional welding procedures. Researchers such as Dragatogiannis [35] have shown through their studies that mechanical properties such as hardness, yield strength, elongation, elastic modulus, and ultimate tensile strength (UTS) can be significantly improved by dissimilar welding aluminium alloys reinforced with titanium carbide nanoparticles. Pantelis [36] also reported similar results, and Bozkurt pointed out that failures in welded specimens usually happened outside the nugget zone, indicating that this region had a higher tensile strength. Despite this, as demonstrated by Bozkurt's [37] study, there has been a 19% drop in tensile strength within the nugget zone in comparison to the base composite.

Subsequent investigation by Ceschini [38] revealed a 28% reduction in the joints' tensile strength in comparison to the base material when fusing AA 6061 alloy with 20% Al_2O_3 particles. The same author also showed that when AA 7005 with 10% Al_2O_3 was used in another study, the tensile strength and joint efficiency of the original composite material were decreased by 20% and 80%, respectively [39]. Amirizad's [40] research, in contrast, showed that welded joints' tensile properties were improved over base composites due to a reduction in porosity and flaws as well as a uniform distribution of ceramic particles. Table 17.1 shows the ultimate tensile strength, joint efficiency, and process parameters of various welded composites.

TABLE 17.1
Ultimate Tensile Strength, Joint Efficiency, and Process Parameters of Various Welded Composites

Workpiece material	Material thickness (mm)	Welding speed (mm/min)	Rotational speed (rpm)	Ultimate tensile stress (MPa)	Joint efficiency	Reference
AA6061/10% ZrB_2	6	48.8	1155	243	95.8	[41]
AA6061/12% B_4C	6	78	997	208	96.8	[45]
AA6061/12% B_4C	6	80	1000	201	93.4	[42]
AA6061/AIN	6	55	1200	225	93.4	[46]
Al/10% TiB_2	6	30	2000	223–282	79–99	[47]
AA7075/10% Al_2O_3	7	56	800	260	83.8	[48]
AA6061/20% Al_2O_3	7	56	800	329	86.8	[48]
Al + 15% Mg_2Si	6	125	1,120	115	-	[49]
AA 6061/22% Al_2O_3	4	260	880	227	99	[50]
AA 6061/10%SiC	6	45	1100	206	74	[51]
AA2124-T6 + 25 SiC	15	75	550	642.1	86	[52]
AA 2009-T4 + 15% SiC	6	100	800	559	104	[53]

According to Dinaharan's [41] research, there is a negative association with elongation percentage and a positive correlation with tensile strength when the volume fraction of reinforcing particles is increased. According to this study, reinforced samples had a slightly lower efficiency at 95% than unreinforced samples, which had a joint efficiency of about 99%. The pattern was further supported by the similar results that Kalaiselvan and Bahrami's [42, 43] work with various aluminium composites shown. By analyzing how tool offset affects tensile strength, Xiao's [44] study offered a novel viewpoint. The findings demonstrated that the specimens' tension strength with tool offset which resulted in the buildup of an aluminium layer on the retreating side was reduced from 85% to 47% when compared to samples without offset. These many investigations highlight the intricate interactions that occur between the parameters of the procedure, the kind and quantity of reinforcing particles and, as a consequence of this, the mechanical properties of the FSW joints in MMCs. The goal of this field's continuous study and experimentation is to improve welding procedures so that they may more effectively utilize the special benefits of MMCs and guarantee their best performance in applications with high demand.

Hardness measurements perpendicular to the welding direction across different zones of welded MMCs joints give important insights on hardness profile variation. In the literature, two different kinds of hardness distributions have been noted. The initial variety shows the maximum hardness values in the nugget zone, progressively declining toward TMAZ and ultimately declining toward the hardness levels of the base composite. This pattern, which has been shown in several experiments, is thought to result in increased hardness because of the even dispersal and refinement of reinforcing particles within the nugget zone. This has been verified by Nami and others [54, 55], individuals who have noticed a notable rise in the level of hardness in the nugget zone in comparison to the base composite. Additionally, they discovered that the overall joint hardness tends to decrease as the number of welding passes increases. The nugget zone had the highest hardness values, according to Feng, who also noted an overall rise in hardness following post-weld heat treatment (PWHT), albeit the improvement in the nugget zone was not as noticeable as it was in other locations. Wang [56] described a situation in which the nugget zone had a hardness of 150 HV, and the welded sample, following T4 conditioning, had a homogeneous hardness level of about

170 HV throughout the joint. With base composites showing the highest hardness, which lowers through the TMAZ and HAZ before increasing again in the nugget zone, the second profile type is distinguished by a "W"-shaped distribution. The hardness of the nugget zone is generally lower than the base materials. Bozkurt [37] saw this "W"-shaped profile and observed that the welded joint's top and lower sides had different levels of hardness. The main cause of this disparity is the differing rates of thermal expansion of aluminium alloys and reinforcing particles, with the nugget zone expanding more thermally than the HAZ. According to Uzun's [57] studies on the FSW of AA 2124/SiC, the hardness of the HAZ is the lowest, and it progressively increases in the TMAZ and nugget zone, albeit being only marginally harder than the base composite.

Yigezu [58] observed that, while hardness tends to decrease with greater welding speeds, it coincides with increased rotating speed and shoulder diameter. Two zones of lowest hardness were observed in the HAZ by Wang [59] at a welding speed of 50 mm/min. When the welding speed was increased to 200 mm/min, the joint hardness improved overall, but the two low-hardness zones persisted. The two low-hardness zones in the HAZ were reduced to one at a welding speed of 800 mm/min, with the nugget zone having the lowest hardness. A four-flute cylindrical pin profile produced the highest joint hardness, according to Bahrami's [43] investigation into the effects of various pin profiles on hardness. A double pass produced a homogenous microhardness distribution, but further passes did not considerably increase hardness, according to Ahn's [60] research on the impact of repeated welding processes on microhardness distribution.

These findings illustrate the complex interactions that occur between welding settings and how these settings influence the hardness properties of welded MMC joints. They also emphasize how important process optimization is to obtaining the right mechanical qualities.

17.5 ULTRASONIC WELDING OF MMC

The use of ultrasonic welding is quickly becoming indispensable at the forefront of technical and industrial breakthroughs. According to Maria [61], this technique uses vibrational energy to fuse metals and polymers, whether they are similar or not. The heat produced by friction, especially when the two materials come into contact with a sonotrode tip, causes fusion to occur at the interface between them in Figure 17.3. When aluminium is ready to be welded, it frequently contains several particle reinforcement composites, such as SiC, B_4C, and TiB_2. According to Chak et al. [62], each of these composites has special qualities suited to particular procedures and applications. Stir-casting is a preferred in-situ method for integrating these reinforcements because it can improve the particle-matrix interface and provide thermodynamically stable outcomes. Research from Laha et al. [63] to Kumar Sharma et al. [64] supports this exothermic process, which promotes processes that aid in the material's self-stabilization when exposed to frictional heat.

Ultrasonic welding has become a crucial method in the rapidly developing field of industrial innovation because of its effectiveness in combining materials that are comparable or dissimilar, such as metals and polymers. According to research, this process is characterized by the conversion of low-power frequency into mechanical energy at high frequencies via a frequency converter. A converter or piezoelectric transducer is used with a clamping mechanism or pneumatic force to form the ultrasonic welding setup. This transducer is essential for converting the ultrasonic generator's high-frequency electrical power into vibratory motion or energy. After that, the energy is focused on the sonotrode tip at the welding location and directed toward the horn in Figure 17.4. This causes the metal sheets to press against a static anvil, which starts the welding process by generating frictional heat [65].

The effectiveness of the welding process is affected by the introduction of particle reinforcement composites, such as SiC, B_4C, and TiB_2. These composites are specifically designed for aluminium welding readiness. Stir-casting is an in-situ procedure that is commonly used to integrate these reinforcements. It is acclaimed for its capacity to achieve stable thermodynamic outcomes through

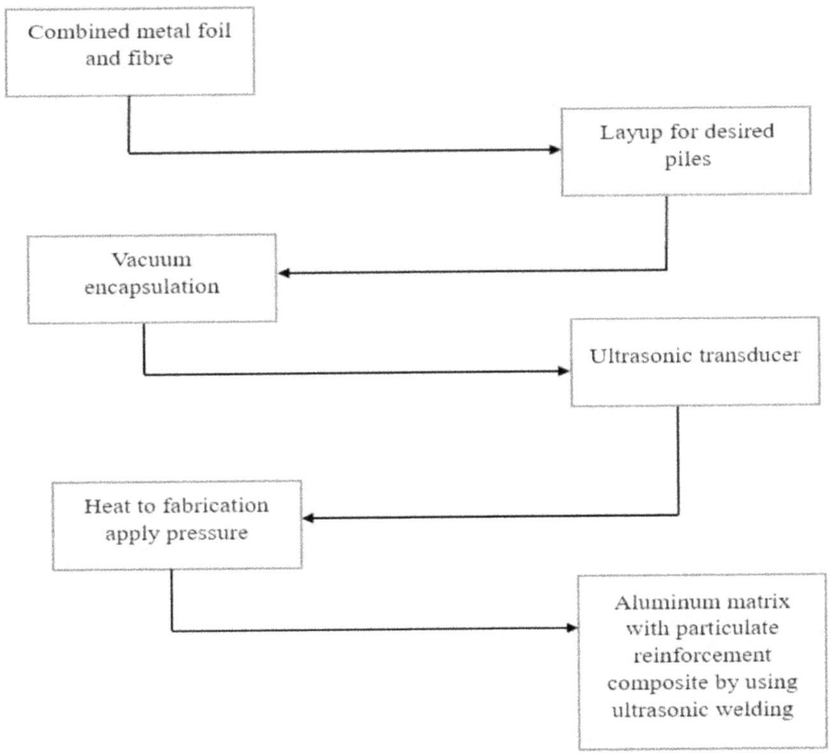

FIGURE 17.3 The structure of the review for understanding the ultrasonic welding of aluminium MMC.

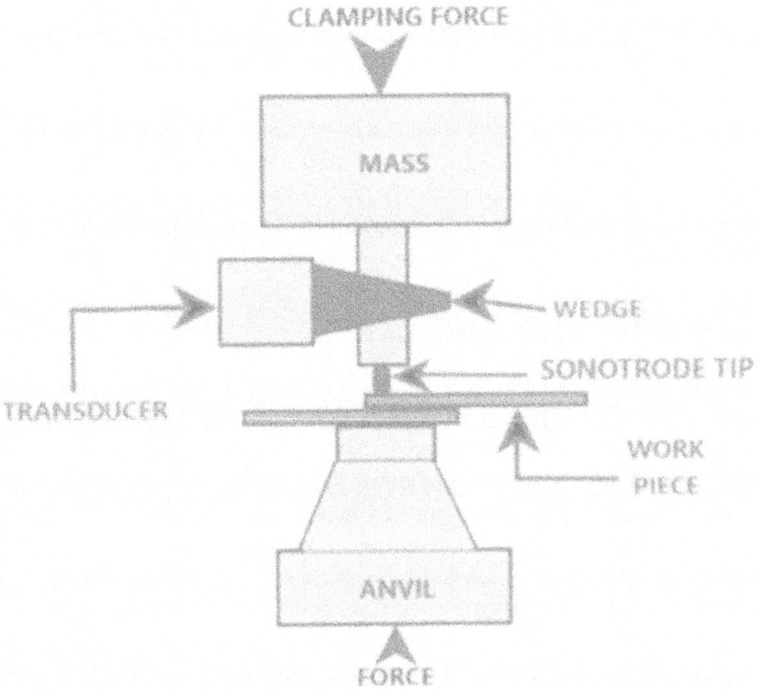

FIGURE 17.4 The ultrasonic welding setup.

exothermic reactions and a carefully controlled particle-matrix interface. This method produces a sensitive particle combination in which the frictional heat created is essential to the self-stabilizing effects [66].

When it comes to ultrasonic welding, the workpiece's thickness is typically restricted to roughly 1 mm, especially when using hard materials. This restriction is essential to shield the workpiece from harm since too much welding pressure might be harmful. One cannot stress the importance of clamping or welding force, and direct comparisons of welding forces between various machines require calibration using a load cell. Furthermore, complexity is added when ceramic fibres, such as silicon carbide, are added to metal matrix composites (MMCs) for structural procedures. Because of the interplay of physical, mechanical, thermal, and chemical qualities as well as the temperature and orientation of the fibres with respect to the welding direction, the integration of these fibres does not always convert into improved outcomes.

Examples of composites where ceramics are used as particle reinforcements within an aluminium metal matrix or as the reinforcement matrix itself are ceramic matrix composites and aluminium metal matrix composites (AMMCs). It's important to pay attention to elements like particle reinforcements that can affect the ultrasonic welding process. The dry friction at the sonotrode tip creates a high frictional force at the welding interface, which quickly raises the surface temperature and promotes metal fusion. But if not enough force is applied during the procedure, the ultrasonic oscillations required for heat creation may not be transmitted [67]. The results of ultrasonic welding are greatly influenced by several elements, which also affect the quality.

Joint Design: Surface machining, appearance, and moldability all depend on the geometric design of the joint, which is crucial. According to Leng et al. [68] lap joints in particular are known to produce the best outcomes.

Frequency Range: The efficiency of the welding process varies with frequency; for plastics, it usually ranges from 15 to 40 kHz, and for metals, it can reach up to 70 kHz. According to Kim et al. [69], the strength of the weld rises with welding time, and lower frequencies such as 20 kHz allow for better aluminium diffusion because of frictional heat than higher frequencies.

Material Type: The selection of a material has a major impact on the results of welding. Because composites like silicon carbide have higher tensile strength than aluminium, reinforced aluminium MMCs have an ideal thickness limit of 1 mm. The measured temperatures at the weld interface should not be higher than the melting point of the aluminium alloy.

Welding process parameters: The final weld can be affected by changes to parameters like amplitude, trigger force, time, and distance. To prevent weld defects and maintain the integrity of the joint, these parameters must be adjusted properly [70]. Comprehending these variables is crucial to refining the ultrasonic welding procedure and guaranteeing that scientific discoveries are effectively utilized to generate sturdy, long-lasting joins for a diverse range of industrial uses.

17.6 FRICTION WELDING OF MMC

Friction welding is a unique technique that is mainly used to fuse incompatible materials together, while it can also be used to combine similar ones. Because it is so simple to weld like materials with conventional techniques, this procedure is particularly preferred for dissimilar materials. In friction welding, one component is spun at a predetermined speed while the other is kept motionless. The spinning portion spins at a predefined speed when the machine is started. Simultaneously, the stationary component is pushed toward the rotating counterpart by an axial force [71]. By applying pressure and generating heat via friction, the process makes it easier for various materials to fuse together. This allows the materials to become flexible and allow for plastic deformation. To guarantee a high-quality weld, the rotation stops after the required length of material has been deformed, and further pressure is then given to the stationary section in Figure 17.5. The amount of friction, the length of the process, the temperature that is produced, and the degree of material displacement are important factors that affect how well friction welding works.

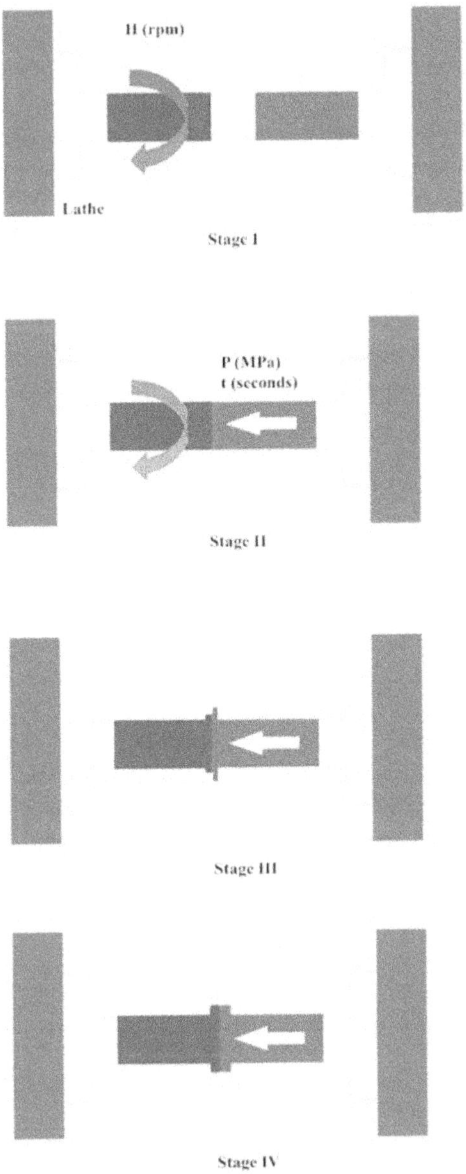

FIGURE 17.5 Stages of friction welding process during dissimilar joints.

Pan et al. [72] investigated the mechanical properties and microstructure of MMC based on aluminium and AISI 304 stainless steel joints that were friction-welded. They observed that intermetallic phases was present at the interface of these different joints. Remarkably, the notch tensile strength of these joints was improved when the rotating speed was increased from 500 to 1,000 revolutions per minute; however, additional speed increases did not affect the strength. In their search for the ideal parameters for friction welding aluminium-based MMCs, Zhou et al. [73] discovered that MMC/Alloy 6061 joints showed noticeably lower notch tensile strengths than other joints and that the fatigue strengths were also inferior to those of the base materials.

The presence of spiral flaws and their impact on the mechanical properties of friction-welded joints between composite materials 6061/Al_2O_3 and aluminium alloy 6061 T6 joints. They discovered that spiral flaws, which originate near the weld's stationary edge and are caused by flow, can have an impact on the quality of the joint. On the other hand, they found that MMC/MMC junctions exhibited noticeably greater tensile strengths than the underlying MMC material when heat-treated post-weld at 280 MPa of friction.

Lin et al. [74] successfully performed friction welding on two composite materials (A7005 aluminium alloy reinforced with Al_2O_3 and SiC) that shared the same matrix but had different types of reinforcement. Particle size and volume % differences between SiC and Al_2O_3 particles affected hardening values, welding strength, and HAZ by affecting the concentration of reinforcement within the zone.

Reddy et al. [75] used a continuous rotation technique to successfully friction weld AISI 304 austenitic stainless steel and AA6061. Fe_2Al_5 production during direct welding resulted in brittle connections; however, this was lessened by electroplating Cu, Ni, and Ag as diffusion barrier interlayers. However, only the silver interlayer successfully stopped Fe_2Al_5 from forming, making the joints stronger and more flexible.

To evaluate the bonding quality of ceramic/metal alloy friction-welded components, Fauzi et al. [76] emphasized the significance of looking at the interface. This demonstrated the vital significance that interface analysis plays in comprehending friction-welded joint integrity.

17.7 CONCLUSIONS

Metal matrix composite joining is still an active field of study in materials science, motivated by the desire to fully realize the potential of these cutting-edge materials in a range of industrial applications. Even though connecting methods that maintain the special qualities of MMCs have gone a long way, further research and development are still needed to solve current problems and open up new avenues for the creation and construction of intricate MMC structures. Leveraging these composites for future technological developments will depend critically on the field's continuing improvement of joining procedures and a greater comprehension of the materials science underpinning MMCs.

REFERENCES

[1] Prabu, S.B., Karunamoorthy, L., Kathiresan, S. and Mohan, B., 2006. Influence of stirring speed and stirring time on distribution of particles in cast metal matrix composite. *Journal of Materials Processing Technology*, *171*, pp. 268–273.

[2] Zhang, L., Duan, Z., Zhu, H. and Yin, K., 2017. Advances in synthesizing copper/graphene composite material. *Materials and Manufacturing Processes*, *32*, pp. 475–479.

[3] Taha, M.A., 2001. Industrialization of cast aluminum matrix composites (AMCCs). *Materials and Manufacturing Processes*, *16*, pp. 619–641.

[4] Kaczmar, J.W., Pietrzak, K. and Włosiński, W., 2000. The production and application of metal matrix composite materials. *Journal of Materials Processing Technology*, *106*, pp. 58–67.

[5] Goni, J., Egizabal, P., Coleto, J., Mitxelena, I., Leunda, I. and Guridi, J.R., 2003. High performance automotive and railway components made from novel competitive aluminium composites. *Materials Science and Technology*, *19*, pp. 931–934.

[6] Liu, J., Khan, U., Coleman, J., Fernandez, B., Rodriguez, P., Naher, S. and Brabazon, D., 2016. Graphene oxide and graphene nanosheet reinforced aluminium matrix composites: Powder synthesis and prepared composite characteristics. *Materials and Design*, *94*, pp. 87–94.

[7] Jayabalakrishnan, D. and Balasubramanian, M., 2018. Eccentric-weave FSW between Cu and AA 6061-T6 with reinforced Graphene nanoparticles. *Materials and Manufacturing Processes*, *33*, pp. 333–342.

[8] Bains, P.S., Sidhu, S.S. and Payal, H.S., 2016. Fabrication and machining of metal matrix composites: A review. *Materials and Manufacturing Processes*, *31*, pp. 553–573.

[9] Mazahery, A. and Ostadshabani, M., 2011. Investigation on mechanical properties of nano-Al 2 O 3-reinforced aluminum matrix composites. *Journal of Composite Materials*, *45*, pp. 2579–2586.

[10] Rathee, S., Maheshwari, S., Siddiquee, A.N. and Srivastava, M., 2018. Distribution of reinforcement particles in surface composite fabrication via friction stir processing: Suitable strategy. *Materials and Manufacturing Processes*, *33*, pp. 262–269.

[11] Rahbari, R.G., Abu Kasim, N.H., Madfa, A.A., Hamdi, M. and Bayat, M., 2011. Porosity reduction model in titanium–hydroxyapatite FGM composites using shrinkage measurement. *Materials Research Innovations*, *15*, pp. s110–s113.

[12] Hu, Q., Zhao, H. and Li, F., 2016. Effects of manufacturing processes on microstructure and properties of Al/A356–B_4C composites. *Materials and Manufacturing Processes*, *31*, pp. 1292–1300.

[13] Verdian, M.M., 2010. Synthesis of TiAl 3-Al_2O_3 composite particles by chemical reactions in molten salts. *Materials and Manufacturing Processes*, *25*, pp. 953–955.

[14] Mao, J.W., Wang, M.M., Wang, L.Q., Xue, X.B., Lu, W.J., Zhang, D., Qin, J.N., Sun, X.J. and Zhu, X.X., 2012. Weld zone characteristic and mechanical performance of in situ titanium matrix composites using gas tungsten arc welding (GTAW). *Science and Technology of Welding and Joining*, *17*, pp. 630–635.

[15] Huang, L., Duan, T., An, Q., Chen, Y., Bai, J., Geng, L. and Lin, S., 2018. Gas tungsten arc welding of network structured titanium matrix composite. *Science and Technology of Welding and Joining*, *23*, pp. 357–364.

[16] Xuan, Z.Z., Gu, X.Y., Zhong, R. and Sun, D.Q., 2010. Microstructures and mechanical properties of tungsten inert gas arc welded magnesium metal matrix composite (TiCp/AZ91D). *Materials Science and Technology*, *26*, pp. 1513–1517.

[17] Prater, T., 2014. Friction stir welding of metal matrix composites for use in aerospace structures. *Acta Astronautica*, *93*, pp. 366–373.

[18] Lionetto, F., Mele, C., Leo, P., D'Ostuni, S., Balle, F. and Maffezzoli, A., 2018. Ultrasonic spot welding of carbon fiber reinforced epoxy composites to aluminum: Mechanical and electrochemical characterization. *Composites Part B: Engineering*, *144*, pp. 134–142.

[19] Ucsnik, S., Scheerer, M., Zaremba, S. and Pahr, D.H., 2010. Experimental investigation of a novel hybrid metal–composite joining technology. *Composites Part A: Applied Science and Manufacturing*, *41*, pp. 369–374.

[20] Li, M., Xiong, X., Ji, S., Hu, W. and Yue, Y., 2021. Achieving high-quality metal to polymer-matrix composites joint via top-thermic solid-state lap joining. *Composites Part B: Engineering*, *219*, p. 108941.

[21] Victor Christy, J., Ismail Mourad, A.-H., Sherif, M.M. and Shivamurthy, B., 2021. Review of recent trends in friction stir welding process of aluminum alloys and aluminum metal matrix composites. *Transactions of Nonferrous Metals Society of China*, *31*, pp. 3281–3309.

[22] Zhang, Z., Gao, Z., Jin, X., Li, S., Qiu, Y., Qiu, D. and Wang, Z., 2024. Joining high volume fraction SiC particle reinforced aluminum matrix composites using Bi_2O_3-B_2O_3-ZnO-BaO-Na_2O glass. *Materials Today Communication*, *38*, p. 108245.

[23] Arenas, J.M., Alía, C., Narbón, J.J., Ocaña, R. and González, C., 2013. Considerations for the industrial application of structural adhesive joints in the aluminium–composite material bonding. *Composites Part B: Engineering*, *44*, pp. 417–423.

[24] Pramanik, A., Basak, A.K., Dong, Y., Sarker, P.K., Uddin, M.S., Littlefair, G., Dixit, A.R. and Chattopadhyaya, S., 2017. Joining of carbon fibre reinforced polymer (CFRP) composites and aluminium alloys—a review. *Composites Part A: Applied Science and Manufacturing*, *101*, pp. 1–29.

[25] Singh, R., Kumar, R., Feo, L. and Fraternali, F., 2016. Friction welding of dissimilar plastic/polymer materials with metal powder reinforcement for engineering applications. *Composites Part B: Engineering*, *101*, pp. 77–86.

[26] Mishra, R.S. and Ma, Z.Y., 2005. Friction stir welding and processing. *Materials Science and Engineering: R: Reports*, *50*, pp. 1–78.

[27] Periyasamy, P., Mohan, B., Balasubramanian, V., Rajakumar, S. and Venugopal, S., 2013. Multi-objective optimization of friction stir welding parameters using desirability approach to join Al/SiCp metal matrix composites. *Transactions of Nonferrous Metals Society of China*, *23*, pp. 942–955.

[28] Gangwar, K. and Ramulu, M., 2018. Friction stir welding of titanium alloys: A review. *Materials and Design*, *141*, pp. 230–255.

[29] Gangwar, K. and Ramulu, M., 2019. Dataset for interpreting the circos figures used in the review of friction stir welding of titanium alloys. *Data Brief*, *22*, pp. 164–168.

[30] Bodunrin, M.O., Alaneme, K.K. and Chown, L.H., 2015. Aluminium matrix hybrid composites: A review of reinforcement philosophies; mechanical, corrosion and tribological characteristics. *Journal of Materials Research and Technology*, *4*, pp. 434–445.

[31] Niu, J., Pan, L., Wang, M., Fu, C. and Meng, X., 2006. Research on laser welding of aluminum matrix composite SiCw/6061. *Vacuum*, *80*, pp. 1396–1399.

[32] Bassani, P., Capello, E., Colombo, D., Previtali, B. and Vedani, M., 2007. Effect of process parameters on bead properties of A359/SiC MMCs welded by laser. *Composites Part A: Applied Science and Manufacturing*, *38*, pp. 1089–1098.

[33] Zhang, X.P., Quan, G.F. and Wei, W., 1999. Preliminary investigation on joining performance of SiCp-reinforced aluminium metal matrix composite (Al/SiCp–MMC) by vacuum brazing. *Composites Part A: Applied Science and Manufacturing*, *30*, pp. 823–827.

[34] Storjohann, D., Barabash, O.M., David, S.A., Sklad, P.S., Bloom, E.E. and Babu, S.S., 2005. Fusion and friction stir welding of aluminum-metal-matrix composites. *Metallurgical and Materials Transactions A*, *36*, pp. 3237–3247.

[35] Dragatogiannis, D.A., Koumoulos, E.P., Kartsonakis, I.A., Pantelis, D.I., Karakizis, P.N. and Charitidis, C.A., 2016. Dissimilar friction stir welding between 5083 and 6082 Al alloys reinforced with TiC nanoparticles. *Materials and Manufacturing Processes*, *31*, pp. 2101–2114.

[36] Pantelis, D.I., Karakizis, P.N., Daniolos, N.M., Charitidis, C.A., Koumoulos, E.P. and Dragatogiannis, D.A., 2016. Microstructural study and mechanical properties of dissimilar friction stir welded AA5083-H111 and AA6082-T6 reinforced with SiC nanoparticles. *Materials and Manufacturing Processes*, *31*, pp. 264–274.

[37] Bozkurt, Y., Uzun, H. and Salman, S., 2011. Microstructure and mechanical properties of friction stir welded particulate reinforced AA2124/SiC/25p–T4 composite. *Journal of Composite Materials*, *45*, pp. 2237–2245.

[38] Ceschini, L., Boromei, I., Minak, G., Morri, A. and Tarterini, F., 2007a. Microstructure, tensile and fatigue properties of AA6061/20 vol.% Al2O3p friction stir welded joints. *Composites Part A: Applied Science and Manufacturing*, *38*, pp. 1200–1210.

[39] Ceschini, L., Boromei, I., Minak, G., Morri, A. and Tarterini, F., 2007b. Effect of friction stir welding on microstructure, tensile and fatigue properties of the AA7005/10 vol.% Al2O3 composite. *Composites Science and Technology*, *67*, pp. 605–615.

[40] Amirizad, M., Kokabi, A.H., Gharacheh, M.A., Sarrafi, R., Shalchi, B. and Azizieh, M., 2006. Evaluation of microstructure and mechanical properties in friction stir welded A356+15%SiCp cast composite. *Materials Letter*, *60*, pp. 565–568.

[41] Dinaharan, I. and Murugan, N., 2012. Effect of friction stir welding on microstructure, mechanical and wear properties of AA6061/ZrB_2 in situ cast composites. *Materials Science and Engineering: A*, *543*, pp. 257–266.

[42] Kalaiselvan, K., Dinaharan, I. and Murugan, N., 2014. Characterization of friction stir welded boron carbide particulate reinforced AA6061 aluminum alloy stir cast composite. *Materials and Design*, *55*, pp. 176–182.

[43] Bahrami, M., Dehghani, K. and Besharati Givi, M.K., 2014. A novel approach to develop aluminum matrix nano-composite employing friction stir welding technique. *Materials and Design*, *53*, pp. 217–225.

[44] Xiao, B.L., Wang, D., Bi, J., Zhang, Z. and Ma, Z.Y., 2010. Friction stir welding of SiCp/Al composite and 2024 Al alloy. *Materials Science Forum*, *638–642*, pp. 1500–1505.

[45] Kalaiselvan, K. and Murugan, N., 2013. Role of friction stir welding parameters on tensile strength of AA6061–B4C composite joints. *Transactions of Nonferrous Metals Society of China*, *23*, pp. 616–624.

[46] Ashok Kumar, B. and Murugan, N., 2014. Optimization of friction stir welding process parameters to maximize tensile strength of stir cast AA6061-T6/AlNp composite. *Materials and Design*, *57*, pp. 383–393.

[47] Battina, N.M., Vanthala, V.S.P. and Chirala, H.K., 2021. Influence of tool pin profile on mechanical and metallurgical behavior of friction stir welded AA6061-T6 and AA2017-T6 tailored blanks. *Engineering Research Express*, *3*, p. 035026.

[48] Cavaliere, P., Cerri, E., Marzoli, L. and Dos Santos, J., 2004. Friction stir welding of ceramic particle reinforced aluminium based metal matrix composites. *Applied Composite Materials*, *11*, pp. 247–258.

[49] Qin, Q., Zhao, H., Li, J., Zhang, Y. and Su, X., 2020. Microstructure and mechanical properties of friction stir processed Al-Mg2Si alloys. *Transactions of Nonferrous Metals Society of China*, *30*, pp. 2355–2368.

[50] Minak, G., Ceschini, L., Boromei, I. and Ponte, M., 2010. Fatigue properties of friction stir welded particulate reinforced aluminium matrix composites. *International Journal of Fatigue*, *32*, pp. 218–226.

[51] Periyasamy, P., Mohan, B. and Balasubramanian, V., 2012. Effect of heat input on mechanical and metallurgical properties of friction stir welded AA6061-10% SiCp MMCs. *Journal of Materials Engineering and Performance*, *21*, pp. 2417–2428.

[52] Cioffi, F., Fernández, R., Gesto, D., Rey, P., Verdera, D. and González-Doncel, G., 2013. Friction stir welding of thick plates of aluminum alloy matrix composite with a high volume fraction of ceramic reinforcement. *Composites Part A: Applied Science and Manufacturing*, *54*, pp. 117–123.

[53] Wang, D., Xiao, B.L., Wang, Q.Z. and Ma, Z.Y., 2014b. Evolution of the microstructure and strength in the nugget zone of friction stir welded SiCp/Al-Cu-Mg composite. *Journal of Materials Science & Technology*, *30*, pp. 54–60.

[54] Nami, H., Adgi, H., Sharifitabar, M. and Shamabadi, H., 2011. Microstructure and mechanical properties of friction stir welded Al/Mg2Si metal matrix cast composite. *Materials and Design*, *32*, pp. 976–983.

[55] Feng, A., Xiao, B. and Ma, Z., 2008. Effect of microstructural evolution on mechanical properties of friction stir welded AA2009/SiCp composite. *Composites Science and Technology*, *68*, pp. 2141–2148.

[56] Wang, D., Xiao, B.L., Wang, Q.Z. and Ma, Z.Y., 2013. Friction stir welding of SiCp/2009Al composite plate. *Materials and Design*, *47*, pp. 243–247.

[57] Uzun, H., 2007. Friction stir welding of SiC particulate reinforced AA2124 aluminium alloy matrix composite. *Materials and Design*, *28*, pp. 1440–1446.

[58] Yigezu, B.S., Venkateswarlu, D., Mahapatra, M.M., Jha, P.K. and Mandal, N.R., 2014. On friction stir butt welding of Al + 12Si/10 wt%TiC in situ composite. *Materials & Design, (1980–2015)*, *54*, pp. 1019–1027.

[59] Wang, D., Wang, Q.Z., Xiao, B.L. and Ma, Z.Y., 2014a. Achieving friction stir welded SiCp/Al-Cu-Mg composite joint of nearly equal strength to base material at high welding speed. *Materials Science and Engineering: A*, *589*, pp. 271–274.

[60] Ahn, B.-W., Choi, D.-H., Kim, Y.-H. and Jung, S.-B., 2012. Fabrication of SiCp/AA5083 composite via friction stir welding. *Transactions of Nonferrous Metals Society of China*, *22*, pp. s634–s638.

[61] Vlad, M., 2007. *Ultrasonic Welding of Aluminum: A Practical Study in Consistency, Part Marking and Control Modes*. Iowa State University, Digital Repository, Ames.

[62] Chak, V., Chattopadhyay, H. and Dora, T.L., 2020. A review on fabrication methods, reinforcements and mechanical properties of aluminum matrix composites. *Journal of Manufacturing Processes*, *56*, pp. 1059–1074.

[63] Laha, T., Agarwal, A., McKechnie, T. and Seal, S., 2004. Synthesis and characterization of plasma spray formed carbon nanotube reinforced aluminum composite. *Materials Science and Engineering: A*, *381*, pp. 249–258.

[64] Kumar Sharma, A., Bhandari, R., Aherwar, A. and Pinca-Bretotean, C., 2020. A study of fabrication methods of aluminum based composites focused on stir casting process. *Materials Today Proceedings*, *27*, pp. 1608–1612.

[65] Truckenmüller, R., Cheng, Y., Ahrens, R., Bahrs, H., Fischer, G. and Lehmann, J., 2006. Micro ultrasonic welding: Joining of chemically inert polymer microparts for single material fluidic components and systems. *Microsystem Technologies*, *12*, pp. 1027–1029.

[66] Daoud, A., 2005. Microstructure and tensile properties of 2014 Al alloy reinforced with continuous carbon fibers manufactured by gas pressure infiltration. *Materials Science and Engineering: A*, *391*, pp. 114–120.

[67] Harichandran, R. and Selvakumar, N., 2016. Effect of nano/micro B_4C particles on the mechanical properties of aluminium metal matrix composites fabricated by ultrasonic cavitation-assisted solidification process. *Archives of Civil and Mechanical Engineering*, *16*, pp. 147–158.

[68] Leng, X., Yang, W., Zhang, J., Ma, X., Zhao, W. and Yan, J., 2018. High-performance joining technology for aluminium matrix composites using ultrasonic-assisted brazing. *Materials Science and Technology*, *34*, pp. 660–663.

[69] Kim, W.H., Kang, E.J. and Park, D.S., 2017. Evaluation of welding performance of 20 kHz and 40 kHz ultrasonic metal welding. *IOP Conference Series Materials Science Engineering*, *248*, p. 012013.

[70] Sharma, P., Dabra, V., Sharma, S., Khanduja, D., Sharma, N., Sharma, R. and Saini, K., 2019. Microstructure and properties of AA6082/(SiC+Graphite) hybrid composites. *Refractories and Industrial Ceramics*, *59*, pp. 471–477.

[71] Sahin, M., 2009. Joining of stainless-steel and aluminium materials by friction welding. *The International Journal of Advanced Manufacturing Technology*, *41*, pp. 487–497.

[72] Pan, C., Hu, L., Li, Z. and North, T.H., 1996. Microstructural features of dissimilar MMC/AISI 304 stainless steel friction joints. *Journal of Materials Science*, *31*, pp. 3667–3674.

[73] Zhou, Y., Zhang, J., North, T.H. and Wang, Z., 1997. The mechanical properties of friction welded aluminium-based metal–matrix composite materials. *Journal of Materials Science*, *32*, pp. 3883–3889.

[74] Lin, C.B., Chou, I.-C. and Ma, C.L., 2002. Manufacturing and friction welding properties of particulate reinforced 7005 Al. *Journal of Materials Science*, *37*, pp. 4645–4652.

[75] Reddy, M.G., Rao, S.A. and Mohandas, T., 2008. Role of electroplated interlayer in continuous drive friction welding of AA6061 to AISI 304 dissimilar metals. *Science and Technology of Welding and Joining*, *13*, pp. 619–628.

[76] Ahmad Fauzi, M.N., Uday, M.B., Zuhailawati, H. and Ismail, A.B., 2010. Microstructure and mechanical properties of alumina-6061 aluminum alloy joined by friction welding. *Materials and Design*, *31*, pp. 670–676.

18 Advanced Forming Methods and Processing Condition of Metal Matrix Composites

Sharath Peramenahalli Chikkegowda, Shrishail Basaprabhu Sollapur, and Bharathkumar Bidare Rangaswamy

18.1 INTRODUCTION

Metal matrix composites, or MMCs, are specially designed materials in which a metallic matrix is reinforced with fibres or particulates to improve the mechanical and physical properties of the material. As a result of the incorporation of fibre or particulate reinforcements into the metallic matrix, the fully fledged composite exhibits enhanced strength, stiffness, and wear resistance compared to the unreinforced matrix. This chapter discusses different methods used to fabricate aluminium-based MMCs with particulate reinforcements, the advantages and disadvantages, as well as on their ability to produce tailored microstructures and improve material properties. It will allow for optimizing the manufacturing process and identifying potential areas for improving MMCs specifically designed for selected applications [1].

Among the various processing techniques for particulate-reinforced MMCs, powder metallurgy stands out as a prominent method. This approach entails blending metal powders with reinforcing particulates, followed by compacting the mixture within a die to form a green compact. Subsequently, the green compact undergoes sintering, a process that facilitates the bonding of the particles with the metal matrix, yielding a strong and homogeneous composite material. Another method is stir casting, which involves dispersing the reinforcing particles in the molten metal matrix using mechanical stirring. Additionally, spray deposition is a method that uses high-velocity gas to propel the metal matrix and reinforcing particles onto a substrate. This is advantageous for the production of coatings or components with complex geometries, owing to the excellent bonding achieved between the reinforcing particles and the matrix [2].

18.2 BASICS OF DEFORMATION PROCESS

Traditionally, deformation techniques such as rolling and extrusion, as well as consolidation techniques such as quick solidification and ultrasonic stirring, have been employed to control secondary phase dispersion or spatial distribution. It has been discovered that plastic deformation-induced accumulative strain input can constantly increase the homogeneity of composite materials, improving their attributes as a result [3]. Secondary particles in composite materials create unique microstructural distributions and enhanced strain hardening through reverse stress and dislocation hardening. High densities of geometrically needed dislocations (GNDs) at interphase contacts generate significant back stress, particularly at particle-matrix interfaces where plastic strain gradients are pronounced.

A schematic representation of MMCs' quasi-static stress-strain behaviour is shown in Figure 18.1. Metal matrix composites (MMCs) exhibit elastic-plastic behaviour where the metallic matrix deforms plastically while ceramic reinforcement remains elastic until failure. While MMCs offer improved strength, stiffness, and strain hardening compared to non-reinforced materials, their reinforcement phase experiences higher internal stress and fails at lower external stress levels, leading to earlier strain-based failure.

DOI: 10.1201/9781003499718-18

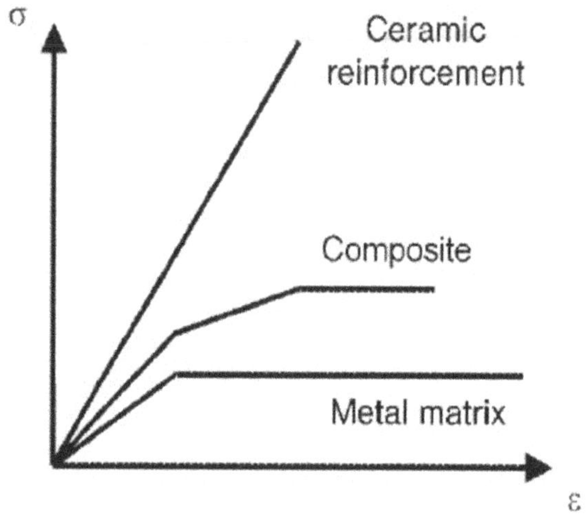

FIGURE 18.1 A schematic illustration of the link between the mechanical behaviour of the individual components and the deformation behaviour of composites. Reproduced with permission from John Wiley and Sons publisher [4].

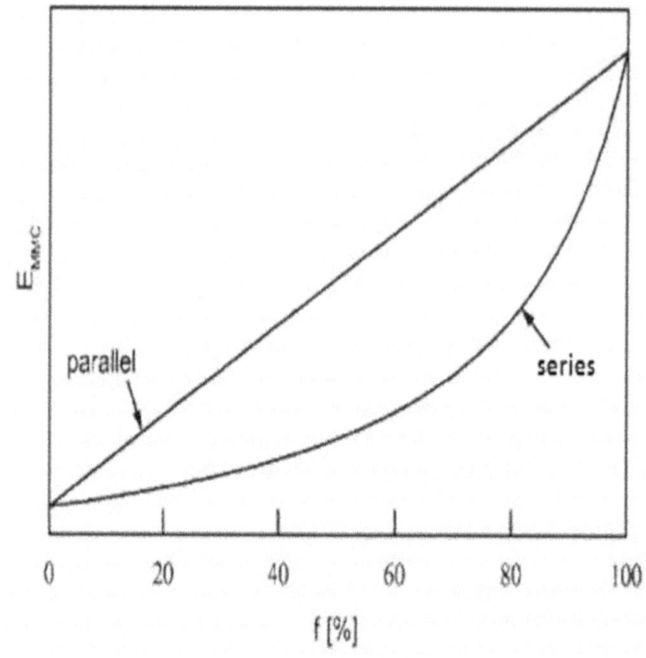

FIGURE 18.2 Demonstrates the effect of the reinforcement volume fraction (f) on the Young's modulus of the composite. Reproduced with permission from John Wiley and Sons publisher [4].

The assessment of Young's modulus [E], an indispensable characteristic for composites, is facilitated by a streamlined and easily implementable procedure [4]. This technique takes into account the arrangement of components and their parts, distinguishing between parallel and serial reinforcing arrangements.

The graphical representation in Figure 18.2 showcases the variation in the elastic modulus of the composite as a function of the reinforcement volume fraction (f), highlighting two limiting scenarios. Composites incorporating continuous reinforcements (long fibres) are characterized by a greater elastic modulus when compared to those reinforced with discontinuous reinforcements (short fibres or particles).

18.3 MICROSTRUCTURE AND MECHANICAL PROPERTIES OF MMC

Manufacturing metal matrix composites (MMCs) presents challenges in optimizing process parameters, managing voids, and achieving uniform reinforcement distribution within the matrix. Key factors affecting MMC quality include wettability between reinforcement and matrix, porosity control, and particle distribution homogeneity, while sustainable manufacturing requires consideration of material degradation and reuse through life cycle assessment. Table 18.1 shows the comparison of different forming processes in metal matrix composites with their parameters and shows the end result.

18.3.1 Extrusion of MMC

Extrusion is the most commonly used method for particle-reinforced MMCs due to its simplicity, efficient reinforcement dispersion, high productivity, cost-effectiveness, and ability to improve microstructure and mechanical properties.

The extrusion process for metal-matrix composites is similar to unreinforced aluminium in speed, exterior quality, and extrusion ratio but requires slightly lower velocity and marginally higher pressure. Davies et al. [35] identified five critical problems in particle-reinforced MMC extrusion: 1) stick-slip tearing, 2) particle fracture, 3) particle distribution homogeneity, 4) porosity, and 5) recrystallization. The stick-slide tearing phenomenon in particle fracture is induced by hydrostatic pressure during extrusion and local stress from applied strain at low temperatures, with temperature significantly affecting the matrix's stress value during the process.

Kang et al. investigated AA6061 with Al_2O_3 short fibres for hot extrusion and its related mechanical properties produced from the compo-casting method [36]. Microstructural analysis showed that increasing the extrusion ratio led to more fibre breakage and improved fibre alignment. Hot-extruded MMCs outperformed the matrix metal in mechanical properties, except for elongation at failure, with extrusion temperature (400–550°C) having minimal impact. Hot extrusion with a constant-strain-rate (CSR) die, where the die radius (*r*) varies with axial position (*z*), is described by [37].

$$r(z) = \left(\frac{r_0^2}{(\lambda - 1)z / L + 1} \right)^{1/2} \tag{18.1}$$

Here, r_0 represents the radius at the die entry, λ denotes the extrusion ratio, and L symbolizes the die length. This study also discovered that the essential extrusion ratio for optimal strength is 5.44. Lastly, fractographs of the extruded MMC tensile specimens showed that the matrix behaves ductile.

Sharma et al. [16] studied solid stir extrusion process for the synthesis of Al-graphene MMC cables. Figure 18.3 shows the process and schematic representation and extruded Al MMC rods.

The figure collectively visualizes the starting materials, the SSE process, the extruded MMC products, and the reference planes for microstructural analysis, providing a comprehensive overview of this solid-state manufacturing technique.

TABLE 18.1
Comparison of Various Deformation Processes in MMCs

	Extrusion process			
References	**Processing parameters**	**Materials**	**Extrusion parameter**	**Mechanical properties**
[5]	Forward extrusion at RT	AA6063 with 5% and 10 % SiCp	12:8 extrusion ratio	Higher silicon carbide particle (SiCp) content reduces the friction factor and affects the required extrusion load.
[6]	Extrusion at RT	A356 with 10% Al_2O_3 (23 μm) along 1, 3 and 5% MoS_2 (5 μm)	12.25, 16, 20.25 extrusion ratio	Reinforcing particles enhance wear resistance through improved bonding while increasing MoS2 content creates better lubrication layers and improved wear properties.
[7]	Friction stir back extrusion	AA7075 with 11% SiC (4 μm)	Rotary billet at 1,400 rpm, feed rate 1.2 mm/s and axial force 15.38 KN	The composite samples demonstrated a substantial rise in their strength and hardness properties.
[8]	Extrusion at 170°C	ZK60-5% SiC whiskers	Varies w.r.t average distance between whiskers	Testing from 25–350°C showed anisotropic structure enhanced both ductility and strength.
[9]	Extrusion at 480°C	Al 6061-6 and 9% BN	3.06 extrusion ratio	Higher boron nitride content enhances hardness and tensile strength but reduces ductility.
[10]	Extrusion at 350°C	Mg–5%Sb (1, 3, and 5 %) SiC	12:1 extrusion ratio	The extrusion process fractures the Mg_3Sb_2 intermetallic compound and improves UTS from 120 to 244 MPa.
[11]	Extrusion at 500°C	AlMg5/1%SiC/ (0, 5, 10, and 15%) piezoelectric Pb-La-(Zr-Ti)-O	25:1 extrusion ratio	Obtained homogenous and dense MMC with improved damping properties.
[12]	Extrusion at 400–450°C	AA7075 with 4.26% Mg, 0.85% Gr, Zn, and Ti	3 mm/min ram rate	Extrusion improves bond strength and reduces surface defects, leading to better flexural strength and tribological properties.
[13]	Cold extrusion	Al6061 with 8% TiB_2	1.2656 extrusion ratio	MMC achieved 179.04% higher UTS compared to a base metal, while cold extruded pure Al 6061 showed only a 66.67% improvement.
[14]	Twist extrusion	AZ91-2% SiC-1.5% B_4C	54° Slope angle	Tensile strength increases to 26.6% and hardness value increases to 38.5% after extrusion
[15]	Extrusion at 380°C	AZ91-3B_4C-3SiC	12:1 extrusion ratio	While uniformly dispersed B_4C and SiC particles enhanced properties through PSN and twinning, micron-sized particles created stress points without improving strength or ductility.
[16]	Solid stir extrusion	Al-graphene	600 rpm tool rotation rate and feed rate 200 mm/min	Al-graphene MMC cables treated with SSE showed a positive synergy between EC and strength.

TABLE 18.1 *(Continued)*
Comparison of Various Deformation Processes in MMCs

References	Processing parameters	Materials	Extrusion process Extrusion parameter	Mechanical properties
Rolling process				
[17]	Rolled at elevated temperature	A359-30%SiC	88% deformation	Hot rolling disrupts the interdendritic eutectic silicon network in the as-cast matrix, influencing the composite's mechanical response.
[18]	Rolled at RT	2014 Al-15 % SiC	Applied load 800 N and 1,300 N	A work hardening maximum at ~175 μm aligns with a pseudo-macro stress minimum.
[19]	Rolled at 400°C	Al6061-10% Zr-3% B	90% reduction with each pass and strain rate of $7x10^{-3}$ s^{-1}	Composite exhibited improved wear resistance and lower friction coefficients.
[20]	ARB and CARB	Al-15% Al_2O_3	50% reduction	CARB specimens outperformed ARB composites in hardness, yield stress, tensile strength, and ductility.
[21]	Non-iso rolling of specimens heated to 1,000°C	Ti-14.25%Mo-5% TiB_2	45% thickness reduction	Yield strength was quite similar, and tensile elongation did not exceed 2%.
[22]	Accumulative roll bonding at 400°C up to 4 passes	Al-Al_2O_3	75% reduction	Dispersed alumina clusters enhanced the composite's strength and toughness.
[23]	Rolling at 150°C and 200°C	Mg-Ta-Al layered MMC	70% reduction and rolling rate 0.17 m s-1	UTS and YS reached 413 and 263 MPa respectively and interface bond strength substantially increased to 54–76 MPa.
Forging process				
[24]	Forging at 594°C	Al_4TiC, Al_4Fe_3C, Al_4MO_2C, and Al_4WC	Two forging steps with 90° rotations, followed by reheating at 594°C for 15 minutes.	Al MMC with carbide reinforcement showed superior hardness and tensile strength under water cooling.
[25]	Hot and cold forged	Al6061-10% SiC	Hot forging at 475 ± 10°C, deformation ratio 4:1	The hardness of 115 BHN was observed for cold forged and tensile strength of 357 MPa for hot forged sample
[26]	Forging at 500°C	AA7075-Cu-(15% and 30%) ZnO, 10%-Nb2Al and SiC	Forging after sintering at 600°C for 45 min	YS and UTS increased with reduced ZnO content, while impact resistance decreased.
[27]	Forging at 1,100°C	Fe-0.4C-1.1Si-4.7Mn-5.1Cr-1.3 Mo-10% carbide	Graphite spray is used as a lubricant and the material used is alloy steel 42CrMo4	The fabricated tools showed enhanced abrasive wear resistance, similar to the wear protection mechanism in wear-resistant metal matrix composites.
[28]	Forging at 532.34°C	AA7075–1% Al_2O_3	Forge and hold for 60 min duration	Increasing the forging temperature and holding time resulted in higher UTS and ductility.

(Continued)

TABLE 18.1 *(Continued)*
Comparison of Various Deformation Processes in MMCs

References	Processing parameters	Materials	Extrusion process Extrusion parameter	Mechanical properties
[29]	Forging at 1,100°C	TRIP steel	Die temperature at 200°C, Boron nitride is used as a lubricant	The yield strength of the composite increased with higher ZrO_2 content, but at 15% reinforcement, it showed more brittle fracture behaviour and reduced strain at break.
[30]	Forging at 450°C	Al6061–(2, 4, and 6)% WC	Reduction ratio 20, 40 and 60%	Forged composites exhibited much-reduced wear rates compared to unforged composites.
Friction stir processing of MMC				
[31]	Melt temp 990°C, preheated ceramic particles to 650°C	AA6061-B_4C and SiC	A spindle speed ranging from 300 to 500 rpm, a stirring time of 10 minutes, and a powder feed rate between 0.75 and 1.0 grams per second	The maximum UTS was around 172.8 MPa for 10% SiC and 3% B4C. Larger strength, better wear characteristics, and hardness are all influenced by a larger SiC content.
[32]	The temperature of melt is 650°C	LM25–5% SiC	Speed of 1,000 rpm, travel speeds of 20–60 mm/min, D/d ratio of 3, axial force of 3 kN, pin length of 7.2 mm, pin dia. of 5.5–7.2 mm	Results indicated excellent wear resistance at a tool travel speed of 40 mm/min, attributed to the significant microhardness value achieved.
[33]	Solution heat treated with precipitation hardening of aluminium alloy	Al-20% B_4C	25 mm corrugated shoulder, a three-edge threaded pin, 1.5-degree tool inclination, and H11 tool steel in the hardened condition	Al-B_4C with a thickness of up to 7 mm were produced, and their mechanical properties did not deteriorate after irradiation.
[34]	Cold spray-Al_2O_3 coating with low pressure	Al-20% Al_2O_3	Threaded pin with 3.3 mm dia. and length of 2.8 mm. A concave shoulder with 10 mm dia. and tool inclination angle of 2 degrees.	The specimens from the third run of the higher friction stir processing exhibited poorer electrochemical performance due to the worsened interface of internal deposits.

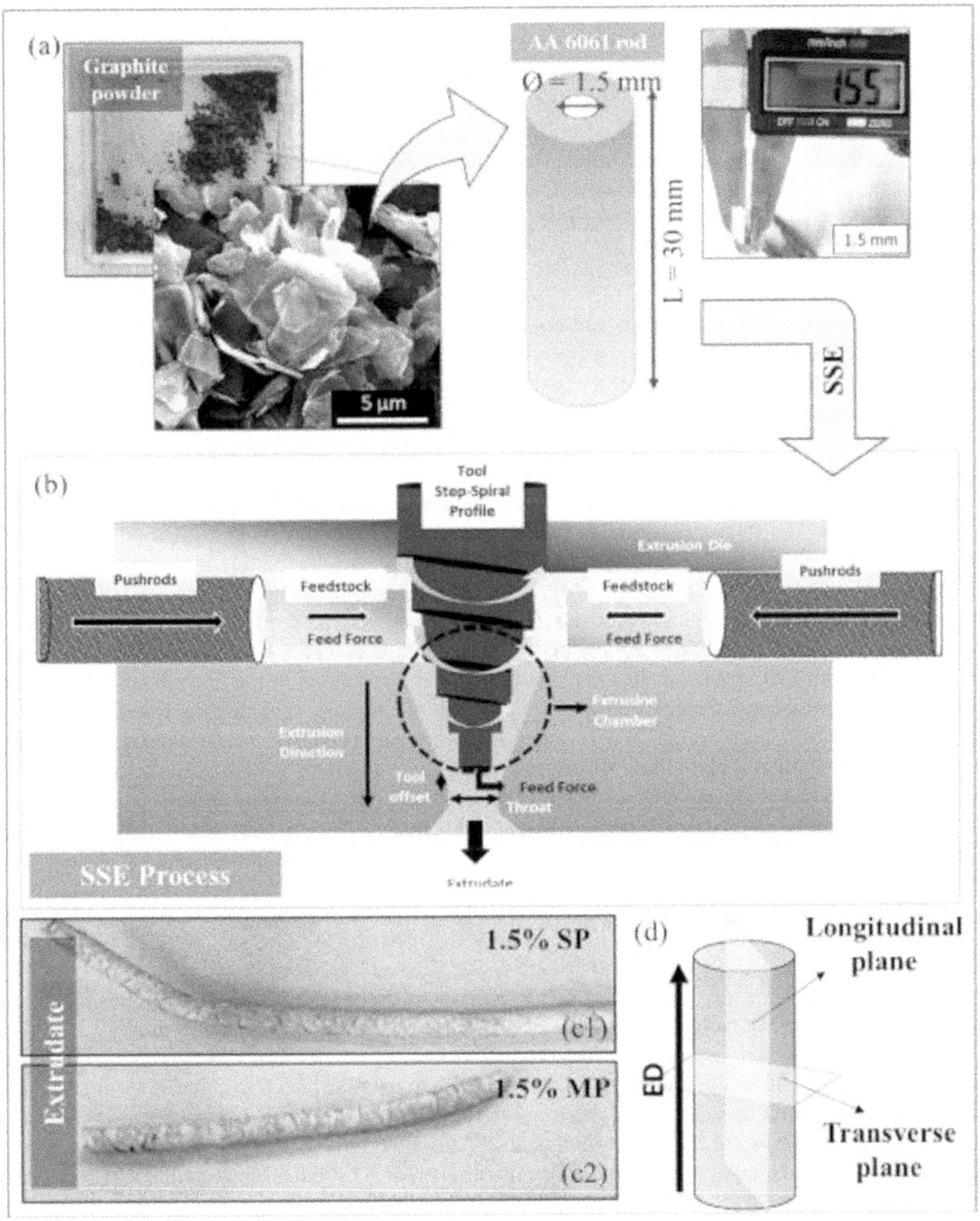

FIGURE 18.3 (a) Visualization of graphite powder and feed rods for the solid-state extrusion (SSE) process, including optical and SEM-BSE images; (b) illustration of the solid-state extrusion (SSE) process with a schematic representation; (c1) and (c2) comparison of Al metal matrix composite (MMC) extruded rods obtained through single-pass (SP) and multiple-pass (MP) processing, with visual analysis; (d) visual representation of the longitudinal and transverse planes of an extruded rod. Reproduced with permission from Elsevier publisher [16].

In the solid-state extrusion process, graphite exfoliation from constant shearing led to the uniform dispersion of multilayered graphene within the aluminium phase. The MP sample showed better graphene distribution than the SP sample. Interfacial bonding between graphene and the Al matrix was partially successful, reducing strength and slightly lowering electrical conductivity compared to the base material (BM). Multilayer graphene increased alloy strength by 54%, with a 2.3% decrease in conductivity. Graphene also significantly reduced grain size, and the desired strength and conductivity in the extruded Al-graphene MMC were achieved through strong texture, solute in solid solution, and fine SiC particle dispersion.

18.3.2 Rolling of MMC

MMCs have been rolled in both hot and cold conditions to enhance particle bonding and to consolidate powders after mixing. Rolling has two effects on distribution: it minimizes the banding of particles perpendicular to the direction of extrusion and breaks up big aggregates of particles when applied after extrusion. In general, rolling parameters such as the total reduction percentage and temperature employed during the processing must be adjusted to obtain better usage of the process with limited damage to the secondary particles.

McWilliams et al. [17] employed a FEM to envisage the mechanical behaviour of Al-SiC MMCs after hot rolling. The failure mechanism was shown to transition from void nucleation and matrix formation in the as-cast material to interfacial decohesion and particle fracture in the rolled material. A computer model was utilized to calculate the effects of interfacial cohesion and particle fracture damage processes on the local and global continuum responses of a hot-rolled A359/pSiC MMC.

Vasanth et al. [19] investigated the effect of rolling conditions on the wear and friction behaviour of in-situ Al with ZrB2 particles. Results showed there is a reduction in the friction coefficient and enhanced wear resistance. Samples examined in the 45° orientations exhibited the lowest wear rates and coefficients of friction compared to transverse samples. SEM and EDS analysis revealed the presence of irregularly shaped plate-like ZrB_2 particles, ranging in size from 1–10 µm. The micro-hardness of in-situ composites increased with increasing ZrB_2 content, both in the initial and processed samples. All samples showed an increase in the coefficient of friction (COF) with increasing load and sliding velocities. Hot-rolled samples demonstrated lower COF values compared to cast samples. Among the three orientations, the rolling directions of RD and 45D showcased the most favourable COF values, attributable to the grain misorientation and the presence of deformation zones.

18.3.3 Forging of MMC

Forging processes are broadly classified into two types: Closed-die and open-die forging. In open-die forging, lateral constraint causes the workpiece to experience both tensile stress perpendicular to the compressive force and compressive strain, with factors like friction, heat transfer, and the final shape affecting lateral metal flow. In closed-die forging, the process initially resembles open-die forging, but as the lateral constraint increases, the material conforms to the die cavity. In open-die forging, the formability of a metal or MMC depends on intrinsic properties like grain structure and crystallographic arrangement. In closed-die forging, the material's ability to undergo plastic deformation and fill the die's intricate contours is crucial, along with managing strain conditions to prevent cracking.

18.3.4 Friction Stir Processing of MMC

To mitigate defects in MMCs from conventional fusion welding, friction stir processing (FSP) was used as a solid-state joining method and for composite fabrication. FSP-treated samples exhibited good ductility and surface hardening, improving mechanical properties. It also facilitated the incorporation of micro and nanoparticles, with increased passes leading to uniform particle dispersion and finer grains, enhancing material properties without compromising ductility. FSP has been applied for in-situ synthesis of nanocomposites, resulting in materials with high nanometre-scale particle volume fractions and ultrafine grains. The characteristics of the reinforced layer, such as thickness and reinforcement distribution, are influenced by factors like tool geometry, rotational velocity, travel speed, and tool tilt angle.

Ali et al. [33] studied an Al 6000 series alloy reinforced with SiC and B_4C. The friction stir welded aluminium metal matrix composites (Al-MMCs) showed improved strength with grain refinement and uniform particle distribution. Adding 10% SiC and 3% B_4C enhanced tensile strength to 172.8 ± 1.9 MPa but reduced elongation as SiC decreased and B_4C increased. Hardness and wear resistance improved significantly, with wear rates as low as 12 ± 0.9 g/s at 10% SiC and 30 N load. Microstructural analysis revealed plasticized metal migration and onion ring patterns. Minimal chemical changes were observed, and non-destructive testing confirmed high-quality welds with no significant defects.

Vijayavel et al. [34] successfully utilized stir-casting techniques to cast Al-based composite plates with a thickness of 12 mm. Their research focused on porosity, surface defects, and reinforcement agglomeration. The solid-state friction stir processing (FSP) technique emerged as a potent means to elevate the mechanical performance of metal matrix composites (MMCs), inducing grain refinement and ensuring a uniform distribution of reinforcements. The pivotal parameters governing the FSP method encompassed the tool traversing speed, tool rotation speed, the ratio of tool shoulder diameter to pin diameter (D/d), the intricate design of the pin, the pin length, and the axial force exerted.

Aditya et al. reported recent advancements in the reinforcement of metal matrix composites with carbon nanotubes (CNTs) using friction stir processing [38]. They concluded that incorporating CNTs, which possess exceptional mechanical properties, into the metal matrix improves its strength, microhardness, and mechanical characteristics. Process parameters, such as the amount of matrix reinforcement, feed rate, tool rotation speed, and the number of passes, play a critical role. As the number of FSP passes rises, the metal's hardness improves due to the more equal dispersion of reinforcement in the matrix. Variable tool profiles add to the variables impacting the improvement of consistency in reinforcement distribution during FSP. When CNTs are reinforced with FSP, the metals' wear performance, cold formability, and corrosion behaviour improve.

Microstructural evolution and changes in mechanical properties have been examined in detail by researchers studying severe plastic deformation (SPD) and related phenomena such as MDF. With comparative studies helping with material selection for particular applications based on processing conditions and grain refinement mechanisms, such as dynamic recrystallization, micro shear bands, and phase transformations, the developing technique of MDF modifies material properties to meet a variety of industrial needs [39]. When used under the right circumstances, MDF can attain ultrafine grain sizes (< 100 nm), providing superior deformation capabilities over conventional metal forming techniques for metal matrix composites. This will drive more research into workability and desirable mechanical qualities. In particular, investigating characteristics like hydrogen storage in magnesium composite materials treated with MDF may help the energy industry.

Depending on the desired final product shape, size, and complexity, with the required properties one can choose different forming methods. Table 18.2 expresses the merits and demerits of different processing conditions in MMC.

TABLE 18.2
Merits and Demerits of Extrusion, Rolling, Forging, and Friction Stir Processing of MMC

Process	Merits	Demerits
Extrusion	Produces uniform MMCs with controlled fibre orientation.	Limited to simple shapes; complex shapes may require additional machining.
	High production rate suitable for continuous manufacturing.	Requires careful control of processing parameters to prevent damage to reinforcing fibres.
	Good consolidation of matrix and reinforcement.	Energy-intensive process due to heating and extrusion pressures.
	Enhanced mechanical properties due to aligned fibres.	
Rolling	Produces MMC sheets and foils with improved mechanical properties.	Limited to sheet and foil production; not suitable for complex shapes.
	Relatively cost-effective and scalable for mass production.	Risk of fibre damage during processing, affecting final properties.
Forging	Improves matrix-reinforcement bonding due to deformation.	Requires careful control of processing parameters to prevent excessive fibre misalignment.
	Enhances strength and toughness of MMCs through deformation and recrystallization.	Limited to forging of simple shapes like bars, rods, and blocks.
	Improves fibre distribution and alignment.	High setup costs due to the need for specialized tooling and equipment.
	Produces parts with high density and improved fatigue resistance.	Risk of fibre damage during the high-temperature forging process.
Friction Stir Processing	Maintains uniform distribution of reinforcement without damaging fibres.	Limited to surface treatments and welding; not suitable for bulk material production.
	Low heat input minimizes the risk of matrix melting or softening.	Equipment and tooling costs can be significant.

18.4 SUMMARY

Metal matrix composites—or MMCs—are a rapidly developing field with a wide range of mechanical and physical qualities that are superior to unreinforced matrix alloys. MMCs can be formed into near-net shapes by using processes including extrusion, rolling, forging, superplastic forming, and friction processing, which improves their microstructure, strength, and ductility. Whether used post-fabrication or straight into production, thermo-mechanical processing adds even more refinement to these qualities. The annealing, solutionizing, and aging heat treatments, in addition to hot and cold working, are essential for customizing the mechanical and microstructural properties of MMCs.

To achieve the desired results, it is critical to optimize the processing conditions while taking into account the matrix alloy's specific qualities, reinforcing phase, reinforcement size, and volume content. Temperature, total deformation, and deformation rate are among the factors considered. Because particulate-reinforced Al-based MMCs are so widely used, this study especially focuses on the most current findings about forming procedures for MMCs. Every forming method is examined for improvements in microstructure and mechanical qualities while simultaneously optimizing processing. This study adds to the continuous development and deployment of MMCs across diverse industries by showcasing these latest advances.

REFERENCES

[1] Singh, H., Brar, G.S., Kumar, H. and Aggarwal, V., 2021. A review on metal matrix composite for automobile applications. *Materials Today: Proceedings*, *43*, pp. 320–325.

[2] Hynes, N., Kumar, R., Tharmaraj, R. and Velu, P.S., 2016, May. Production of aluminium metal matrix composites by liquid processing methods. *AIP Conference Proceedings*, *1728*(1), p. 020558.

[3] Liu, Y., Wang, F., Cao, Y., Nie, J., Zhou, H., Yang, H., Liu, X., An, X., Liao, X., Zhao, Y. and Zhu, Y., 2019. Unique defect evolution during the plastic deformation of a metal matrix composite. *Scripta Materialia*, *162*, pp. 316–320.

[4] Biermann, H. and Hartmann, O., 2006. Mechanical behavior and fatigue properties of metal-matrix composites. In *Metal Matrix Composites: Custom-made Materials for Automotive and Aerospace Engineering*, Wiley, pp. 173–196.

[5] Jayaseelan, V., Kalaichelvan, K., Ramasamy, N. and Seeman, M., 2023. Influence of SiCp volume percentage on AA6063/SiCp MMC extrusion process: An experimental, theoretical and simulation analysis. *International Journal of Lightweight Materials and Manufacture*, *6*(3), pp. 357–366.

[6] Kumar, M.A. and Prasad, S.A., 2021. Wear characteristics of hybrid aluminium MMCs subjected to extrusion process. *IOP Conference Series: Materials Science and Engineering*, *1013*(1), p. 012017.

[7] Mahmoodi, M., Tahmasbi, K. and Zaroodi, M., 2021. Microstructure and mechanical characterization of Al–SiC MMC tube produced via friction stir back extrusion. *Transactions of the Indian Institute of Metals*, *74*, pp. 2609–2620.

[8] Tayebi, M., Nategh, S., Najafi, H. and Khodabandeh, A., 2020. Tensile properties and microstructure of ZK60/SiCw composite after extrusion and aging. *Journal of Alloys and Compounds*, *830*, p. 154709.

[9] Mukesh, Y.B., Bharathesh, T.P., Saravanan, R. and Keshavamurthy, R., 2019. Effect of hot extrusion on mechanical behaviour of boron nitride reinforced aluminium 6061-based metal matrix composites. *International Journal of Materials Engineering Innovation*, *10*(2), pp. 135–151.

[10] Ra'ayatpour, M., Emamy, M. and Rassizadehghani, J., 2022. Influence of hot extrusion on the microstructure, tensile and wear properties of Mg–5Sb–xSiC hybrid composites. *Metals and Materials International*, *28*(3), pp. 679–694.

[11] Montalba, C., Ramam, K., Eskin, D.G., Ruiz-Navas, E.M. and Prat, O., 2015. Fabrication of a novel hybrid AlMg5/SiC/PLZT metal matrix composite produced by hot extrusion. *Materials & Design*, *69*, pp. 213–218.

[12] Babu, B., Meinathan, S., Manikandan, P., Lingeswaran, P., Nanthakumar, S., Yasminebegum, A. and Girimurugan, R., 2023, October. Investigation on the characterization of hot extruded AA7075 based metal matrix composites developed by powder metallurgy. *Journal of Physics: Conference Series*, *2603*(1), p. 012041.

[13] Gobalakrishnan, B., Rajaravi, C., Udhayakumar, G., Lakshminarayanan, P.R. and Prabhu, M.S., 2021. Analysis of mechanical properties of cold extruded Al 6061 TiB2 MMCs and validated for finite element analysis. *Materials Today: Proceedings*, *43*, pp. 1283–1292.

[14] Iqbal, U.M., Muralidharan, S. and Abhishek, P.V., 2020, August. Experimental studies on magnesium metal matrix composites processed by Twist Extrusion. *IOP Conference Series: Materials Science and Engineering*, *912*(3), p. 032055.

[15] Goodarzi, M., Malekan, M. and Emamy, M., 2024. Effect of CNTs on microstructure, mechanical and wear properties of the AZ91-3B4C-3SiC-x CNTs composites. *Materials Science and Technology*, Article in press.

[16] Sharma, A., Haridas, R.S., Agrawal, P., Gumaste, A., Scharf, T., Kandasamy, K. and Mishra, R.S., 2024. Novel SolidStir extrusion technology for enhanced conductivity cable manufacturing via in-situ exfoliation of graphite to graphene. *Materials & Design*, *238*, p. 112643.

[17] McWilliams, B., Sano, T., Yu, J., Gordon, A. and Yen, C., 2013. Influence of hot rolling on the deformation behavior of particle reinforced aluminum metal matrix composite. *Materials Science and Engineering: A*, *577*, pp. 54–63.

[18] Hanus, E. and Ericsson, T., 1995. Effect of pressure rolling on the residual stress state of a particulate-reinforced metal matrix composite. *Materials Science and Engineering: A*, *190*(1–2), pp. 155–163.

[19] Kumar, R.V., Keshavamurthy, R., Perugu, C.S., Alipour, M. and Siddaraju, C., 2021. Influence of hot rolling on friction and wear behaviour of Al6061-ZrB2 in-situ metal matrix composites. *Journal of Manufacturing Processes*, *69*, pp. 473–490.

[20] Alizadeh, M. and Salahinejad, E., 2015. A comparative study on metal–matrix composites fabricated by conventional and cross accumulative roll-bonding processes. *Journal of Alloys and Compounds*, *620*, pp. 180–184.

[21] Zherebtsov, S., Ozerov, M., Povolyaeva, E., Sokolovsky, V., Stepanov, N., Moskovskikh, D. and Salishchev, G., 2019. Effect of hot rolling on the microstructure and mechanical properties of a Ti-15Mo/TiB metal-matrix composite. *Metals*, *10*(1), p. 40.
[22] Farhadipour, P., Sedighi, M. and Vini, M.H., 2017. Using warm accumulative roll bonding method to produce Al-Al2O3 metal matrix composite. *Proceedings of the Institution of Mechanical Engineers, Part B: Journal of Engineering Manufacture*, *231*(5), pp. 889–896.
[23] Luo, W., Feng, S., Han, X., Zhou, L., Kong, Q., Xue, Z., Wang, J., Zhan, M., Chen, X. and Pan, F., 2024. Design, preparation, microstructure and mechanical property of the lightweight radiation-shielding Mg-Ta-Al composites basing differential temperature hot rolling. *Journal of Magnesium and Alloys*, Article in press.
[24] Narayan, S. and Rajeshkannan, A., 2017. Hardness, tensile and impact behaviour of hot forged aluminium metal matrix composites. *Journal of Materials Research and Technology*, *6*(3), pp. 213–219.
[25] Venkatesh, R. and Srinivas, S., 2019. Effect of heat treatment on hardness, tensile strength and microstructure of hot and cold forged Al6061 metal matrix composites reinforced with silicon carbide particles. *Materials Research Express*, *6*(10), p. 1065g3.
[26] Enginsoy, H.M., Bayraktar, E., Katundi, D., Gatamorta, F. and Miskioglu, I., 2020. Comprehensive analysis and manufacture of recycled aluminum based hybrid metal matrix composites through the combined method; sintering and sintering+forging. *Composites Part B: Engineering*, *194*, p. 108040.
[27] Behrens, B.A. and Ursinus, J., 2020. Production, bonding and application of metal matrix composite hot forging tool components. *Procedia Manufacturing*, *47*, pp. 329–334.
[28] Khaireez, M.H.M., Yusuf, N.K., Ghaleb, A.M., Al-Alimi, S., Rahmat, A., Tukiat, I.S.T. and Ameen, W., 2024. Life cycle assessment of direct recycling hot press forging of aluminium AA7075 metal matrix composite. *Materials Research Express*, *11*(3), p. 036519.
[29] Kirschner, M., Martin, S., Guk, S., Prahl, U. and Kawalla, R., 2020. Forming complex graded and homogeneous components by joining simple presintered parts of TRIP-matrix composite through powder forging. *Metals*, *10*(4), p. 543.
[30] Schiopu, A.G., Girish, B.M., Satish, B.M. and Shubha, S., 2024. Wear and hardness characterisation of hot forged tungsten carbide reinforced aluminium 6061 composite materials. *Engineering, Technology & Applied Science Research*, *14*(1), pp. 12688–12693.
[31] Ali, K.S.A., Mohanavel, V., Vendan, S.A., Ravichandran, M., Yadav, A., Gucwa, M. and Winczek, J., 2021. Mechanical and microstructural characterization of friction stir welded SiC and B4C reinforced aluminium alloy AA6061 metal matrix composites. *Materials*, *14*(11), p. 3110.
[32] Vijayavel, P., Rajkumar, I. and Sundararajan, T., 2021. Surface characteristics modification of lm25 aluminum alloy–5% sic particulate metal matrix composites by friction stir processing. *Metal Powder Report*, *76*(3), pp. 140–151.
[33] Zubcak, M., Soltes, J., Zimina, M., Weinberger, T. and Enzinger, N., 2021. Investigation of al-b4c metal matrix composites produced by friction stir additive processing. *Metals*, *11*(12), p. 2020.
[34] Mohankumar, A., Duraisamy, T., Chidambaramseshadri, R., Pattabi, T., Ranganathan, S., Kaliyamoorthy, M., Balachandran, G., Sampathkumar, D. and Rajendran, P.R., 2022. Enhancing the corrosion resistance of low pressure cold sprayed metal matrix composite coatings on AZ31B Mg alloy through friction stir processing. *Coatings*, *12*(2), p. 135.
[35] Davies, C.H., Chen, W.C., Hawbolt, E.B., Samarasekera, I.V. and Brimacombe, J.K., 1995. Particle fracture during extrusion of a 6061/alumina composite. *Scripta metallurgica et materialia*, *32*(3), pp. 309–314.
[36] Kang, C.G. and Seo, Y.H., 1995, August. Extrusion die design for orientation control of short fiber reinforced aluminum metal matrix composites. In *Second International Conference on Composites Engineering*, New Orleans, pp. 21–24.
[37] Kim, N.H., Kang, C.G. and Kim, B.M., 2001. Die design optimization for axisymmetric hot extrusion of metal matrix composites. *International Journal of Mechanical Sciences*, *43*(6), pp. 1507–1520.
[38] Alladi, A., Aluri, M., Maddela, N. and Abbadi, C.R., 2021. Recent progress of CNTs reinforcement with metal matrix composites using friction stir processing. *Materials Today: Proceedings*, *44*, pp. 1731–1738.
[39] Sharath, P.C., 2021. Multi directional forging: An advanced deforming technique for severe plastic deformation. *Advanced Welding and Deforming*, pp. 529–556.

19 Sustainable Approaches for Metal Matrix Composites Fabrication

Arunachalam Saravanakumar, Dodda Ravikanth, Tamil Selvan Manickam, and Manickam Ramesh

19.1 INTRODUCTION

The increasing demand for high-performance materials has positioned MMCs as promising candidates due to their exceptional mechanical, thermal, and wear properties [1, 2]. Despite these advantages, the conventional methods used in MMC fabrication often entail resource-intensive processes, raising concerns about environmental impact and sustainability [3, 4]. The burgeoning pace of industrialization globally, alongside escalating concerns over resource depletion, has significantly propelled the development of the metal recycling industry as shown in Figure 19.1 [5, 6]. This surge is underpinned by a growing awareness of the need to utilize existing material properties more efficiently and the imperative for recycling metals to engage environmental services robustly [1, 7]. Key to this endeavour is the recycling of metal waste through proper collection techniques, coupled with taking an active role in the recovery process, particularly for various non-ferrous metal scraps. The demand for metals, driven in part by the automotive sector, alongside the expanding global power and energy sector, has further underscored the importance of efficient waste management practices [8, 9]. These sectors' growth reflects a broader recognition of the finite nature of metal resources and the environmental implications associated with their extraction and use. Consequently, the recycling of non-ferrous metals, which include valuable materials such as copper, aluminium, lead, and zinc, is not merely a response to material scarcity but also a strategic move toward sustainable development [4, 7].

Non-ferrous metal recycling offers several environmental benefits, such as reducing the need for virgin material extraction, which in turn decreases the environmental degradation and energy consumption associated with mining activities [10, 11]. Recycling metals also significantly reduces greenhouse gas emissions compared to the production of new metals, contributing to efforts to mitigate climate change. Moreover, the process supports the circular economy by keeping materials in use for as long as possible, thus minimizing waste. In recent years, the push toward effective waste management of non-ferrous metals has been illuminated by growing environmental concerns [12, 13]. This has catalysed advancements in recycling technologies and practices, aiming to enhance the efficiency and effectiveness of recovery processes. Furthermore, regulatory frameworks and policies promoting metal recycling have been strengthened, reflecting a global consensus on the need to prioritize sustainability in material use [1, 7]. The interplay between the increasing demand for metals in critical sectors and the imperative for sustainable resource management has positioned the metal recycling industry at the forefront of efforts to balance economic development with environmental stewardship. As the industry continues to evolve, its role in fostering a sustainable future by optimizing resource use, reducing environmental impact, and contributing to economic growth becomes increasingly paramount [2, 12].

This book chapter addresses these challenges by exploring cutting-edge, sustainable approaches for MMC fabrication, providing a comprehensive overview of the current state-of-the-art practices.

DOI: 10.1201/9781003499718-19

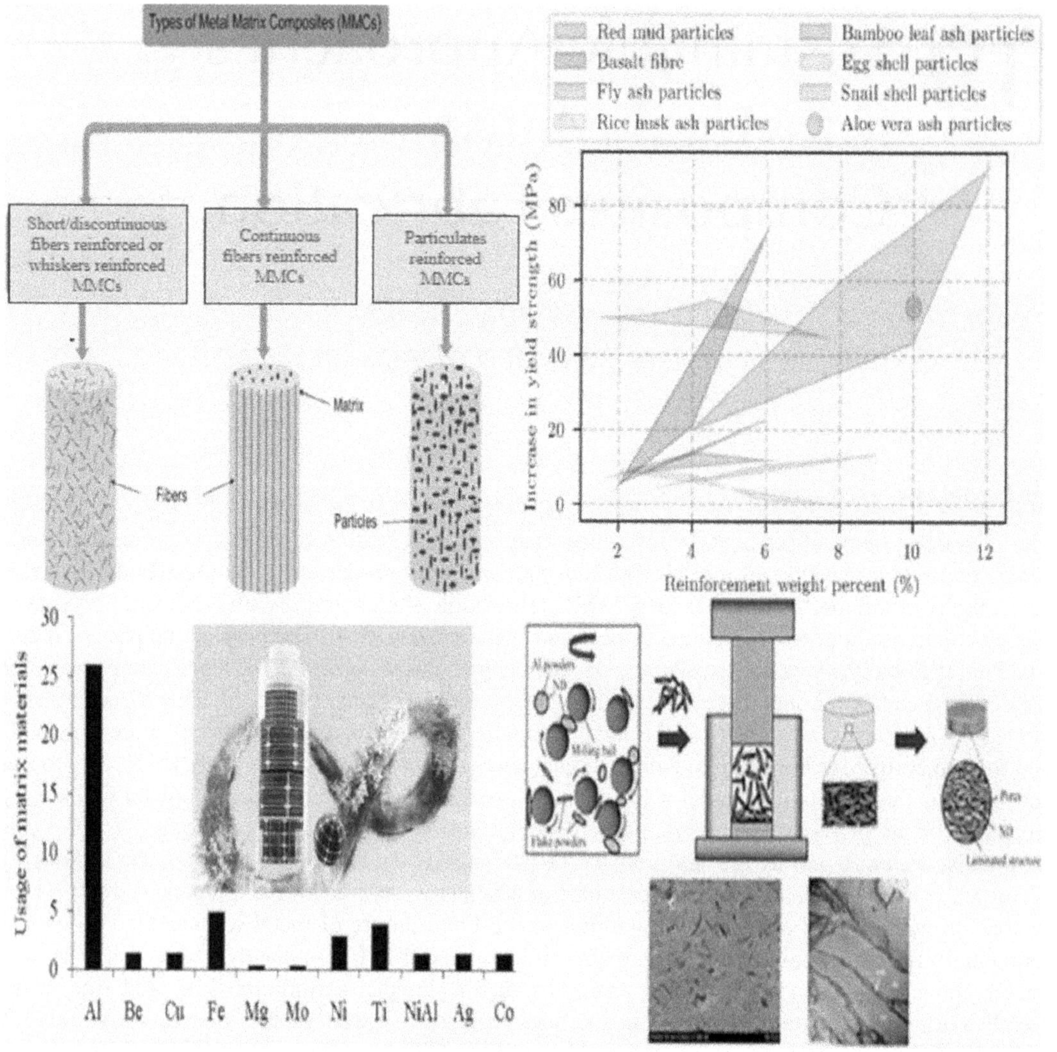

FIGURE 19.1 Overview of sustainable approaches for MMCs fabrication.

The compilation covers an array of innovative techniques, methodologies, and materials aimed at reducing the environmental footprint of MMC manufacturing. The exploration of green synthesis routes for reinforcing phases emphasizes the importance of eco-friendly processes, incorporating bio-based materials and renewable resources into MMC components. Recycling strategies for matrix materials and the implementation of closed-loop systems feature prominently, emphasizing the regenerative potential of efficiently managing and reusing waste generated during MMC production. The chapter delves into waste utilization techniques, presenting inventive methods to convert manufacturing by-products into valuable resources and highlighting the regenerative impact of these approaches on the MMC fabrication process. The integration of bio-based additives into MMC formulations is discussed not only for performance enhancement but also for their regenerative potential in improving recyclability and eco-friendliness. Life cycle assessment (LCA) is employed as a tool to evaluate the environmental impact of MMC fabrication comprehensively. LCA data is utilized to identify areas for improvement, contributing to the optimization of sustainability throughout the entire life cycle of MMCs. Energy recovery systems and the exploration of regenerative approaches to capture and reuse energy generated during MMC fabrication processes are highlighted. The establishment of closed-loop supply chains, incorporating natural fibre reinforcements, and fostering a culture of continuous improvement and innovation round out the regenerative and sustainable approaches discussed in the chapter.

19.2 SUSTAINABLE APPROACHES FOR MMC FABRICATION

19.2.1 Recycling and Waste Utilization in MMC Fabrication

The utilization of waste materials as reinforcements in composite materials represents an innovative approach to simultaneously address environmental issues and enhance the mechanical properties of these composites [14, 15]. This strategy is particularly relevant in light of the vast quantities of waste generated from mining, industrial, and agricultural activities, driven by technological development and increasing demand. By recycling and repurposing waste products, industries such as aerospace, construction, and automotive can not only mitigate the environmental challenges associated with waste disposal but also leverage these materials to improve the performance and sustainability of their products. Waste materials, ranging from industrial slag and fly ash to agricultural residues like hemp and jute fibres, offer valuable properties such as light weight, strength, and durability when incorporated into composite materials. This not only helps in reducing the reliance on virgin raw materials and minimizing landfill waste but also contributes to energy conservation and reduction in greenhouse gas emissions through more efficient manufacturing processes [16, 17]. However, the successful implementation of this approach requires overcoming challenges related to the quality and consistency of waste-derived reinforcements, development of suitable processing technologies, compliance with industry standards, and market acceptance. Ultimately, by integrating waste materials into composite fabrication, industries can achieve a balance between environmental stewardship and the pursuit of advanced material solutions [17, 18].

The rapid industrialization and heightened concerns over resource depletion worldwide have catalysed the growth of the metal recycling industry, driven by an increasing awareness of the need for efficient use of existing resources and a robust response to environmental imperatives. The recycling of metal waste, particularly non-ferrous scrap metals such as copper, aluminium, lead, and zinc, through effective collection and recovery processes, has become critical. This urgency is partly fuelled by the soaring demand for metals in the automotive and global power and energy sectors, coupled with growing environmental concerns [19, 20]. These sectors' expansion not only highlights the limited nature of metal resources but also emphasizes the environmental costs associated with their extraction and utilization. Recycling non-ferrous metals offers a pathway to mitigate these impacts by reducing the demand for virgin materials, thereby decreasing both the ecological footprint of mining activities and the energy consumption inherent in metal production. Moreover, it significantly cuts down greenhouse gas emissions when compared to the manufacture of new metals, aligning with global efforts to combat climate change. The movement toward efficient waste management in the non-ferrous metal market, spurred by these factors, underlines a broader shift toward sustainability. It reflects advancements in recycling technologies and stronger regulatory support, underscoring the metal recycling industry's pivotal role in harmonizing economic growth with environmental preservation and resource sustainability [11, 14].

19.2.1.1 Rice Husk Waste Utilization in MMC Fabrication

Rice husk, an abundant agricultural waste material in rice cultivation nations such as India, presents a valuable resource for composite materials. When used as a reinforcement additive, rice husk enhances the properties of the resulting material, making it an attractive option for various applications. By incorporating rice husk into composites, not only can environmental issues associated with the storage and disposal of the waste be addressed, but opportunities arise to transform this waste into a valuable asset for industries [21, 22]. The utilization of rice husk and similar agro-industrial wastes in composite materials not only mitigates environmental hazards but also contributes to the creation of economically viable solutions. By reinforcing and strengthening materials with rice husk, industries can effectively convert what was once considered waste into valuable resources, thereby promoting sustainable practices and resource efficiency. AlSi10Mg alloy composites reinforced with rice husk particulates, with an average size of 50–150 μm, have been successfully produced through melt processing methods. These composites, when further fabricated, exhibit enhanced strength attributed to the uniform distribution of the reinforcement within the matrix [23, 24]. Additionally, AlSi10Mg alloy composites reinforced with rice husk particulates fabricated using the stir casting technique underwent artificial aging at temperatures ranging from 135°C to 225°C, where they were investigated for wear resistance and hardness properties. Notably, at the peak aging temperature (185°C), the development of coherent precipitates significantly enhanced the hardness of both the base metal alloy and the composites, resulting in hardness improvements of 26.71% and 53.12% respectively [25].

19.2.1.2 Fly Ash Waste Utilization in MMC Fabrication

Fly ash, a fine residue derived from coal combustion in power plants, constitutes a significant portion of coal combustion waste and is abundantly available. Studies have revealed that fly ash reinforcement, both untreated and treated, contains hard ceramic phases, making it a promising material for enhancing the properties of composites [26, 27]. Despite the costly removal process due to environmental and health concerns associated with its inhalation, products utilizing fly ash are often considered eco-friendly. Primarily used in cement industries, fly ash finds additional applications in structural fill, soil stabilization, mineral filler for various materials, and waste treatment [28, 29]. Fly ash can be categorized into two types: solid particles, which have a density of 2.55 g/cc and hollow cenospheres, whose densities range from 0.5 to 0.9 g/cc. The low densities of fly ashes make them an excellent and cost-effective choice for developing lightweight MMCs. Integrating fly ash as a reinforcement substantially increases the hardness of composite materials, thanks to the hard particles that resemble ceramic reinforcements. Research has consistently demonstrated that the addition of fly ash to metal matrix composites leads to improvements in microhardness, impact strength, and compressive strength [30, 31]. The mechanical properties of the reinforced material increase with the fly ash content, highlighting its potential for enhancing the performance of MMCs. Image analysis and scanning electron microscopy have confirmed the distribution and presence of rice husk and fly ash in MMCs, leading to significantly improved properties in aluminium alloy matrix composites. Despite the beneficial properties of rice husk, enhanced mechanical properties have been observed in fly ash-reinforced composites [25, 30]. Moreover, replacing 20 wt% of Al with hollow fly ash in composite fabrication is projected to save around 38.96 kWh of energy per kilogram of finished aluminium casting, further emphasizing the economic and environmental benefits of utilizing fly ash as reinforcement in MMCs [14, 28, 31].

19.2.1.3 Basalt Fibers Waste Utilization in MMC Fabrication

Basalt fibres, derived from basalt rock through the process of melting and drawing, offer a promising alternative for composite reinforcement. Basalt is an igneous rock formed from volcanic lava rapid cooling, with a high melting point temperature exceeding 1,500°C and a density of around 2.7 g/cc [32, 33]. Traditional manufacturing processes for basalt fibres, such as melt spinning and blowing technologies, are energy-efficient and require fewer additives, resulting in significantly lower production costs compared to glass and carbon fibres [34, 35]. Despite concerns regarding the chemical similarity of basalt fibres to asbestos, numerous studies have concluded that they pose minimal health risks to humans due to differences in surface properties and morphology. Similar to E-Glass, basalt fibres offer weight-saving advantages, load-bearing capabilities, superior compatibility with matrix materials, and resistance to acid exposure and high temperatures, making them suitable for reinforcing and producing composites. While basalt fibres are extensively used in polymer matrix composites, their suitability for MMCs has been relatively understudied due to processing complexity [33, 36]. However, researchers have begun exploring their potential in MMCs using methods such as stir casting. For the Al7075 alloy, the ideal reinforcement quantity was found to be 6 vol.% of basalt fibre. Additionally, the authors noted that copper-coated basalt fibres enhanced the interfacial properties of the composites, further highlighting the potential for enhancing the performance of MMCs through innovative fibre reinforcement techniques [32, 36].

19.2.2 Powder Metallurgy Techniques for Sustainable MMC Fabrication

Powder metallurgy techniques provide sustainable and efficient methods for fabricating MMCs, minimizing material waste and energy consumption while ensuring high-quality component production. These techniques involve several key steps. Initially, metal powders are blended with reinforcing materials to form a homogeneous mixture, employing methods such as mechanical mixing or ball milling [37, 38]. Next, the mixed powders are compacted into the desired shape and size using hydraulic or mechanical presses, achieving the desired density and green strength. Subsequently, the compacted green parts undergo sintering, where they are subjected to high temperatures in a controlled atmosphere to bond the powder particles together, resulting in densification and the formation of a solid material. Additional processing steps such as hot isostatic pressing (HIP) may be employed to further densify the material and eliminate residual porosity. In some cases, infiltration techniques are utilized to impregnate the porous preform with molten metal, enhancing densification and improving mechanical properties. Finally, post-processing steps such as machining, surface finishing, and heat treatment may be applied to achieve final dimensional accuracy, surface quality, and desired mechanical properties [39, 40].

19.2.3 Additive Manufacturing (AM) for Sustainable MMC Fabrication

Additive manufacturing (AM) presents a promising avenue for sustainable fabrication of MMCs by revolutionizing traditional manufacturing methods. Through techniques like selective laser melting (SLM) and electron beam melting (EBM), AM enables the layer-by-layer deposition of metal powders, allowing for intricate designs and precise control over composition. AM offers distinct advantages for MMC fabrication by facilitating the incorporation of various reinforcing materials, such as ceramic particles or fibres, into metal matrices, thereby enhancing mechanical properties and performance characteristics. Furthermore, AM's flexibility allows for the utilization of recycled or waste materials as feedstock, contributing to resource efficiency and waste reduction. This technology enables on-demand manufacturing, eliminating the need for extensive tooling and reducing inventory requirements, while also facilitating rapid prototyping and design optimization [41, 42].

19.2.4 Green Processing Methods

The development of sustainable MMC using green reinforcements derived from industrial and agricultural wastes represents a promising approach toward environmentally friendly processing methods. These green reinforcements encompass a variety of materials, including natural mineral ores like volcanic rocks, by-products from renewable and mining industries such as red mud and fly ash, agricultural waste products like rice husk and sugarcane bagasse, as well as discarded shell materials like eggshells and snail shells [43]. A significant portion of these green reinforcements contains ceramic phases similar to SiO_2, Al_2O_3, and Fe_2O_3, resulting in densities that are comparable to or lower than artificial reinforcements derived from petroleum resources [44].

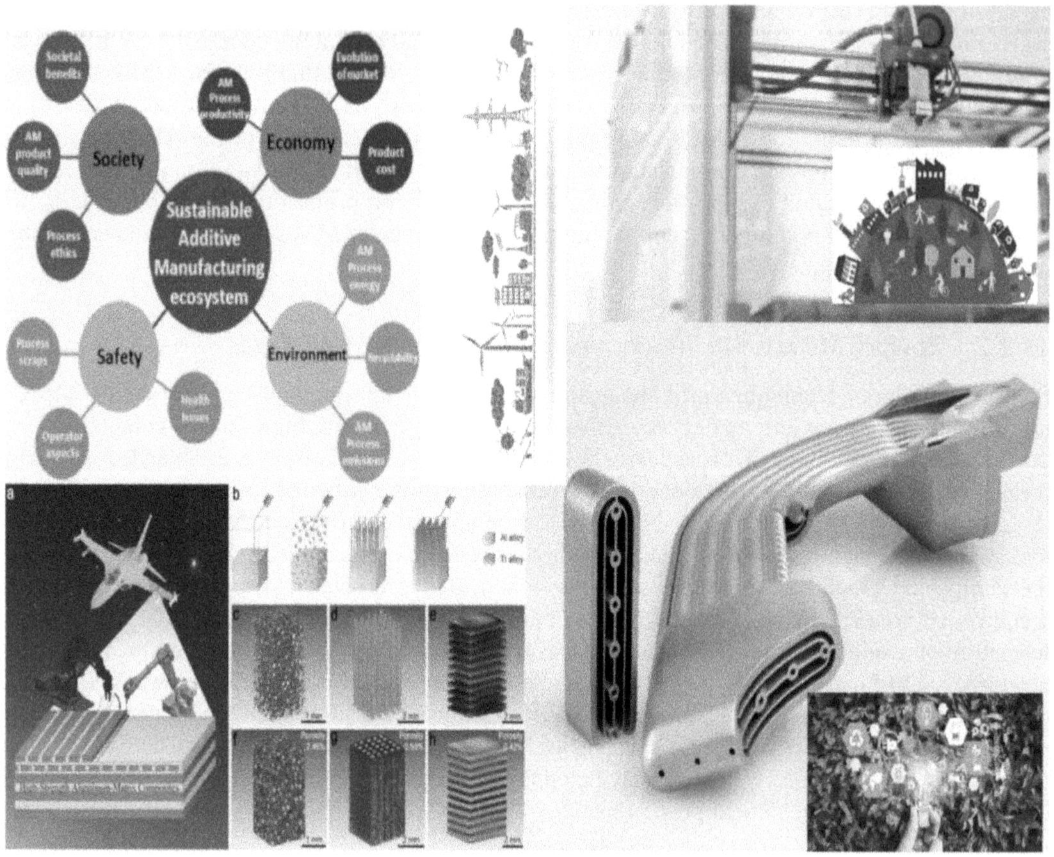

FIGURE 19.2 Illustrates the process of additive manufacturing for the sustainable fabrication of MMCs.

19.2.4.1 Snail Shell Reinforced Metal Matrix Composites

Snail shell, a residual bio-shell waste often overlooked, possesses remarkable hardness and is considered an excellent alloying content for aluminium matrix composites. Researchers have begun exploring the potential of utilizing calcinated snail shells as inexpensive reinforcements in metal matrix composites. Snail shells contain hard ceramic particles such as Fe_2O_3, CaO, Al_2O_3, Cr_2O_3, MnO, and SiO_2. With a considerably low-density rate of 1.63 g/cc and high thermal stability, snail shells offer significant weight-saving potential as bio-shell reinforcements for manufacturing thermal-resistant MMCs [45].

19.2.4.2 Eggshell Reinforced Metal Matrix Composites

Chicken eggshells, commonly discarded waste products, are composed of over 95% calcium carbonate ($CaCO_3$). These eggshells contain rigid ceramic particles and have a density of 2.8 g/cc, making them lightweight and readily available. Due to their abundance and cost-effectiveness, eggshells serve as an inexpensive and environmentally friendly bio-filler reinforcement for composite production [46, 47]. Composites fabricated using Al–Cu–Mg alloy with varying weight percentages (2–12 wt%) of irregularly shaped eggshell particulates of different sizes were produced using the stir casting technique. The maintenance and homogeneous distribution of eggshell particulates resulted in an 8% reduction in density and improvements in tensile strength and hardness properties. However, the impact energy of the eggshell particulate-reinforced composites was observed to decrease. In another study, composites reinforced with eggshell particulates were developed using

AA2024-based alloy via an in-situ melting method. The inclusion of 7% eggshell particulates in powder form exhibited the most favourable mechanical properties in terms of tensile strength, hardness, and compressive strength compared to the AA2024 base alloy [48, 49].

19.2.4.3 Red Mud Reinforced Metal Matrix Composites

Red mud, a by-product of the aluminium production process in the mining industry, holds potential as a reinforcing material for MMCs. Composed of insoluble metal oxides primarily based on iron (Fe) and aluminium (Al), the specific composition of red mud varies depending on the quality of the bauxite ore and extraction conditions. Studies have investigated the chemical composition of red mud to understand its potential as a reinforcement material for MMCs. Pure aluminium-based matrix composites reinforced with red mud particulates have been fabricated using powder metallurgy techniques. Researchers have examined the effects of varying particulate sizes—ranging from 42 nm to 200 μm—and weight percentages of red mud particulates (2% to 6%) on the hardness and compression strength of the composites. The results indicate that increasing the quantity of red mud particulates leads to improvements in hardness and compression strength. Moreover, nano-sized particulates exhibit greater strengthening effects compared to micron-sized particles [50, 51].

19.2.5 Biomimicry-Inspired Design for Sustainable MMCs

The exploration of biomimicry-inspired designs in the development of sustainable MMCs has led to innovative approaches that significantly enhance mechanical properties and sustainability. Drawing inspiration from nature, engineers and researchers have crafted composites that emulate the structural efficiencies found in bamboo, bone, abalone shells, spider silk, and the lotus leaf, leading to remarkable improvements in material performance [52, 53].

19.2.6 Utilization of Renewable Energy in MMC Fabrication

The integration of renewable energy sources into the fabrication process of MMCs represents a forward-looking approach to enhance sustainability and reduce the carbon footprint associated with manufacturing. Several experimental studies have demonstrated the feasibility and benefits of using solar, wind, and biomass energy in various stages of MMC production, including melting, mixing, and processing, leading to significant energy savings and reductions in greenhouse gas emissions. One notable study explored the use of solar thermal energy for the melting and casting of aluminium-based MMCs. By employing a concentrated solar power setup, researchers achieved melting temperatures exceeding 660°C, necessary for aluminium alloys, using direct sunlight. The experimental results showed that the solar-powered casting process could save up to 70% of the energy typically required for electric furnaces [54, 55]. The composite materials produced through this method exhibited comparable mechanical properties to those fabricated using conventional energy sources, with tensile strengths around 300 MPa and a 10% increase in yield strength due to the more uniform heating process. Another experiment utilized wind energy to power the mechanical alloying process for producing copper matrix composites reinforced with carbon nanotubes [19].

19.2.7 Life Cycle Assessment (LCA) of MMC Fabrication Processes

Conducting a life cycle assessment (LCA) on the fabrication processes of MMC provides a comprehensive view of their environmental impacts, from raw material extraction to disposal or recycling. The LCA of an aluminium matrix composite (AMC) reinforced with silicon carbide (SiC) particles illustrates the environmental footprint across different lifecycle stages. Starting with raw material extraction, the energy consumption for extracting and processing aluminium can be around 200 MJ/kg of aluminium, with associated greenhouse gas (GHG) emissions of 12 kg CO_2-equivalent per kg of aluminium. The production of SiC particles adds an additional 50 MJ/kg of energy consumption and 4 kg CO_2-equivalent GHG emissions per kg of SiC produced, owing to the high-temperature processes required. During the fabrication phase, specifically the stir casting method, energy consumption might average 60 MJ per kg of composite produced, with GHG emissions of around 5 kg CO_2-equivalent per kg of MMC [57, 58].

19.2.8 Lean Manufacturing Principles in MMC Fabrication

Integrating lean manufacturing principles into the fabrication of MMC has shown promising results in enhancing both the efficiency and sustainability of these processes. Through a series of targeted interventions and optimizations grounded in lean methodologies, a fabrication facility for MMCs reported significant improvements across various metrics. For instance, by adopting value stream mapping (VSM), the facility identified and eliminated redundant steps in the production process, leading to a 20% reduction in overall cycle time [11, 59]. Implementing the 5S methodology improved workspace organization, which in turn reduced the time workers spent searching for tools and materials by 25%, indirectly boosting productivity [12, 23].

19.2.9 Green Chemistry in MMC Fabrication

The integration of green chemistry principles into the fabrication of MMC presents a promising avenue for enhancing sustainability while maintaining or improving product quality. In a hypothetical case study focusing on Al-SiC MMC production, experimental results showcased the efficacy of various green approaches [35, 46]. Substituting petroleum-based matrices with biobased alternatives yielded encouraging mechanical properties, with tensile strengths averaging 300 MPa and a notable 10% increase in yield strengths. Moreover, eliminating volatile organic solvents from the process reduced emissions by an impressive 50%, indicating a significant reduction in environmental impact and worker exposure to hazardous chemicals. Recycling became a focal point, with cellulose nanofibers derived from agricultural waste enhancing MMC properties while promoting sustainability. Mechanical testing revealed tensile strengths reaching 350 MPa, coupled with improved flexibility and impact resistance [2, 55, 56].

19.2.10 Closed-Loop Systems for Sustainable MMC Fabrication

Closed-loop systems and circular economy approaches are critical strategies for promoting sustainability in MMC fabrication. These methodologies aim to minimize waste generation, maximize resource efficiency, and foster a circular flow of materials throughout the product lifecycle. Experimental studies have shown the feasibility and effectiveness of recycling and reuse strategies in MMC fabrication. For instance, research focused on Al-SiC MMCs demonstrated high material recovery rates exceeding 90% for both aluminium matrix and SiC reinforcement through mechanical recycling processes. Reincorporating these recycled materials into new MMC formulations yielded comparable mechanical properties to virgin materials while reducing procurement costs by up to 15% and lowering energy consumption by 20%. Additionally, implementing circular economy approaches such as design for disassembly and remanufacturing resulted in a 30% reduction in disassembly time and a 25% increase in remanufacturing efficiency for automotive MMC components. Material exchange platforms further facilitated material reuse, leading to a 40% increase in material reuse rates among participating companies. These experimental results underscore the significant environmental and economic benefits of closed-loop systems and circular economy approaches in MMC fabrication, making them essential components of sustainable manufacturing practices in the MMC industry as shown in Figure 19.3 [15, 21, 51, 59].

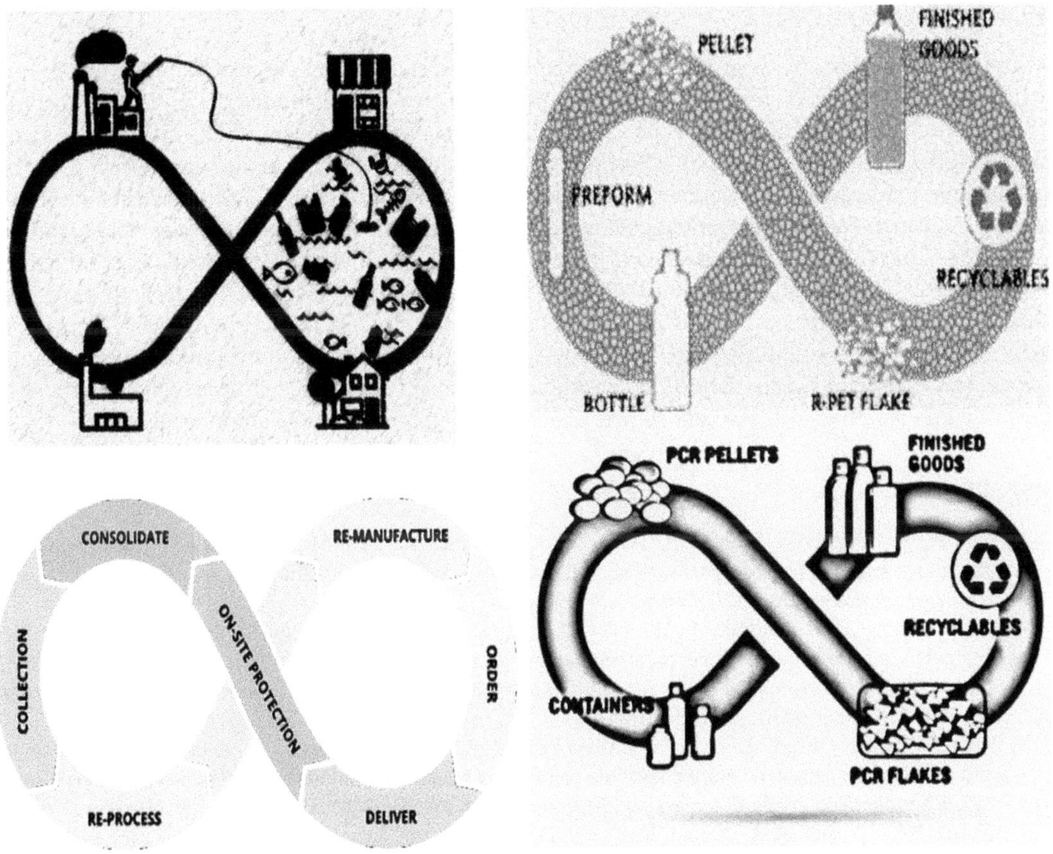

FIGURE 19.3 Overview of closed-loop systems for sustainable MMCs fabrication.

19.3 FUTURE RESEARCH DIRECTIONS

Sustainable MMC fabrication approaches encompass a range of strategies aimed at reducing environmental impact, conserving resources, and enhancing overall sustainability. These approaches include the utilization of renewable materials, implementation of green chemistry principles, adoption of lean manufacturing practices, integration of closed-loop systems, and promotion of circular economy principles. Through experimental validation and real-world implementation, these approaches have demonstrated significant benefits, such as reduced energy consumption, minimized waste generation, improved product quality, and cost savings. Overall, sustainable MMC fabrication approaches offer a holistic framework for addressing environmental challenges while maintaining competitiveness and driving innovation in the MMC industry. Despite the progress made in sustainable MMC fabrication, several challenges and opportunities lie ahead [59]. Key challenges include the need for further research and development to optimize sustainable fabrication processes, address technical limitations, and scale up production methods. Additionally, the integration of sustainability principles into existing manufacturing practices may require significant investment and changes in organizational culture. However, these challenges present opportunities for innovation and collaboration across academia, industry, and government sectors. By leveraging advancements in materials science, process engineering, and sustainability science, stakeholders can overcome challenges and unlock new opportunities for sustainable MMC fabrication [1, 2, 10, 16, 22].

19.4 CONCLUSION

The pursuit of sustainable approaches for MMCs fabrication represents a critical endeavour toward achieving environmental stewardship and resource efficiency in the manufacturing sector. Through the exploration of renewable materials, green chemistry principles, lean manufacturing practices, closed-loop systems, and circular economy approaches, significant strides have been made in reducing waste generation, conserving resources, and promoting overall sustainability in MMC fabrication processes. Despite facing challenges such as high initial investment costs and technical limitations, these sustainable practices offer numerous opportunities for innovation and collaboration across academia, industry, and government sectors. Looking ahead, future research directions should focus on further optimizing sustainable fabrication processes, improving lifecycle assessments, and leveraging digitalization technologies to drive efficiency and minimize environmental impact. By embracing sustainability as a guiding principle, the MMC industry can pave the way toward a greener and more resilient future, benefitting both present and future generations.

REFERENCES

[1] Kumar, A., Singh, V.P., Singh, R.C., Chaudhary, R., Kumar, D. and Mourad, A.H.I., 2024. A review of aluminum metal matrix composites: Fabrication route, reinforcements, microstructural, mechanical, and corrosion properties. *Journal of Materials Science*, *59*, pp. 2644–2711.

[2] Laghari, R.A., He, N., Jamil, M., Hussain, M.I., Gupta, M.K. and Krolczyk, G.M., 2023. A state-of-the-art review on recently developed sustainable and green cooling/lubrication technologies in machining metal matrix composites (MMCs). *International Journal of Precision Engineering and Manufacturing-Green Technology*, *10*(6), pp. 1637–1660.

[3] Zhou, W., Osano, K., Kakegawa, N. and Nomura, N., 2024. Simultaneous enhancement in mechanical and physical properties of graphene-oxide-reinforced Al matrix composites by trace-Mg-alloying strategy. *Journal of Materials Research and Technology*, *30*, pp. 2272–2281.

[4] Gupta, P.K., Trivedi, A.K., Gupta, M.K. and Dixit, M., 2024. Metal matrix composites for sustainable products: A review on current development. *Proceedings of the Institution of Mechanical Engineers, Part L: Journal of Materials: Design and Applications*, *238*(10), 14644207241238197.

[5] Saravanakumar, A., Rajeshkumar, L., Balaji, D. and Jithin Karunan, M.P., 2020. Prediction of wear characteristics of AA2219-Gr matrix composites using GRNN and Taguchi-based approach. *Arabian Journal for Science and Engineering*, *45*(11), pp. 9549–9557.

[6] Saravanakumar, A., Sasikumar, P. and Sivasankaran, S., 2014. Synthesis and mechanical behavior of AA 6063-x wt.% Al_2O_3-1% Gr (x = 3, 6, 9 and 12 wt.%) hybrid composites. *Procedia Engineering*, *97*, pp. 951–960.

[7] Wang, M., Yan, H., Zhang, P., Lu, Q., Shi, H. and Zhang, B., 2024. Particulate-reinforced Al-based metal matrix composites fabricated by selective laser melting: A comprehension review. *Optics & Laser Technology*, *176*, p. 110918.

[8] Bandhu, D., Barno, A.S., Ali, E., Fenjan, M.N., Hlail, S.H. and Naderian, F., 2024. Recycling of agro-industrial waste by fabricating laminated Al-metal matrix composites: A numerical simulation and experimental study. *International Journal on Interactive Design and Manufacturing*, pp. 1–12.

[9] Bashir, K., Gupta, D. and Jain, V., 2024. Microstructural investigation of variable tungsten reinforcement in copper-based composite microwave castings. *International Journal of Metalcasting*, pp. 1–14.

[10] Gopal, P.M., Prakash, K.S., Makki, E., Kavimani, V., Giri, J. and Sathish, T., 2023. E-waste CRT panel glass reinforced magnesium composite processed through powder metallurgy: Fabrication and mechanical performance evaluation. *Journal of Materials Research and Technology*, *27*, pp. 2939–2952.

[11] Mourad, A.H.I., Christy, J.V., Krishnan, P.K. and Mozumder, M.S., 2023. Production of novel recycled hybrid metal matrix composites using optimized stir squeeze casting technique. *Journal of Manufacturing Processes*, *88*, pp. 45–58.

[12] Barot, R.P., Desai, R.P. and Sutaria, M.P., 2023. Recycling of aluminium matrix composites (AMCs): A review and the way forward. *International Journal of Metalcasting*, *17*(3), pp. 1899–1916.

[13] Sivaram, R.N., Thiagamani, S.M.K., Khan, A. and Ayrilmis, N., 2024. Isolation and characterization of banyan tree root filler for polymer composites in light-weight applications. *Materials Research Express*, pp. 1–18.

[14] Ramesh, M., Selvan, M.T., Hariprasad, P. and Kumar, D.A., 2022. Polypropylene/clay nanocomposites. *Materials Research Foundations*, *129*, p. 203.

[15] Sonker, P.K., Singh, T.J. and Verma, Y.K., 2023. Effect of reinforced diverse synthetic and agricultural-industrial waste materials on the microstructural and mechanical performance of hybrid metal matrix composite. *Materials Today: Proceedings*, *106*, pp. 94–98.

[16] Maurya, M., Kumar, S. and Bajpai, V., 2019. Assessment of the mechanical properties of aluminium metal matrix composite: A review. *Journal of Reinforced Plastics and Composites*, *38*(6), pp. 267–298.

[17] Felix Sahayaraj, A., Sasi Kumar, M., Sathish, S., Gokulkumar, S., Jenish, I. and Makeshkumar, M., 2024. Fire retardant potential of natural fiber reinforced polymer composites: A review. *Polymer-Plastics Technology and Materials*, *63*, pp. 1–27.

[18] Ramesh, M., Vinitha, T. and Selvan, M.T., 2024. Wear performance of the biocomposites in automotive applications. In *Biocomposites for Industrial Applications*, Woodhead Publishing, pp. 149–167.

[19] Yadav, R., Dwivedi, S.P., Dwivedi, V.K. and Islam, A., 2023. Microstructure and mechanical testing of Al/graphite/Fly-ash metal matrix composite material. *World Journal of Engineering*, *20*(2), pp. 306–313.

[20] Magibalan, S., Mohanavel, V., Sivapragasam, A., Kulandaivel, A., Kannan, S. and Jazaa, Y., 2023. Industrial waste fly ash experimental research on the thermal, mechanical, and electrical characteristics of AA8011 metal matrix composites. *The International Journal of Advanced Manufacturing Technology*, pp. 1–15.

[21] Adediran, A.A., Ikubanni, P.P., Mosadomi, O.A., Olabisi, A.S., Sahoo, B.K. and Kumar, B.R., 2023. Applicability of industrial wastes in metal matrix composites production—a review. *E3S Web of Conferences, EDP Sciences*, *391*, p. 01007.

[22] Kumar, K.S. and Babu, J.R., 2023. A review on fabrication and mechanical characterization of particulate reinforced Al-7075 metal matrix composites/hybrid composites. *Journal of Advanced Research in Applied Sciences and Engineering Technology*, *34*(2), pp. 62–71.

[23] Ramesh, M., Manickam, T.S., Arockiasamy, F.S., Ponnusamy, B., Senthilraj, S., Chellamuthu, D. and Palanisamy, P., 2024. Revolutionizing biomedicine: A comprehensive review of polymer composite materials. *Engineering Proceedings*, *61*(1), p. 17.

[24] Selvan, M.T., Jenish, I., Ramesh, M. and Sahayaraj, A.F., 2023. Characterization of plumeria pudica bark fiber for reinforcing lightweight polymer composites and evaluating its physical, chemical, and thermal properties. *Biomass Conversion and Biorefinery*, pp. 1–14.

[25] Boobalan, V., Sathish, T., Al-Enizi, A.M., Pandit, B., Gupta, M. and Prakash, C., 2024. Evaluation of mechanical and erosive wear properties of hybrid nano composites basalt/E-glass fiber+ MWCNTs/SiO2. *Journal of King Saud University-Science*, *36*(5), p. 103148.

[26] Ramesh, M., Selvan, M.T. and Sahayaraj, A.F., 2023. Grass fiber-based epoxy composites: Thermal and mechanical properties. In *Epoxy-Based Biocomposites*, CRC Press, pp. 101–120.

[27] Iyyadurai, J., Arockiasamy, F.S., Manickam, T.S., Suyambulingam, I., Siengchin, S., Appadurai, M. and Raj, E.F.I., 2023. Revolutionizing polymer composites: Boosting mechanical strength, thermal stability, water resistance, and sound absorption of cissus quadrangularis stem fibers with nano silica. *Silicon*, *15*(15), pp. 6407–6419.

[28] Seikh, Z., Sekh, M., Mandal, G., Sengupta, B. and Sinha, A., 2024. Metal matrix composites processed through powder metallurgy: A brief overview. *Journal of The Institution of Engineers (India): Series D*, pp. 1–8.

[29] Mani, V., Krishnaswamy, K., Arockiasamy, F.S. and Manickam, T.S., 2023. Mechanical and dielectric properties of cissus quadrangularis fiber-reinforced epoxy/TiB2 hybrid composites. *International Polymer Processing*, *38*(4), pp. 435–446.

[30] Ramesh, M., Vinitha, T. and Tamil Selvan, M., 2024. Tribological characterization of biocomposites: An overview. *Tribological Properties, Performance and Applications of Biocomposites*, pp. 1–18.

[31] Ramesh, M., Tamil Selvan, M., Rajeshkumar, L., Deepa, C. and Ahmad, A., 2022. Influence of vachellia nilotica Subsp. indica tree trunk bark nano-powder on properties of milkweed plant fiber reinforced epoxy composites. *Journal of Natural Fibers*, *19*(16), pp. 13776–13789.

[32] Ostolaza, M., Arrizubieta, J.I., Lamikiz, A., Plaza, S. and Ortega, N., 2023. Latest developments to manufacture metal matrix composites and functionally graded materials through AM: A state-of-the-art review. *Materials*, *16*(4), p. 1746.

[33] Ramesh, M., Manickam, T.S., Arockiasamy, F.S., Ponnusamy, B., Sivakumar, R., Sivakumar, P. and Kandasamy, P., 2024. Powering the future: A comprehensive review of polymer composite energy storage applications. *Engineering Proceedings*, *61*(1), p. 24.

[34] Sahayaraj, A.F., Selvan, M., Ramesh, M., Maniraj, J., Jenish, I. and Nagarajan, K.J., 2024. Extraction, purification, and characterization of novel plant fiber from tabernaemontana divaricate stem to use as reinforcement in polymer composites. *Biomass Conversion and Biorefinery*, pp. 1–15.

[35] Singh, C., Khanna, V. and Singh, S., 2023. Sustainability of microwave heating in materials processing technologies. *Materials Today: Proceedings*, *73*, pp. 241–248.

[36] Nagpal, Y., Sharma, R., Sharma, N. and Tyagi, R.K., 2023. Fabrication of AA2024/SiC/eggshell reinforced hybrid green aluminium matrix composite by stir casting route. *Materials Today: Proceedings*, *78*, pp. 726–730.

[37] Şap, S., 2024. Machining and energy aspect assessment with sustainable cutting fluid strategies of Al–12Si based hybrid composites. *International Journal of Precision Engineering and Manufacturing-Green Technology*, *11*(1), pp. 33–53.

[38] Ramesh, M., Niranjana, K. and Tamil Selvan, M., 2024. Wear and friction behavior of biocomposites fabricated through additive manufacturing. *Tribological Properties, Performance and Applications of Biocomposites*, pp. 219–246.

[39] Felix Sahayaraj, A., Jenish, I., Tamilselvan, M., Muthukrishnan, M. and Kumar, B.A., 2022. Mechanical and morphological characterization of sisal/kenaf/pineapple mat reinforced hybrid composites. *International Polymer Processing*, *37*(5), pp. 581–588.

[40] Sadeghi, B., Cavaliere, P. and Shabani, A., 2023. Design strategies for enhancing strength and toughness in high performance metal matrix composites: A review. *Materials Today Communications*, p. 107535.

[41] Cao, J., Li, F., Yang, Q., Zhan, K., Yang, Z., Wang, Z. and Zhao, B., 2023. A review on interfacial structure optimization and its mechanism on the properties of carbon reinforced metal-matrix composites. *Composite Interfaces*, *30*(5), pp. 543–583.

[42] Ramesh, M., Selvan, M.T., Chandran, P.R., Hariprasad, P., Arunkumar, A., Kumar, C.R. and Kumar, R.P., 2023. An experimental investigation on the effect of lithium and alkali resistant glass fiber on aluminum AA5383 fabricated by the stir-casting process. *Materials Today: Proceedings*, pp. 1–10.

[43] Ramesh, M., Deepa, C., Rajeshkumar, L., Tamil Selvan, M. and Balaji, D., 2021. Influence of fiber surface treatment on the tribological properties of calotropis gigantea plant fiber reinforced polymer composites. *Polymer Composites*, *42*(9), pp. 4308–4317.

[44] Sadeghi, B. and Cavaliere, P.D., 2023. Reviewing the integrated design approach for augmenting strength and toughness at macro-and micro-scale in high-performance advanced composites. *Materials*, *16*(17), p. 5745.

[45] Ramesh, M., Selvan, M.T. and Niranjana, K., 2022. Thermal characterization and hygrothermal aging of lignocellulosic agave cantala fiber reinforced polylactide composites. *Polymer Composites*, *43*(9), pp. 6453–6463.

[46] Manickam, T., Iyyadurai, J., Jaganathan, M., Babuchellam, A., Mayakrishnan, M. and Arockiasamy, F.S., 2023. Effect of stacking sequence on mechanical, water absorption, and biodegradable properties of novel hybrid composites for structural applications. *International Polymer Processing*, *38*(1), pp. 88–96.

[47] Wang, K., Fang, N., Huang, Z., Xu, J., Zhang, S. and Qin, J., 2023. A physics-informed knowledge distillation model with spatial–temporal attention for energy consumption pre-assessment in sustainable additive manufacturing. *Sustainable Energy Technologies and Assessments*, *60*, p. 103537.

[48] Chaurasiya, S. and Singh, G., 2024. Evaluating the environmental footprint: Life cycle assessment of metal matrix composites manufactured by the stir casting process. *Process Integration and Optimization for Sustainability*, *8*(1), pp. 135–142.

[49] Ali, M., Alahmari, S.D., Abdelmohsen, S.A., Alanazi, M.M., Al-Sehemi, A.G., Abdullah, M., Aman, S. and Farid, H.M.T., 2024. Fabrication of MnSe/WSe_2 nanohybrid electrode prepared through hydrothermal method for supercapacitor applications. *Ceramics International*, *50*(4), pp. 6931–6940.

[50] Lostado-Lorza, R., Corral-Bobadilla, M., Íñiguez-Macedo, S. and Somovilla-Gómez, F., 2023. Characterization, LCA and FEA for an efficient ecodesign of novel stainless steel woven wire mesh reinforced recycled aluminum alloy matrix composite. *Journal of Cleaner Production*, *411*, p. 137380.

[51] Chaurasiya, S. and Singh, G., 2023. Life cycle assessment of nanocomposite manufactured using ultrasonic stir casting. *Journal of Materials Science*, *58*(12), pp. 5298–5318.

[52] Iyyadurai, J., Arockiasamy, F.S., Manickam, T., Rajaram, S., Suyambulingam, I. and Siengchin, S., 2023. Experimental investigation on mechanical, thermal, viscoelastic, water absorption, and biodegradability behavior of sansevieria ehrenbergii fiber reinforced novel polymeric composite with the addition of coconut shell ash powder. *Journal of Inorganic and Organometallic Polymers and Materials*, *33*(3), pp. 796–809.

[53] Ramesh, M., Tamil Selvan, M. and Niranjana, K., 2022. Hygrothermal aging, kinetics of moisture absorption, degradation mechanism and their influence on performance of the natural fibre reinforced composites. In *Aging Effects on Natural Fiber-Reinforced Polymer Composites: Durability and Life Prediction*, Springer, pp. 257–277.

[54] Lee, T., Jeong, W., Chung, S., Kim, J.J., Lee, D., Kim, J. and Ryu, H.J., 2023. Role of B4C in the directed energy deposition of aluminum alloy 6061/B_4C composite using core-shell powder for crack annihilation, pore mitigation, and grain refinement. *Additive Manufacturing*, *72*, p. 103622.

[55] Debnath, B., Shakur, M.S., Bari, A.M. and Karmaker, C.L., 2023. A bayesian best–worst approach for assessing the critical success factors in sustainable lean manufacturing. *Decision Analytics Journal*, *6*, p. 100157.

[56] Ramesh, M., Deepa, C., Tamil Selvan, M. and Reddy, K.H., 2022. Effect of alkalization on characterization of ripe bulrush (typha domingensis) grass fiber reinforced epoxy composites. *Journal of Natural Fibers*, *19*(3), pp. 931–942.

[57] Ramesh, M., Deepa, C., Selvan, M.T., Rajeshkumar, L., Balaji, D. and Bhuvaneswari, V., 2021. Mechanical and water absorption properties of calotropis gigantea plant fibers reinforced polymer composites. *Materials Today: Proceedings*, *46*, pp. 3367–3372.

[58] Ramesh, M., Rajeshkumar, L., Deepa, C., Tamil Selvan, M., Kushvaha, V. and Asrofi, M., 2022. Impact of silane treatment on characterization of ipomoea staphylina plant fiber reinforced epoxy composites. *Journal of Natural Fibers*, *19*(13), pp. 5888–5899.

[59] Babu, J.S., S. Murthy, V.S., Kumar, U.P., Sathiyamoorthy, V. and Saravanakumar, A., 2024. Novel spinifex littoreus fibre and sugarcane biosilica on mechanical, wear, time dependent and water absorption behaviour of epoxy structural composite. *Biomass Conversion and Biorefinery*, *14*(3), pp. 4017–4028.

20 New Insights into Metal Matrix Composite Recycling and Reclamation

Raja Thiyagarajan, Shenbaga Velu Pitchumani, Senthil Kumar Marikannan, Venkatachalam Gopalan, and Sakthi Selvarasu

20.1 INTRODUCTION

Materials that are recyclable, lightweight, and have tremendous performance are in huge demand across all industries. One of the greatest choices is engineered materials known as metal matrix composites or MMCs [1]. The automotive, aviation, and military sectors currently substitute several traditional materials with MMCs. In most cases, a metallic alloy matrix will include either continuous or dispersed fibres, whiskers, or particles. Because of these reinforcements, the composite gains properties that are impossible to achieve with monolithic alloys. The performance of MMCs is superior at higher temperatures compared to their base metal equivalents. Aluminium, magnesium, and titanium alloys are often used as base matrices in producing MMCs [2, 3]. The incorporation of carbides, nitrides, borides, and oxides further strengthens these metals. In contemporary material science and engineering, recycling MMCs is essential because it fosters technical innovation, economic efficiency, and environmental sustainability. Metal matrix composites, combined with reinforcing fibres or particles, provide better strength, stiffness, heat conductivity, and wear resistance compared to conventional materials [4–6]. The construction, automotive, electronics, and aerospace industries use these cutting-edge materials extensively. To manage the end-of-life of these materials, efficient recycling techniques have become more and more necessary as the use of MMCs has increased. Several crucial viewpoints can be used to explain the significance of recycling MMCs.

20.1.1 Sustainability of the Environment

Lowering the environmental impact is one of the main advantages of recycling MMCs. The manufacturing of virgin metals and composite materials requires significant energy resources, often involving environmentally harmful extraction and processing techniques. For example, deforestation, soil erosion, and water contamination could result from bauxite mining for aluminium production. Recycling MMCs significantly reduces the need for virgin materials, protects natural resources, and lowers the energy and carbon footprint involved in the extraction, processing, and production of raw materials [7]. By recycling MMCs, industries can greatly reduce the environmental effects of their operations and support global sustainability goals.

20.1.2 Preservation of Resources

Rare or hard-to-find metals, such as titanium or rare earth elements used in certain high-performance applications, are frequently found in MMCs. By preserving raw materials and lowering dependency on limited natural resources, recycling these materials contributes to the maintenance of a sustainable supply chain. This is especially important because rising industries and technological

 DOI: 10.1201/9781003499718-20

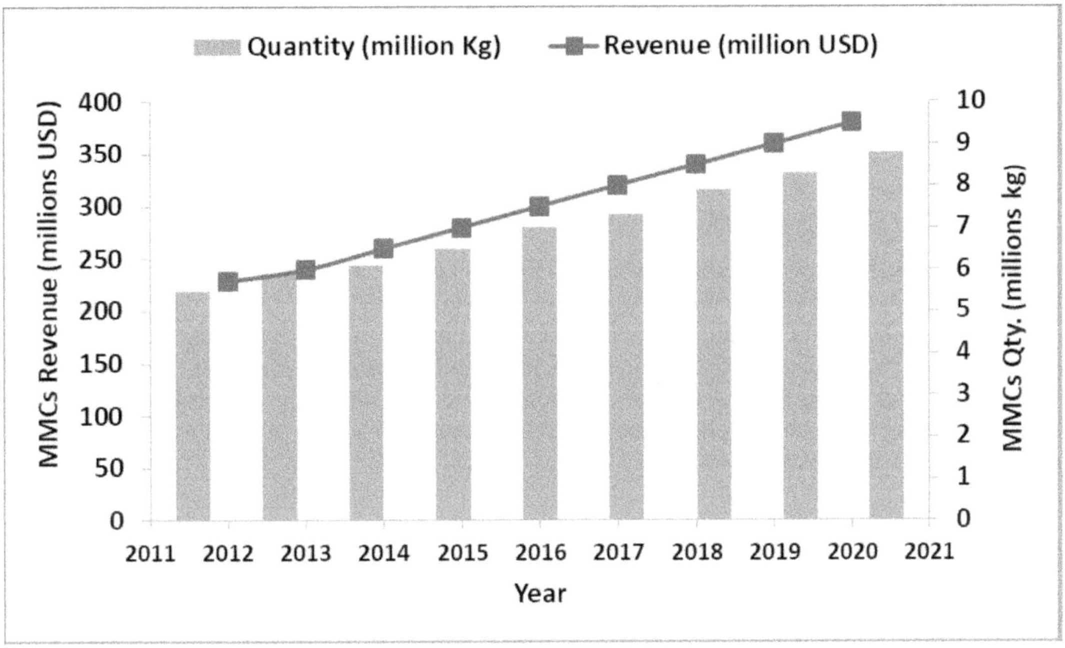

FIGURE 20.1 Global demand for MMCs and their income [8].

improvements are driving up the global demand for these materials [9, 10]. The materials found in MMCs can be recovered and repurposed through recycling, prolonging the useful life of these priceless resources and encouraging the development of a more circular economy.

20.1.3 Economic Gains

There are also major financial benefits to MMC recycling. By offering a less expensive supply of raw materials than the extraction and processing of new metals, it can, in the first place, reduce material prices. Manufacturers can save money by using recycling technologies, which typically use less energy and have lower operating costs than the manufacturing of new materials. Recycling MMCs can also lower trash disposal costs and possibly even turn a profit by selling recycled materials [11]. Industries can improve their competitiveness in the market and economic efficiency by adopting recycling.

20.1.4 Innovation in Materials and Technology

Material science and engineering innovation is stimulated by recycling MMCs. New or enhanced MMCs can be created by processing and engineering recycled materials to fulfil particular performance requirements [12, 13]. This promotes the development of recycling technologies and procedures while also expanding the application possibilities of MMCs. By improving material recovery, separation, and reintegration efficiency and effectiveness, recycling innovations can produce high-quality recycled MMCs with attributes that are on par with or better than those of their virgin equivalents.

20.1.5 Difficulties and Prospects for the Future

Notwithstanding the obvious advantages, recycling MMCs presents several difficulties, such as the problem of guaranteeing the quality and integrity of recovered components and separating the metal matrix from the reinforcing materials. Ongoing studies and technical developments are tackling these issues, nevertheless. For instance, the recyclability of these materials can be increased by the development of increasingly complex separation methods and the design of MMCs with recycling in mind [14, 15]. Furthermore, it is expected that advances and investments in MMC recycling technologies would be further stimulated by the increased emphasis across industries on sustainability and the principles of the circular economy.

20.2 CHALLENGES IN RECYCLING OF MMCs

Recycling metal matrix composites (MMCs), which are engineered by embedding reinforcement materials like fibres or particles into a metal base, faces numerous hurdles due to their sophisticated structures and the distinct characteristics of their components. These composites, celebrated for their enhanced mechanical, thermal, and electrical capabilities, are extensively used across various sectors, including aerospace, automotive, and construction. However, the very features that make MMCs advantageous also complicate their recycling process. The following phenomenon are the reason for the challenges in recycling of MMCs.

20.2.1 Complex Material Compositions

The intricate makeup of MMCs, combining metals with ceramic, carbon-based, or other metallic reinforcements, introduces significant recycling challenges [16]. The divergent properties of these components necessitate bespoke recycling methods for each, complicating the recovery process. Separating these materials efficiently, without compromising their quality, poses a considerable challenge.

20.2.2 Extraction of Reinforcement Materials

Extracting the reinforcing materials from the metal matrix is a critical step in recycling MMCs, yet it presents a formidable challenge. Conventional melting methods suitable for homogeneous metals fall short for MMCs, requiring more sophisticated physical or chemical techniques for separation [17].

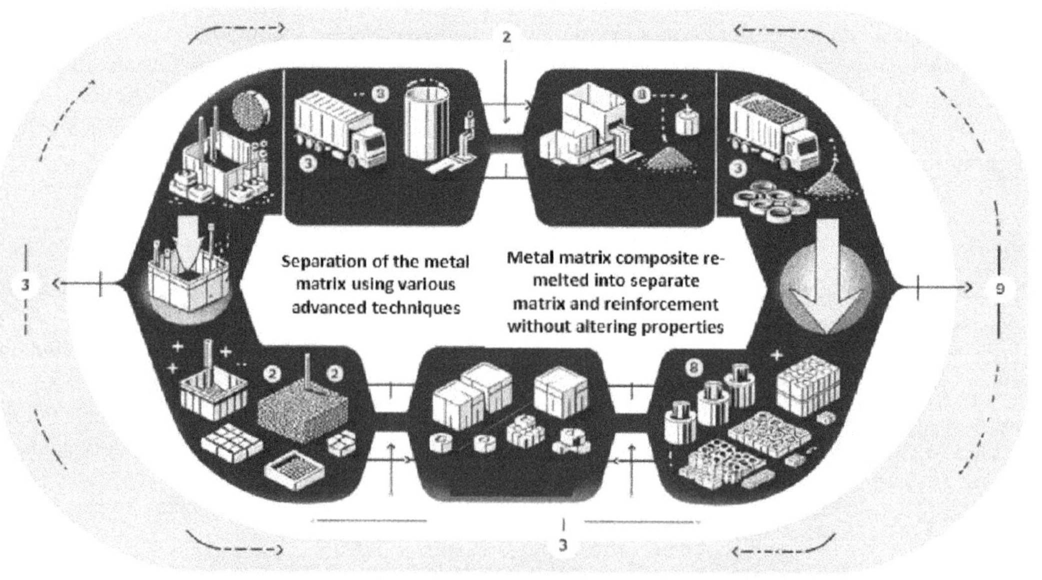

FIGURE 20.2 Recycling process of MMCS.

These methods, however, can be resource-intensive and might degrade the quality of the reinforcement, limiting its reuse.

20.2.3 Retention of Material Qualities

A significant concern in recycling MMCs is maintaining the original properties of the materials for their subsequent reuse. Processes such as remelting can cause oxidation, and mechanical handling may damage the structural integrity of reinforcements [18]. Developing recycling techniques that preserve these properties is a key technical obstacle.

20.2.4 Economic Considerations

The financial aspects of recycling MMCs also warrant attention. The expenses involved in collecting, sorting, and processing scrap MMCs need to be weighed against the potential value of the reclaimed materials [8]. The specialized nature of MMC recycling can incur higher costs compared to traditional metal recycling, potentially diminishing the economic appeal of recycling efforts.

20.2.5 Environmental Considerations

Though recycling is typically environmentally favourable, the specific processes for recycling MMCs can have adverse environmental effects if not managed properly [19]. Techniques like chemical leaching can generate hazardous wastes, necessitating careful disposal, and the high energy demands of some recycling methods may counteract the economic limits.

20.2.6 Technological and Research Development Needs

Ongoing technological innovation is vital to overcoming the recycling challenges posed by MMCs. There are significant gaps in research, particularly in understanding how various composite materials behave during recycling. Progress in separation technologies, processing techniques, and the design of MMCs for easier recycling is crucial for surmounting these obstacles.

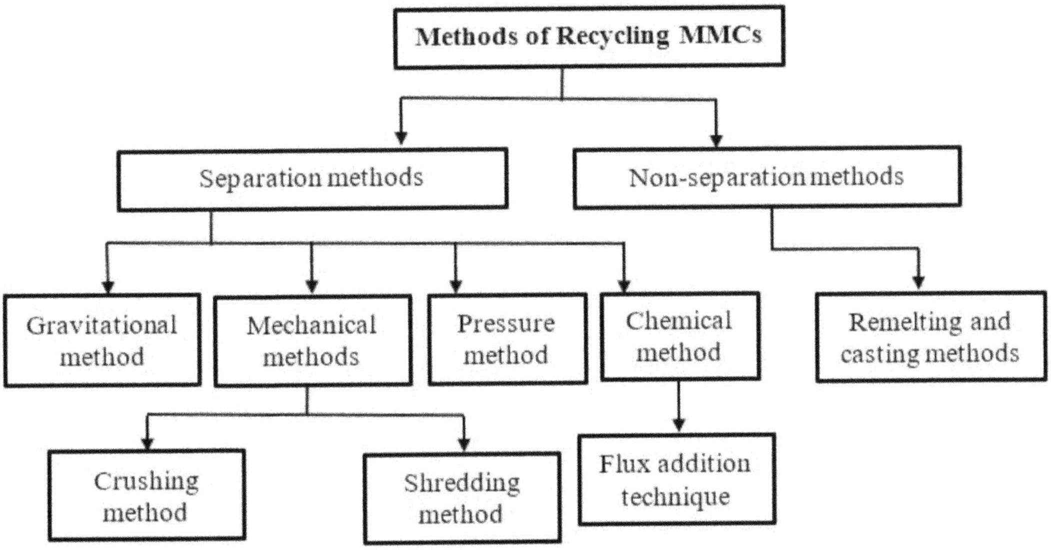

FIGURE 20.3 The recycling techniques used for AMCs.

In summary, while recycling metal matrix composites involves complex challenges, it is a crucial process for achieving sustainable material management and environmental protection. Tackling these challenges necessitates a collaborative, multidisciplinary effort that spans material science, manufacturing technologies, and sustainable development strategies [20]. As the utilization of MMCs expands, so too does the importance of crafting efficient, cost-effective, and eco-friendly recycling methodologies.

Two basic approaches are primarily used for recycling MMCs.

a. MMC recycling with separation process
b. MMC recycling without separation process

20.3 MMC RECYCLING WITH SEPARATION PROCESS

20.3.1 Mechanical Method and Challenges

Recovering and reusing materials is achieved by the recycling of MMCs through the separation of matrix and reinforcements. The majority of material separation processes involve some combination of mechanical, chemical, pressure-driven, or gravity forces. Mechanical procedures for separation include shredding and crushing. Mechanical grinding or ball milling procedures are used to transform the generated AMCs into the form of powder in tearing and crushing techniques. Next, a cyclone separator or sieve processes are used to separate the phases. Shishkin A. et al. [21] separated the Al/WB composites using a milling disintegrator, and they assessed the grindability of recycled powder. It was identified that the mechanical recycling method has been most effectively used for the fibre reinforced metal matrix composites.

20.3.1.1 Challenges

Several obstacles arise when mechanical techniques are used to recycle metal matrix composites (MMCs). The intricate composition of MMCs which include metal matrices (such as magnesium or aluminium) reinforced with ceramic particles, fibres, or whiskers is the root cause of these difficulties. Physical operations including shredding, grinding, and separation are at the basis of mechanical recycling technologies. It is challenging to evenly treat MMCs since they comprise many materials with differing mechanical characteristics. Ceramic reinforcements, such as Al_2O_3 and SiC, etc., are less effective in recycled materials because of their size reduction and fragility to damage from mechanical procedures such as grinding or milling. The strong bonding between matrix and reinforcement requires high energy to separate them. The mechanically separated matrix results in contamination with the ceramic particles which causes degradation of the quality of material. In addition, wear and tear with high energy consumptions are the major drawbacks in the mechanical separation process.

20.3.2 Gravitational Method and Challenges

For MMCs with a greater density variation among the matrix and reinforcements, the phases are separated using the gravity or density separation technique. In this technique, the produced composites are heated again in the furnace until they reach a molten condition. The first step is to rapidly agitate the mixture to eliminate contaminants and slag. The next step is to keep the slurry at a higher temperature than its melting point for an extended period of time without mixing it. This will cause the particles to settle to the bottom of the crucible as a result of their greater densities. Afterwards, the phases are separated by either transferring the metal matrix phase out from the furnace or letting the composite slurry harden there. Sun N. et al. [16] studied the supergravity-induced separation technique in order to recycle MMC waste. The conceptual design of supergravity apparatus to deal with industrial-scale quantities of scrap MMCs was developed.

20.3.2.1 Challenges

Recycling metal matrix composites (MMCs) using gravitational methods poses several challenges. Effective gravitational separation needs a large matrix-reinforcement particle density difference. When this difference is small, separation is inefficient. The smaller particles have the potential to obstruct separation procedures that depend on centrifugal or gravitational forces. And gravitational approaches may not separate evenly distributed reinforcing particles from the matrix. While separating materials, unintended chemical reactions may occur, potentially changing the recycled materials' characteristics.

20.3.3 Chemical Method and Challenges

The chemical method uses a variety of chemical fluxes to dewet the material, with the goal of reducing the energy at the surface of the reinforcements relative to the matrix. The interfacial connection between the matrix materials and reinforcements is broken when fluxes are included. A steady stream of argon gas is used to include several fluxes into the molten composite slurry, including dolomite, borax, potassium chloride, the mineral fluorite, and limestone. Incorporating fluxes into the process while maintaining a steady flow of bubbled argon causes the top layer to develop sludge, which consists of contaminants and dewetted reinforcing content. Sano H et al. [17], used a variety of combinations, including $CaCl_2$-NaCl, NaCl-KCl, NaCl-KCl-KF, and $CaCl_2$-NaCl-KF, to extract the aluminium from Al/SiC composites.

20.3.3.1 Challenges

It is challenging to find a single chemical technique that can effectively separate all components due to the heterogeneous nature. Matrix metals are reactive with common chemicals, which may cause undesirable by-products to develop and make separation and purification more difficult. Significant environmental and safety issues are associated with the dissolving process's usage of powerful acids, alkalis, or other dangerous compounds. Dealing with huge quantities or very reactive substances might drive up the cost of chemicals processes used in MMCs recycling.

20.3.4 Pressure Method

Filters are used under pressured circumstances to separate particles from Al in the pressure-driven process. This process involves applying pressure to a slurry of melt composites, which allows the matrix to flow through a filter while the particles settle to the bottom. Sun N. et al. [16] used centrifugation and a carbon fibre filter (whose pores are smaller than the particles) to separate 30 gm of aluminium-SiC composites into their individual components. In addition, the residue contains the Al_4C_3 component in addition to SiC reinforcement. Mizumoto et al. separated Al and SiC particles employing a nozzle filtering technique. This allowed the molten metal to flow through the tip of the nozzle by continually applying pressurized argon gas to the metal. The particles aggregate and create a skeleton-like structure as a result of the combined action of applied pressure and gravity; as a result, only the matrix flows through the nozzle, with the particles being separated within. The findings showed that although the highest collected metal yield exceeds 60%, it increases as gas pressure increases. It has been noted by the investigators that this technology is unsuitable for reclaiming bulk composite.

Researchers segregated aluminium alloys and reinforcements employing mechanical and chemical techniques. Melt purity and cleanliness are crucial for future uses of recovered aluminium alloys. Following recycling, it is crucial to ensure that the recovered alloys are of high quality and that there is strict supervision over their decontamination. Experts have previously used a variety of methods, including the K mould approach, X-ray radiography and computed tomography, reduced pressure test, and spectral analysis, to assess the purity and quality of melt for metal alloys. Further study is needed to increase melt quality and purity of retention during numerous recycling processes.

In conclusion, the following are the primary limitations of MMC recycling and separation:

a. Lower return on investment
b. The procedure is more complicated and costlier
c. Modifications that might be made to the matrix concerning its quality and cleanliness
d. Problems with the retrieved reinforcements due to contaminants

20.4 MMC RECYCLING WITHOUT SEPARATION PROCESS

20.4.1 Remelting SiC-Reinforced MMCs

Klasik A et al. [22] studied the findings from microstructural analyses and fatigue life tests of A359 matrix composites augmented with SiC particles and exposed to numerous remelting processes. There is little correlation between the frequency of remelting and the dispersion of eutectic precipitates inside the combined matrix. The squeeze casting technique should be used to do numerous remelting in order to avoid shrinkage porosity. The results of the A359 composite achieve the greatest results by numerous remelting. Braszczyńska-Malik K. N. [23] investigated the microstructural characterization of casted magnesium-SiC composite. The SEM microstructure indicated the good bonded SiC among the aluminium matrix. And also, the homogeneous dispersion was observed in Figure 17.5.

Mandal D. and Viswanathan S. [24] revealed the form of interface compounds of aluminium composite after remelting. The experimental findings show that as the remelted composite plate goes from top to bottom, the volume proportion of SiC particles drops. Remelting the SiC particle causes it to break into Si and C, according to microstructure investigations and EDX analyses performed on the interface of the remelted composite. At the interface encompassing the SiC particle, the Si-rich phase and carbides (Al_4C_3) are detected.

Sharma A. and Ahn B. [25] found that remelting, holding, and casting are effective methods for recycling Al matrix alloy from composite trash into molten metal alloy and SiC particles.

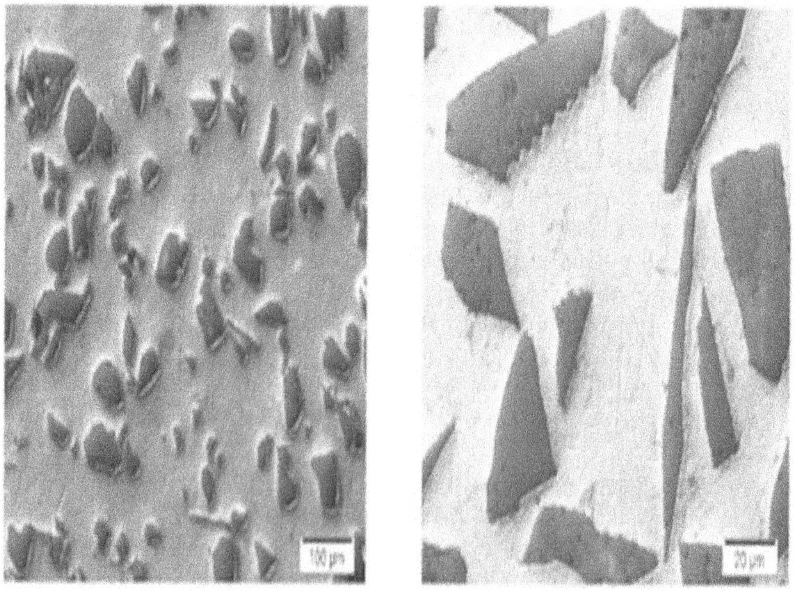

FIGURE 20.4 **a)** SEM Microstructure of SiC reinforced Mg matrix composite **b)** SEM microstructure with higher magnification [23].

FIGURE 20.5 **a)** A359 alloy droplets formed, and SiC particles separated and moved toward the salt phase **b)** High-magnification Al droplets surrounded by SiC particles [26]

Approximately 50–77% of the aluminium alloy matrix may be recovered from the composite trash including SiC particles of varying particle size and weight fraction. A higher concentration of particles and a greater proportion of weight result in a shorter period of time required to reach the maximum recovery of the aluminium alloy. The SiC particles separated from the A359 aluminium matrix using molten salt of NaCl-KCl-NaF. The separated SiC mixed with the molten salt as a separate layer and the aluminium alloy droplets formed are shown in Figure 20.5. The higher magnification image identified the droplets surrounded with the SiC particles.

20.4.2 Al_2O_3-Reinforced MMCs Remelting

The investigators have examined the effects of repeated melting on the features and properties of alumina reinforced metal matrix composites. Pietrzak K. et al. [27] performed the several straight remeltings of composites made from the A359 metal and 20% alumina particles. The investigated composite is fully free of gas-shrinkage porosity when using the squeeze casting technique after each remelting, in contrast to when using standard gravity casting for each remelting, which results in gas shrinkage porosity after the initial remelting [28, 29]. Tongxiang [30] investigated the behaviour of alumina/Al composite using the remelting process. The reaction remelting process was carried out using the CuO. The findings also indicated that in the Al_2O_3/Al composite, the fracture mechanisms shifted from a particle fracture to interfacial debonding, which was mainly caused by a rise in interfacial bonds between the in situ Al_2O_3 particles and aluminium.

Zhong et al. [31], examined the reactions that occur at the interface of as-cast composites and composites that were remelted at 800°C for periods of up to four hours using Al_2O_3 particles and 5083 Al-Mg alloy. The results show that when the remelting time increases, the reaction product changes. Following a 4-hour remelting process, MgO is converted to $MgAl_2O_4$, and the area between the two surfaces becomes filled with huge crystals of $MgAl_2O_4$, each measuring around 1 mm in circumference. These changes show that the amount of magnesium in the matrix controls the final product type.

20.4.3 Other Reinforced MMCs Remelting

Researchers also attempted to learn more about the properties of Al/CuO and Al/carbon AMCs after various remelting processes [30]. It was assessed that the AMCs' behaviour during five separate remelts of Al/CuO metal matrix composites. The findings showed that the Al_2O_3 and Al2Cu intermetallics are formed when CuO particulates react chemically with the matrix over the remelting process. According to their findings, a greater negative Gibbs free energy value led to the creation of the Al_2O_3 phase after remelting. Fan T. et al. [32], investigated the carbon fibre reinforced aluminium alloy with enhanced mechanical properties. Additionally, the results of the research show that it is rather appropriate to utilize the waste composite materials as charge material for future composite casting manufacturing. Experimental findings did not support the claim that remelting composites reduces their characteristics. Authors show that the original composite manufacturing method, together with the remelting and casting procedures, may be carefully selected to maintain a suitable level of composite characteristics. The production of unwanted intermetallics as a result of chemical interactions between reinforcements and certain matrix components is the main disadvantage of recycling AMCs without separation. Additionally, the production of undesired intermetallics directly impacts the properties, features, and quality of MMCs due to their brittle nature.

20.5 OBSTACLES AND POTENTIAL SOLUTIONS

To regulate the dissociation of particles and limit the development of unwanted intermetallics during the process of remelting, the researchers have introduced coated particles. The researchers noted that the presence of a coating on particles leads to a decrease in chemical reactions. Nevertheless, the prevention of unwanted intermetallic compounds is not entirely achieved even with repeated remelting. Furthermore, the investigation was conducted by injecting a reactive addition, namely titanium, during the remelting process. Nevertheless, the desired outcomes were not attained as a consequence of the existence of harmful intermetallics during repeated remelting. Hence, it is necessary to expand the efforts to suppress and separate the harmful intermetallic compounds that are produced during the process of remelting. Previously, scientists have noted that when aluminium alloys are combined with substances like titanium dioxide, boric acid, potassium titanium fluoride, potassium fluoroborate, potassium fluorozirconate, etc., it leads to the formation of beneficial compounds such as titanium carbide, boride, and zirconium boride. This, in turn, enhances the mechanical and tribological characteristics of the original alloy.

20.6 CONCLUSIONS

The growing need for energy savings and environmental concerns, including environmentally conscious production and a green economy, has made metal matrix composite recycling an urgent matter. This research provides a comprehensive analysis of the literature on MMC recycling. Typically, MMC recycling involves either separating the particles and matrix or doing it without. Primary findings from this investigation are as follows:

- Separating metal matrix composites for recycling has challenges such as lower purity and low metal extraction. Results show a maximum retrieved matrix metal output of 60%. Additionally, the retrieved particles are unusable owing to contaminants and pollution
- Investigations should focus on the purity and quality of the melt in order to facilitate future uses of the recovered matrix materials
- In the process of without separation, MMCs are directly melted and casted. The formation of unwanted intermetallics as a result of particle fragmentation is the main obstacle in this approach

- The creation of harmful intermetallics causes an obvious reduction in hardness, maximum tensile strength, and wear following each remelting, according to investigators. Researchers have shown that the weight percentage of intermetallics increases with each consecutive remelting. The Al_4C_3 and $MgAl_2O_4$ intermetallic compounds are formed by repeated remelting of silicon carbide and alumina reinforced aluminium matrix composites (AMCs), respectively.
- The incorporation of additives during the remelting process of composites reinforced with SiC and Al_2O_3 is a method that may be used to inhibit the occurrence of intermetallics that are undesirable or damaging.

Within this framework, further research may concentrate on improving the remelting process of AMCs by adding reactive additives, which can inhibit the synthesis of undesirable compounds while facilitating the creation of desirable compounds and intermediates.

REFERENCES

[1] Baran, I., Cinar, K., Ersoy, N., Akkerman, R. and Hattel, J.H., 2017. A review on the mechanical modeling of composite manufacturing processes. *Archives of Computational Methods in Engineering*, *24*, pp. 365–395.
[2] Shuaib, N.A., Sultan, A.A.M., Ismail, S.O., Samat, A.A., Omar, N.W.Y., Azmi, A.I. and Mativenga, P.T., 2021. Recycling of composite materials. In *Advances in Machining of Composite Materials*. Engineering Materials, Cham: Springer, pp. 527–552.
[3] Bulei, C., Todor, M.P., Kiss, I. and Alexa, V., 2019. Aluminium matrix composites: Processing of matrix from re-melted aluminium wastes in micro-melting station. *IOP Conference Series: Materials Science and Engineering*, p. 012049.
[4] Kudyba, A., Akhtar, S., Johansen, I. and Safarian, J., 2021. Aluminum recovery from white aluminum dross by a mechanically activated phase separation and remelting process. *JOM*, *73*, pp. 2625–2634.
[5] Yusuf, N.K., Medi, A.S., Lajis, M.A., Chan, B.L. and Shamsudin, S., 2021. Mechanical properties of direct recycling metal matrix composite (MMC-AlR) AA7075 aircraft aluminium alloy. *International Journal of Integrated Engineering*, *13*, pp. 89–94.
[6] Friedrich, B., 2003. Recycling of MMCs and aluminium foam. In *European Metallurgical Conference*, pp. 108–114.
[7] Ling, P.S. and Gupta, M., 2000. Recycling of aluminium based metal matrix composite using disintegrated melt deposition technique. *Materials Science and Technology*, *16*, pp. 568–574.
[8] Ajay Kumar, P., Rohatgi, P. and Weiss, D., 2020. 50 Years of foundry-produced metal matrix composites and future opportunities. *International Journal of Metalcasting*, *14*, pp. 291–317.
[9] Mironovs, V., Usherenko, Y., Boiko, I. and Kuzmina, J., 2022. Recycling of aluminum-based composites reinforced with boron-tungsten fibres. *Materials*, *15*, p. 3207.
[10] Yang, Y., Boom, R., Irion, B., van Heerden, D.J., Kuiper, P. and de Wit, H., 2012. Recycling of composite materials. *Chemical Engineering and Processing: Process Intensification*, *51*, pp. 53–68.
[11] Gergel, Y.V., Deg, H.P. and Clyne, T.W., 2000. Recycling of MMCs and production of metallic foams. *Comprehensive Composite Materials*, pp. 797–820.
[12] Nishida, Y., 2001. Recycling of metal matrix composites. *Advanced Engineering Materials*, *3*, pp. 315–317.
[13] Nishida, Y., Izawa, N. and Kuramasu, Y., 1999. Recycling of aluminum matrix composites. *Metallurgical and Materials Transactions A*, *30*, pp. 839–844.
[14] Ravi, K.R., Pillai, R.M., Pai, B.C. and Chakraborty, M., 2007. Separation of matrix alloy and reinforcement from aluminum metal matrix composites scrap by salt flux addition. *Bulletin of Materials Science*, *30*, pp. 393–398.
[15] Schuster, D.M., Skibo, M.D., Bruski, R.S., Provencher, R. and Riverin, G., 1993. The recycling and reclamation of metal-matrix composites. *JOM*, *45*, pp. 26–30.
[16] Sun, N., Wang, Z., Guo, L., Wang, L. and Guo, Z., 2020. Efficient separation of reinforcements and matrix alloy from aluminum matrix composites by supergravity technology. *Journal of Alloys and Compounds*, *843*, p. 155814.

[17] Sano, H., Kato, S., Motomura, T. and Fujisawa, T., 2005. Closed recycling process for Al-based composite materials. *Materials Transactions*, *46*, pp. 3067–3072.
[18] Jalilvand, M.M., Bagh, N.T., Akbarifar, M. and Divandari, M., 2020. A new insight to dynamic oxidation of molten metals by the parametric study of oxide/metal/oxide sandwich formation. *International Journal of Metalcasting*, *14*, pp. 949–961.
[19] Prasadh, S., Suresh, S., Ratheesh, V., Wong, R. and Gupta, M., 2021. Biocompatibility of metal matrix composites used for biomedical applications. *Encyclopedia of Materials: Composites*, *1*, pp. 475–501.
[20] Krauklis, A.E., Karl, C.W., Gagani, A.I. and Jørgensen, J.K., 2021. Composite material recycling technology—state-of-the-art and sustainable development for the 2020s. *Journal of Composites Science*, *5*, p. 28.
[21] Shishkin, A., Mironov, V., Goljandin, D. and Lapkovsky, V., 2012. Recycling of Al-W-B composite material. *Key Engineering Materials*, *527*, pp. 143–147.
[22] Klasik, A., Maj, M., Pietrzak, K., Wojciechowski, A. and Sobczak, J., 2016. Fatigue life and microstructure after multiple remelting of A359 matrix composites reinforced with SiC particles. *Archives of Metallurgy and Materials*, *61*, pp. 2123–2128.
[23] Braszczyńska-Malik, K.N., 2021. Types of component interfaces in metal matrix composites on the example of magnesium matrix composites. *Materials (Basel)*, *14*, p. 5182.
[24] Mandal, D. and Viswanathan, S., 2013. Effect of re-melting on particle distribution and interface formation in SiC reinforced 2124 Al matrix composite. *Materials Characterization*, *86*, pp. 21–27.
[25] Sharma, A. and Ahn, B., 2018. Recycling of aluminum alloy from Al-Cu metal matrix composite reinforced with SiC particulates. *Korean Journal of Materials Research*, *28*, pp. 691–695.
[26] Madarasz, D., Budai, I. and Kaptay, G., 2011. Fabrication of SiC-particles-shielded Al spheres upon recycling Al/SiC composites. *Metallurgical and Materials Transactions A*, *42*, pp. 1439–1443.
[27] Pietrzak, K., Klasik, A., Makowska, K., Sobczak, J.J., Wojciechowski, A., Rudnik, D. and Sienicki, E., 2016. Structural determinants of the mechanical properties of A359 + Al_2O_3 composites after multiple remelting. *Transactions of Foundry Research Institute*, *57*, pp. 87–112.
[28] Pietrzak, K., Sobczak, N., Sobczak, J.J., Klasik, A., Kudyba, A., Wojciechowski, A. and Sienicki, E., 2017. Stability of metal-ceramic slurry after multiple re-melting of A359 aluminum alloy based composite reinforced with Al_2O_3 particles. *Transactions of Foundry Research Institute*, *57*, pp. 69–84.
[29] Pietrzak, K., Klasik, A., Maj, M., Sobczak, J. and Wojciechowski, A., 2018. Microstructure and fatigue life of the A359 alloy reinforced with Al_2O_3 after multiple remelting. *Archives of Foundry Engineering*, *18*, pp. 39–44.
[30] Fan, T., Zhang, D., Yang, G., Shibayanagi, T. and Naka, M., 2003. Fabrication of in situ Al_2O_3/Al composite via remelting. *Journal of Materials Processing Technology*, *142*, pp. 556–561.
[31] Zhong, W.M., L'Esperance, G. and Suéry, M., 2002. Effect of current Mg concentration on interfacial reactions during remelting of Al-Mg(5083)/Al_2O_3p composites. *Materials Characterization*, *49*, pp. 113–119.
[32] Łagiewka, M. and Konopka, Z., 2015. Properties of AlSi9Mg alloy matrix composite reinforced with short carbon fibre after remelting. *Archives of Foundry Engineering*, *15*, pp. 39–42.

Index

A

accumulative roll bonding, 109, 259
additive manufacturing, 60, 66–75, 97, 111, 271
aluminium matrix composites, 18, 222, 287
applications, 9, 30, 60, 85, 132, 155, 208

B

binder jetting, 67, 115–117, 127

C

case study, 62
challenges, 10, 28, 61, 76, 97, 133, 232, 282, 285
classification, 4, 197
cloud computing, 181, 183, 185
corrosion, 165

D

degassing, 28–29, 146
diffusion bonding, 54, 58–59, 92, 126
directed energy deposition, 67, 72, 127
drilling, 211, 217

E

electrical discharge machining, 222, 225, 232
equal channel angular pressing, 6, 25
evolution, 2, 121, 181
extrusion, 126, 257, 258, 264

F

forging, 94, 259, 262, 264
forming, 255–264
friction stir
    processing, 52, 187, 260, 261, 264
    welding, 6, 169, 242
functionally graded materials (FGM), 21, 57, 72, 138

G

grinding, 56, 213, 218–219

I

Industry 4.0, 181–193
intelligent MMC, 121–135
Internet of Things, 181, 186

J

joining, 60, 126, 169, 240–250

L

life cycle assessment, 268, 273
liquid metallurgy, 35, 47
liquid state technique, 15, 36, 109, 152

M

machining, 196, 208, 222
magnesium matrix composites, 20, 198, 224
mechanical alloying, 55, 81, 273
milling, 187, 202, 211, 216
    ball milling, 55, 56, 81, 271, 284
    reactive milling, 56

P

physical vapour deposition, 6, 91
post-processing, 16, 26, 37, 55, 116, 225
powder bed fusion, 67, 71, 98
powder metallurgy, 5, 52, 92, 122, 125, 222, 271

R

reclamation, 280
recycling, 135, 188, 267, 269, 280–288
robotics, 186, 189

S

safety, 31, 123, 135
selective laser melting (SLM), 60, 79–83, 111, 115, 127
solid-state techniques, 5, 51, 58, 92, 109, 262
spark plasma sintering, 54, 58, 97
stir casting, 16, 28–31, 36, 40, 158
surface roughness, 215, 218, 227, 229
sustainability, 132–133, 191, 269, 274, 280

T

tribology, 47, 138

U

ultrasonic, 41, 95, 204, 241, 246–248

W

wear, 129, 146, 201, 218

For Product Safety Concerns and Information please contact our EU
representative GPSR@taylorandfrancis.com
Taylor & Francis Verlag GmbH, Kaufingerstraße 24, 80331 München, Germany

www.ingramcontent.com/pod-product-compliance
Ingram Content Group UK Ltd.
Pitfield, Milton Keynes, MK11 3LW, UK
UKHW050353071126
477531UK00011B/599
* 9 7 8 1 0 3 2 8 1 4 0 7 0 *